Food Safety Handbook

Food Safety Handbook

A Practical Guide for Building a Robust Food Safety Management System

International Finance Corporation

WORLD BANK GROUP
THE WORLD BANK IBRD • IDA
IFC International Finance Corporation

1818 H Street NW, Washington, DC 20433
Telephone: 202-473-1000; Internet: www.worldbank.org

1 2 3 4 23 22 21 20

ISBN (paper): 978-1-4648-1548-5
ISBN (electronic): 978-1-4648-1549-2
DOI: 10.1596/978-1-4648-1548-5

Cover photo: A group of women work in a selection line of an olive factory in Lima, Peru. © Christian Vinces / Shutterstock.com. Used with the permission of Christian Vinces / Shutterstock.com. Further permission required for reuse.

Cover and interior design: Kirsten Dennison

The Library of Congress Control Number has been requested.

Contents

Foreword

Every year, millions of people around the world suffer from serious foodborne illnesses.

Food safety is a global issue that is more urgent than ever as the global population races toward a projected 9 billion by 2050. The demand for safe food is growing, pressuring the world's food supply chains, and leaving no room for complacency or slack standards.

In addition to posing health risks, foodborne illnesses can wreak considerable economic damage. *The Safe Food Imperative*, a 2018 World Bank report, estimates that food safety issues cost developing countries a staggering $110 billion in lost productivity and medical treatment in 2016 alone.

A single serious food safety lapse can badly tarnish the brands of restaurants, hotels, and food producers and processors, which can require years of investment and trust-building to repair.

The good news is that most food safety issues are preventable, especially if they are addressed systematically. Businesses that establish rigorous food safety systems are also in a better position to expand and attract investment. In surveys conducted in 2010–18, approximately 27 client companies of the International Finance Corporation (IFC) attributed $478 million in increased sales and $564 million in investment to better food safety practices.

IFC's *Food Safety Handbook: A Practical Guide for Building a Robust Food Safety Management System,* now in its fourth edition, has input from leading industry experts to identify and eliminate problems along the entire food supply chain long before they affect consumers or the bottom line of businesses.

The handbook is a practical instruction manual aimed at business owners who seek to develop or improve a food safety system. Companies of any size, location, or point along the food production chain can apply the handbook's rules and lessons to establish a systematic approach to food safety. The handbook's templates can also be tailored to specific needs.

The handbook has proven successful with food sector businesses in Africa, Asia, and Europe, providing practical information that covers prerequisite programs, the hazard analysis critical control point (HACCP) system, food safety management system documentation, international best practice and legislation, and guidance for top management.

This latest edition, the first since 2016, includes revised food safety standards and legislation, updated tools and techniques for implementing food safety systems, and new information on management's role and responsibilities regarding food safety. This update also features the latest Global Food Safety Initiative (GFSI) benchmarking requirements and the new version of International Organization for Standardization (ISO) 22000:2018, the food safety management standards.

As a global leader supporting sustainable private sector development in emerging markets, IFC takes food safety seriously. This handbook will help small and large businesses in emerging markets and more-developed economies to feel secure knowing the foods they produce, process, serve, store, or ship are always safe to eat.

Tania Lozansky
Senior Manager
Manufacturing, Agribusiness, and Services Advisory
International Finance Corporation

Acknowledgments

This handbook has been prepared and published thanks to the efforts of many dedicated professionals.

The core team, led by Natia Mgeladze, Food Safety Specialist, International Finance Corporation (IFC), included Michael Gerard Kelly of the Kelmac Group, Ireland; Rob Kooijmans, Food Safety Experts, the Netherlands; and Mark Nielsen, Consultant, Small and Medium Enterprises and Value Chain Solutions, IFC Financial Institutions Group.

Many thanks also to the following IFC staff for producing and reviewing the handbook's contents: Sarah Ockman, Yuriy Zvazhenko, Kateryna Onul, Uy Duc Pham, Jacqueline Khanali Njonjo, Jason Hopps, Beverley McDonald, Irina Sarchenko, Deepa Chakrapani, Elaine Reinke, and Maria Paulina Mogollon.

Special acknowledgments go to Steven M. Jaffee, Lead Agricultural Economist, Food and Agriculture Global Practice, World Bank; and Lystra N. Antoine, Program Manager, Food and Agriculture Global Practice, World Bank. Those outside the World Bank who have dedicated time and effort to review the handbook include Delia Grace Randolph, Program Leader (joint) Animal and Human Health, International Livestock Research Institute; Marlynne Hopper, Deputy Head of Standards and Trade Development Facility, World Trade Organization; and Akhila Vasan, Program Manager, Food Safety and Education, Science and Education Foundation, Grocery Manufacturers Association.

The combined efforts of these individuals have helped deliver a document we hope will support food sector businesses around the world for years to come.

This fourth edition of the *Food Safety Handbook* has been generously funded by the World Bank Group's Creating Markets Advisory Window. Thanks are also due to the Austrian Ministry of Finance and the Royal Norwegian Ministry of Foreign Affairs for funding and supporting the first three editions.

Abbreviations

5S program	sort, set in order, shine, standardize, sustain
CAC	Codex Alimentarius Commission
CCP	critical control point
CFR	Code of Federal Regulations (United States)
CU	customs union
DMS	document management system
EAEU	Eurasian Economic Union
EC	European Commission
ELISA	enzyme-linked immunosorbent assay
EU	European Union
FBO	food business operator
FDA	Food and Drug Administration (United States)
FSIS	Food Safety and Inspection Service (United States)
FSMA	Food Safety Modernization Act (United States)
FSMS	food safety management system
FSSC	Food Safety System Certification
GAP	good agricultural practice
GDP	good distribution practice
GFSI	Global Food Safety Initiative
GHP	good hygiene practice
GMP	good manufacturing practice
HACCP	hazard analysis critical control point
HRMS	human resources management system
IFC	International Finance Corporation
ISO	International Organization for Standardization
MRL	maximum residue level
OPRP	operational prerequisite program
PRP	prerequisite program
SMART	specific, measurable, attainable (or achievable), realistic, and time-bound
SOP	standard operating procedure

SWOT	strengths, weaknesses, opportunities, and threats
TACCP	threat assessment and critical control point
TR	technical regulation
TS	technical specification
USDA	U.S. Department of Agriculture
VACCP	vulnerability assessment and critical control point

Note: All dollar amounts are U.S. dollars ($) unless otherwise indicated.

Introduction

The IFC and global food safety

The International Finance Corporation (IFC)—part of the World Bank Group—is the largest global development institution focused on the private sector in emerging markets.[1] It works with more than 2,000 businesses worldwide, using its capital, expertise, and influence to create markets and opportunities in the toughest business environments in the world. In fiscal year 2018/19, it delivered more than $19 billion in long-term financing among developing countries, leveraging the power of the private sector to reduce extreme poverty and boost shared prosperity.

Developing agribusiness

IFC has made agribusiness a priority because of the potential of agribusiness for broad development impacts and an especially strong role in poverty reduction. IFC combines investment and advisory services to support the sector in addressing the growing demand and escalating food prices in an environmentally sustainable and socially inclusive manner. IFC invests across the agribusiness supply chain, from farm to retail, to boost production, increase liquidity, improve logistics and distribution, and expand the access to credit among small farmers. The IFC approach in agribusiness is comprehensive and covers the entire value chain. IFC aims to bring land into sustainable production, enhance productivity by transferring technologies and proven practices, and make the best use of water and other natural resources. As urbanization continues, IFC works to support efficient supply chains to bring safe, affordable food to cities. To help clients prefinance inventories, seeds, fertilizers, and chemicals among farmers, IFC offers working capital facilities. It is helping clients maintain competitiveness, upgrade sanitary and food safety standards, and expand market access. With both the private sector and the public sector, it pursues investments in infrastructure—including in ports, warehouses, cold storage, and telecommunication—that can facilitate trade and reduce costs. To reach small farmers and rural enterprises, particularly in low-income countries, IFC is working with trading companies and financial intermediaries, helping channel financing and advisory services effectively.

The IFC global food safety platform

For more than 15 years, the IFC global food safety platform has provided high-quality professional services to help more than 200 companies apply international food safety standards and adapt sustainable business models.[2] IFC support includes food safety assessments, staff training, and guidance in obtaining international certification. Improved food safety is helping IFC clients meet regional and export market requirements, attract investment, realize cost savings, and strengthen brands. The twin goals of the IFC global food safety platform are healthier balance sheets and high-quality food on plates.

Purpose of the handbook

The IFC *Food Safety Handbook* is designed to enable enterprises in developing markets to reduce key risks in growing a sustainable food business to meet the ever-increasing demands, needs, expectations, and trust of customers, wholesalers, retailers, government food safety regulators, and, ultimately, consumers. IFC has developed the handbook with the support of food industry experts. It is based on Codex Alimentarius requirements and best industry practices and standards.[3]

The handbook provides companies with the expertise to develop, implement, and maintain modern food safety management systems based on hazard analysis critical control point (HACCP) system principles.[4] HACCP aims to identify and prevent potential food safety problems proactively. In simple terms, this means safely handling and storing ingredients and supplies that enter and exit food sector businesses.

The handbook offers an entirely voluntary system to help companies identify gaps in their existing practices and develop more efficient food safety systems. By following the sections relevant to their facilities and business, companies may carry out the following:

- Apply the handbook within any process regardless of production facility size or location and regardless of food safety sophistication
- Develop systemic science-based approaches to food safety management
- Benchmark a food safety system against the best international practice
- Use the handbook as a simple, practical self-service tool, replicating the steps it describes on all production lines as necessary
- Tailor the handbook templates in accordance with enterprise needs

Organization of the handbook

The handbook consists of six chapters. It serves as a clear and informative road map to help companies manage their food safety systems. The purpose of each chapter is as follows:

Chapter 1 offers an overview of the Global Food Safety Initiative (GFSI) and other food safety management system (FSMS) schemes and standards that a food sector business might implement to manage food safety.[5] The schemes and standards highlighted in the handbook are all based on international best practices and recognized by the GFSI.

Chapter 2 provides an overview and the best sources of currently available primary food safety legislation, plus a description of food sector companies and of the roles and responsibilities of various enforcement agencies. The chapter also outlines how companies may demonstrate their legal compliance with food safety requirements. It includes useful links to new food safety amendments and related regulations.

Chapter 3 provides information on the dairy sector prerequisite program (PRP) and HACCP system implementation and two other analysis and critical control point systems (TACCP and VACCP). The methodologies described may be applied to any food products. The chapter consists of two main elements: (1) an overview of PRPs based on the requirements of International Organization for Standardization technical specification (ISO/TS) 22002-1, six examples of PRPs associated with milk processing based on ISO/TS 22002-1, a PRP template that food business operators (FBOs) can use to develop PRPs (editable templates can be found at http://www.ifc.org/foodsafety/handbook/templates), and general information on planning and developing PRPs and (2) general information on the HACCP, including history, principles, and benefits.[6] Preliminary steps for developing an HACCP system are also outlined, and an example is given of a milk-processing HACCP plan based on ISO 22000:2018. In the milk-processing example, two critical control points (CCPs) and one operational prerequisite program (OPRP) are described. Also included is an HACCP plan template and TACCP/VACCP templates to help FBOs develop their own HACCP plans (editable templates can be found at http://www.ifc.org/foodsafety/handbook/templates).

Based on the example of the dairy sector, chapter 4 explains how to establish and develop FSMS procedures and documentation. The chapter provides an overview of a typical FSMS documentation structure or hierarchy, an explanation of the purpose and benefits of a documented FSMS, a description of the various documents needed for an FSMS (such as documents on policies, objectives, procedures, work instructions, food safety plans, specifications, and forms and records), and the control of FSMS documents and general information on record management and retention. The chapter also offers examples of the primary documented procedures an FBO is likely to need, as defined by the various GFSI and other FSMS schemes, and a basic example of a food defense plan, a vulnerability assessment, and a food fraud remediation plan enabling the FBO to meet its food security requirements.

Chapter 5 supplies general information on training and development; an example of a training and development procedure; an example of an FSMS responsibility, training needs analysis, and training plan; an FSMS training and development matrix; and a description of methods followed to evaluate the effectiveness of a training program an FBO might consider implementing as part of its FSMS.

Chapter 6 describes the establishment and development of a food safety policy, methods for demonstrating the commitment of top management, and the resources required to establish, develop, implement, and maintain an effective FSMS. This chapter also includes an example of an FSMS management review procedure and a PowerPoint template that might be used by FBOs to document the results of an FSMS management review, including the evaluation of follow-up activities. The section on nonquality costs will help FBO management understand more fully the costs involved in nonquality.

Figure I.1 provides a visual description of how these chapters fit together.

Figure I.1 The Food Safety House

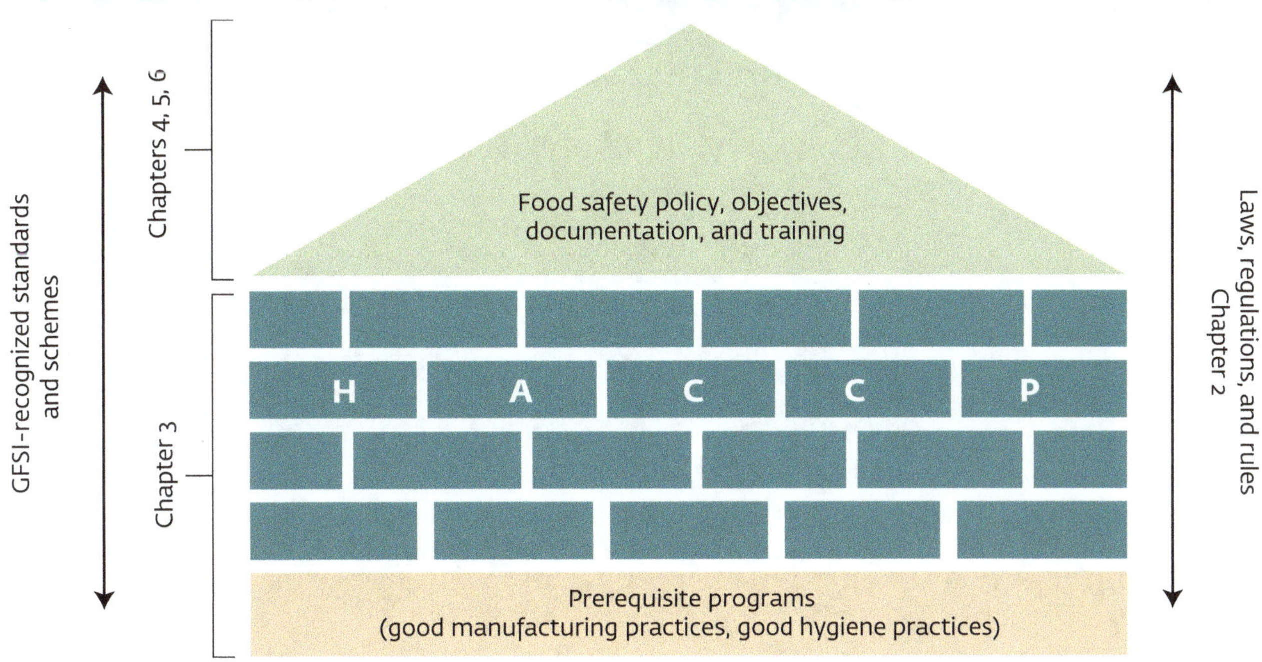

Notes

1. The IFC website is at https://www.ifc.org/wps/wcm/connect/corp_ext_content/ifc_external_corporate_site/home.
2. For more information on IFC food safety advisory services or questions about this handbook, contact Sarah Ockman, Program Lead, at sockman@ifc.org, or consult "Global Food Safety Advisory Program," International Finance Corporation, Washington, DC, http://www.ifc.org/foodsafety.
3. See "Codex Alimentarius: International Food Standards," Joint Food and Agriculture Organization of the United Nations–World Health Organization Food Standards Programme, Rome, http://www.fao.org/fao-who-codexalimentarius/en/.
4. See "Hazard Analysis Critical Control Point (HACCP)," U.S. Food and Drug Administration, Silver Spring, MD, https://www.fda.gov/food/guidance-regulation-food-and-dietary-supplements/hazard-analysis-critical-control-point-haccp.
5. See "Global Food Safety Initiative," Consumer Goods Forum, Levallois-Perret, France, https://mygfsi.com/.
6. To find ISO/TSs and standards, go to the website of the ISO, at https://www.iso.org/home.html, and enter the ISO reference designation, such as ISO/TS 22002-1, in the space provided next to the search icon.

CHAPTER 1

Food Safety Standards and Schemes

Introduction

This chapter provides an overview of voluntary food safety and food quality schemes and standards applicable among food business operators (FBOs).[1] The chapter includes a variety of private and government certification programs and standards for food safety management recognized by the Global Food Safety Initiative (GFSI).

The GFSI is a facilitated collaboration between food safety experts in retail, manufacturing, and food service companies, as well as international organizations, governments, academia, and service providers. It provides leadership and guidance on food safety management systems (FSMSs) in the food supply chain. It is managed by the Consumer Goods Forum, a global parity-based food industry network.[2]

GFSI recognition offers a passport to the global market among both recognized certification program owners and the companies they certify. To be recognized by the GFSI, certification program owners must verify that they meet the GFSI benchmarking requirements, one of the most widely accepted benchmarking schemes across food safety programs (GFSI 2018).

The GFSI benchmarking requirements were created in 2001 by a group of retailers motivated by the need to harmonize food safety standards across the global supply chain. The requirements are frequently updated with input from food safety experts around the world to keep up with food safety trends. They do not constitute a food safety standard in their own right, nor can food businesses be audited or certified against them. Recognized certification program owners are relied on to undertake these roles. In fact, the knowledge possessed by FBOs on these schemes and standards is limited. This is partly caused by the large number of FSMS schemes and standards already in the market.

This chapter offers an overview of food safety certification programs and standards as an aid in helping FBOs consider which FSMS scheme may be most relevant to them, their customers, and consumers.

BRC Global Standards

The BRC Global Standards represent a safety and quality certification program used by over 28,000 certificated suppliers in more than 130 countries. Certification is issued through a worldwide network of accredited certification bodies. The BRC Global Standards are a market-leading global brand that helps build confidence in the supply chain. Its standards for food safety, packaging and packaging materials, storage and distribution, consumer products, agents and brokers, and retail set the benchmark for good manufacturing practices (GMPs) and help provide assurance to customers that products are safe, legal, and high quality (table 1.1).

The BRC Global Standard for Food Safety, issue 8, was published in 2018 (BRCGS 2018). It provides a framework for managing product safety, integrity, legality, and quality and for the operational controls of the associated criteria in food and food ingredient manufacturing, processing, and packing.

This BRC Global Standard focuses on the following:

- Encouraging the development of a product safety culture
- Expanding the requirements for environmental monitoring to reflect the increasing importance of this technique
- Encouraging sites to develop systems for security and food defense
- Adding clarity to the requirements for high-risk, high-care, and ambient high-care production zones
- Providing greater clarity for sites manufacturing pet foods
- Ensuring global applicability and benchmarking for the GFSI

Table 1.1 BRC Global Standards and the GFSI

Standard	GFSI benchmarked?	GFSI scope
Food safety, version 8	Yes	BII, CO, CI, CII, CIII, CIV, K
Packaging and packaging materials, version 5	Yes	I
Storage and distribution, version 3	Yes	G
Agents and brokers, version 2	Yes	FII
Consumer products, version 4	No	
Retail, version 1	No	
Ethical trade and responsible sourcing, version 1	No	
Gluten-free certification program	No	
Plant-based global standard	No	

Sources: BRCGS 2015, 2016a, 2016b, 2016c, 2016d, 2017, 2018, 2019.

Note: The BRC Global Standard for Food Safety, issue 8, is recognized by the GFSI. The letters in the right column refer to GFSI scopes of recognition, as follows: BII = farming of grains and pulses; CO = animal primary conversion; CI = processing of perishable animal products; CII = processing of perishable plant products; CIII = processing of perishable animal and plant products (mixed products); CIV = processing of ambient stable animal and plant products (mixed products); FII = food broker/agent; G = provision of storage and distribution services; I = production of food packaging; K = production of (bio)chemicals and biocultures used as food ingredients.

The BRC Global Standard for Food Safety is divided into nine sections, as follows.

Senior management commitment. Commitment at a senior level is essential in the development of a good food safety culture and is therefore necessary to ensure the effectiveness, application, and ongoing development of food safety systems.

The food safety plan: hazard analysis critical control point (HACCP). Effective hazard and risk analysis enables companies to identify and manage hazards that may pose a risk to the safety, quality, and integrity of their products. The BRC Global Standard requires the development of an effective HACCP program based on the requirements of the internationally recognized Codex Alimentarius system.

The food safety and quality management system. This ensures that companies implement well-documented, systematic management systems that form the basis for the product and process controls necessary to produce safe products, meet customer expectations, and ensure that staff are well trained.

Site standards. This covers the suitability, cleanliness, and control of sites and includes factory conditions, cleaning, equipment, pest control, foreign body controls, food defense, and site security.

Product control. Establishing product controls, such as allergen management, the prevention of food fraud, and product testing, is important in the reliable delivery of safe, authentic products.

Process control. This ensures that the documented HACCP plan is put into operation every day, together with effective procedures to manufacture products consistently to the correct level of quality.

Personnel. Training, protective clothing, and proper hygiene practices are covered in this section.

High-risk, high-care, and ambient high-care production zones. A specific section of the standard deals with products that are susceptible to potential pathogen contamination and therefore require additional controls to ensure product safety.

The requirements for traded products. This is a voluntary section of the standard for sites that purchase and sell food products that would normally fall within the scope of the standard and are stored at site facilities, but that are not manufactured, additionally processed, or packed at the sites being audited.

For more information, see the BRC Global Standards website, at https://www.brcgs.com/, and the GFSI website, at https://mygfsi.com/.

International Featured Standards

The International Featured Standards were established in 2003 and were eventually expanded. The standards are governed by IFS Management, a legally independent company headquartered in Berlin. The nine standards have been developed for and by stakeholders involved in all parts of the supply chain. They are all process standards that help users implement legal provisions on food and product safety by providing uniform guidelines on food and product safety and quality issues. Table 1.2 illustrates relevant standards.

The food standard is a GFSI-recognized standard for auditing food manufacturers. The focus is on food safety and the quality of processes and products. This covers food processing companies and companies that pack loose food products.

The standard applies if products are processed or if there is a hazard of product contamination during primary packing. The standard is important for all food manufacturers, especially those producing private labels, because it includes many requirements related to compliance with customer specifications.

The standard supports production and marketing departments in their efforts at brand safety and quality. The standard has been developed with the full and active involvement of certification bodies, retailers, the food industry, and food service companies.

Table 1.2 International Featured Standards and the GFSI

Standard	GFSI benchmarked?	GFSI scope
Food, version 6.1	Yes	BII, CO, CI, CII, CIII, CIV, K
Logistics, version 2.2	Yes	G
PACsecure, version 1.1	Yes	I
Broker, version 3	Yes	FII
HPC	No	
Wholesale, cash and carry	No	
Global markets food	No	
Global markets HPC	No	
Global markets logistics	No	

Source: IFS Database (International Featured Standards), IFS Management GmbH, Berlin, https://www.ifs-certification.com.

Note: The letters in the right column refer to GFSI scopes of recognition, as follows: BII = farming of grains and pulses; CO = animal primary conversion; CI = processing of perishable animal products; CII = processing of perishable plant products; CIII = processing of perishable animal and plant products (mixed products); CIV = processing of ambient stable animal and plant products (mixed products); FII = food broker/agent; G = provision of storage and distribution services; I = production of food packaging; K = production of (bio)chemicals and biocultures used as food ingredients. HPC = household and personal care. PAC = packaging.

The standard is used to audit food manufacturers on food safety and the quality of processes and products. The list of requirements is organized by the following topics:

- Senior management responsibility
- The quality and food safety management system
- Resource management
- The planning and production process
- Measurements, analysis, and improvements
- Food defense

For more information, see the International Featured Standards website, at https://www.ifs-certification.com/, and the GFSI website, at https://mygfsi.com/.

Food Safety System Certification 22000

Food Safety System Certification (FSSC) 22000 is a certification program for FSMSs and feed safety management systems that are in compliance with the publicly available FSMSs standard International Organization for Standardization (ISO) 22000 (requirements for any organization in the food chain), sector-specific technical specifications (TSs) (BSI Group Publicly Available Specification, ISO/TS, Royal Netherlands Standardization Institute–Netherlands Technical Agreement) for food safety prerequisite programs (PRPs), and additional scheme requirements (figure 1.1, table 1.3). Besides the three components shown in figure 1.1, there's a voluntary FSSC 22000 quality module based on all requirements of ISO 9001. FSSC 22000 published version 5 of its scheme in May 2019.

FSSC 22000 is used to audit and certify the FSMSs of food chain organizations in farming animals for milk, meat, eggs, and honey; farming fish and seafood; processing perishable animal products, such as meat, poultry, eggs, dairy, and fish products; processing perishable plant products, such as fresh fruits and fresh juices, preserved fruits, fresh vegetables, grains, nuts, and pulses; processing perishable animal and plant products (mixed products); processing ambient stable products with a long shelf life at ambient temperature, such as

Figure 1.1 The FSSC 22000 Scheme: Required Components

ISO 22000	ISO 22000 provides a common framework across the supply chain for managing requirements and internal and external communication and for continually improving the system
PRPs	Sector-specific PRPs (ISO/TS and BSI Group Publicly Available Specification)
FSSC 22000	FSSC 22000 adds specific requirements to ensure consistency and integrity and to provide scheme governance and management

Table 1.3 FSSC 22000 Standards and the GFSI

FSSC scope	GFSI benchmarked?	GFSI scope
Food manufacturing (ISO 22000:2018, ISO/TS 22002–1:2009, FSSC 22000 additional requirements: Part II 2.1.4)	Yes	BII, C0, CI, CII, CIII, CIV, DI, K
Food packaging manufacturing (ISO 22000:2018, ISO/TS 22002–4:2013, FSSC 22000 additional requirements: Part II 2.1.4)	Yes	I
Transport and storage (ISO 22000:2018, NEN NTA 8059:2016, FSSC 22000 additional requirements: Part II 2.1.4)	Yes	G
Farming (ISO 22000:2018, ISO/TS 22002–3:2011, FSSC 22000 additional requirements: Part II 2.1.4)	No	
Animal feed production (ISO 22000:2018, ISO/TS 22002–6:2016, FSSC 22000 additional requirements: Part II 2.1.4)	No	
Catering (ISO 22000:2018, ISO/TS 22002–2:2013, FSSC 22000 additional requirements: Part II 2.1.4)	No	
Retail and wholesale (ISO 22000:2018, BSI PAS 221:2013, FSSC 22000 additional requirements: Part II 2.1.4)	No	
Biochemicals	No	
Quality management system (ISO 9001)	No	

Source: FSSC 22000 website, at https://www.fssc22000.com/

Note: The letters in the right column refer to GFSI scopes of recognition, as follows: BII = farming of grains and pulses; C0 = animal primary conversion; CI = processing of perishable animal products; CII = processing of perishable plant products; CIII = processing of perishable animal and plant products (mixed products); CIV = processing of ambient stable animal and plant products (mixed products); DI = production of feed; G = provision of storage and distribution services; I = production of food packaging; K = production of (bio)chemicals and biocultures used as food ingredients. BSI PAS = BSI Group Publicly Available Specification. NEN NTA = Royal Netherlands Standardization Institute–Netherlands Technical Agreement.

canned products, biscuits, snacks, oil, drinking water, beverages, pasta, flour, sugar, and salt; manufacturing (bio)chemical food ingredients, such as vitamin supplements, additives, and biocultures, but excluding pesticides, drugs, fertilizers, and cleaning agents; production of feed and pet food; production of food and feed packaging and packaging materials with direct or indirect contact with food; catering; retail and wholesale; and food transport and storage services.

For more information, see the FSSC 22000 website, at https://www.fssc22000.com/, and the GFSI website, at https://mygfsi.com/.

The PrimusGFS Standards

PrimusGFS is a GFSI-recognized food safety audit scheme with certifications in 20 countries. It is endorsed by more than 7,000 organizations. PrimusGFS is mainly focused on food safety among agricultural products designated for human consumption in the fresh state or after minimum processing. PrimusGFS establishes a series of requirements for managing production, handling, processing, and storage operations to ensure consumer safety (table 1.4). PrimusGFS audits consist of several modules (table 1.5). The applicability of the modules depends on the type of operation being audited. PrimusGFS audits cover, for example, FSMSs, good agricultural practices (GAPs), GMPs, HACCP, and preventive control.

Table 1.4 PrimusGFS Standards and the GFSI

Standard	GFSI benchmarked?	GFSI scope
PrimusGFS, version 3	Yes	BI, BII, BIII, CII, CIII, CIV, G

Source: PrimusGFS website, at http://www.primusgfs.com/.

Note: The letters in the right column refer to GFSI scopes of recognition, as follows: BI = farming of plants (other than grains and pulses); BII = farming of grains and pulses; BIII = primary conversion of plant products; CII = processing of perishable plant products; CIII = processing of perishable animal and plant products (mixed products); CIV = processing of ambient stable animal and plant products (mixed products); G = provision of storage and distribution services.

Table 1.5 PrimusGFS Audit Modules

Module	Operation	Applicability
1	FSMS	All operations
2	Farm	Farm operations
3	Indoor aquaculture	Indoor agricultural operations
4	Harvest crew	Harvest crew operations
5	Facility	All facility operations
6	HACCP	All facility operations
7	Preventive control	Optional, all facility operations

Source: PrimusGFS website, at http://www.primusgfs.com/.

For more information, see the PrimusGFS website, at http://www.primusgfs.com/, and the GFSI website, at https://mygfsi.com/.

The Global Red Meat Standard

The Global Red Meat Standard is a scheme specifically developed for the red meat industry. The standard sets out the requirements for all processes relating to the production of meat and meat products. It focuses on areas critical to achieving the greatest safety and the highest quality. The goal of the standard is to deliver transparency in animal welfare, quality, food safety, and hygiene in factories that slaughter, cut, debone, process, and handle meat and meat products derived from pork, beef, lamb or sheep, goats, and horses.

The standard was developed by the Danish Agriculture and Food Council, in partnership with the council's abattoir members and the Danish Meat Research Institute.[3] It was launched in 2006. The Global Red Meat Standard, version 6, was published in 2018 and benchmarked by the GFSI (table 1.6).

Table 1.6 The Global Red Meat Standard and the GFSI

Standard	GFSI benchmarked?	GFSI scope
Global Red Meat Standard, version 6	Yes	C0, CI

Source: Global Red Meat Standard website, at https://grms.org/.

Note: The letters in the right column refer to GFSI scopes of recognition, as follows: C0 = animal primary conversion; CI = processing of perishable animal products.

The Global Red Meat Standard covers the following processes and products: (1) processes: transport, lairage, slaughtering, evisceration, chilling, cutting, deboning, curing, marinating, mincing, mixing, fermentation, smoking, cooking, packing, chilling, freezing, and storage; (2) products: fresh meat, meat products, meat preparations, mixed products, and edible by-products.

For more information, see the Global Red Meat Standard website, at https://grms.org/, and the GFSI website, at https://mygfsi.com/.

CanadaGAP

CanadaGAP is an onfarm food safety program for companies that produce and handle fruits and vegetables. It is designed to help implement and maintain effective food safety procedures within fresh produce operations. The program has received full Canadian government recognition.

The CanadaGAP global agricultural standards and the associated certification program were developed by the Canadian Horticultural Council, the national industry association for fruit and vegetable producers in Canada, as a means of standardizing and updating onfarm food safety programs.[4] It covers eight crop groupings, revolving around the safe production, storage, and packing of fresh produce. The council participates in the Canadian federal On-Farm Food Safety Recognition Program, which involves comprehensive reviews by provincial and federal governments to ensure the technical soundness of the CanadaGAP standards.[5] The owner of the Scheme is CanAgPlus, a Canadian not-for-profit corporation.

Two manuals have been developed by the horticultural industry and reviewed for technical soundness by Canadian government officials. The first is specific to greenhouse operations (CanadaGAP 2018a). The second is aimed at other fruit and vegetable operations (CanadaGAP 2018b). The manuals are designed for companies implementing GAPs in their production, packing, and storage operations and for repackers and wholesalers implementing GMPs and HACCP programs. The program is also designed for fresh produce brokers implementing best practices in supplier management and product traceability. The following are among the topics covered in the manuals: commodity starter products; premises; commercial fertilizers, pulp sludge, and soil amendments; manure, compost, compost tea, and other products; mulch and row cover materials; agricultural chemicals; agricultural water; equipment; cleaning and maintenance materials; waste management; personnel hygiene facilities; employee training; visitor policy; pest program for buildings; water (for fluming and cleaning); ice; packaging materials; growing and harvesting; sorting, grading, packing, repacking, storing, and brokerage; storage of product; transportation; identification and traceability; deviations and crisis management; and HACCP plan and food safety program maintenance and review. The manuals are based on a rigorous hazard analysis applying the seven principles of the internationally recognized HACCP approach.

Table 1.7 The CanadaGAP Standards and the GFSI

Standard	GFSI benchmarked?	GFSI scope
CanadaGAP, version 7.1, options B, C, and D	Yes	BI, BIII

Source: CanadaGAP website, at https://www.canadagap.ca/.

Note: The letters in the right column refer to GFSI scopes of recognition, as follows: BI = farming of plants (other than grains and pulses); BIII = primary conversion of plant products.

For more information, see the CanadaGAP website, at https://www.canadagap.ca/, and the GFSI website, at https://mygfsi.com/.

GLOBALG.A.P.

GLOBALG.A.P., formerly known as EurepGAP, was launched in 1997 as a retailer initiative rooted in the Euro-Retailer Produce Working Group. Its starting point was an effort to develop standards and procedures for the development of GAPs in conventional agriculture especially by highlighting the importance of integrated crop management and a responsible approach to worker welfare. Over the next 10 years, the initiative spread throughout Europe and beyond. Driven by the impacts of globalization, a growing number of producers and retailers around the globe joined in, gaining the European organization global significance. To reflect both its global reach and its goal of becoming the leading international GAP standard, the name of the organization was changed from EurepGAP to GLOBALG.A.P. in 2007. The scheme is managed by FoodPLUS GmbH, Cologne, Germany.

GLOBALG.A.P. is a private sector body that sets voluntary standards for agricultural product certification around the world. The GLOBALG.A.P. standard is designed to reassure consumers about how their food is produced on the farm. Focal points include food safety and traceability; biodiversity, minimizing the detrimental environmental impacts of farming operations, and reducing the use of chemical inputs; and ensuring a responsible approach to worker health, safety, and welfare and animal welfare. The organization aims to establish one standard for GAPs, with varied product applications capable of interfacing seamlessly with the whole pattern of global agriculture. This includes integrated crop management, integrated pest control, the quality management system, and HACCP.

GLOBALG.A.P. is a prefarmgate standard. The certificate covers the process of generating certified products from farm inputs, such as feed or seedlings, including all related farming activities until the product leaves the farm.

The GLOBALG.A.P. Integrated Farm Assurance Standard covers GAPs in agriculture, aquaculture, livestock, and horticulture production. It also covers additional aspects of the food production and supply chain, such as the chain of custody and compound feed manufacturing. The standard is built on a system of modules that enable producers to obtain certification for several subscopes in one audit. The system consists of the following:

- *General regulations:* These map out the criteria for successful implementation of control points and compliance criteria and set guidelines for the verification and the regulation of the standard.
- *Control points and compliance criteria:* These clearly define the requirements for achieving the quality standard required by GLOBALG.A.P.

The control points and compliance criteria are based on modules consisting of the following:

- The all farm base module is the foundation of all standards. It consists of all the requirements that all producers must fulfill to gain certification.
- The scope module defines clear criteria in the various food production sectors. GLOBALG.A.P. covers three scopes: crops, livestock, and aquaculture.
- The subscope module includes control points and compliance criteria that cover all the requirements for a particular product or the various aspects of the food production and supply chain.

These modules are illustrated in figure 1.2.

Figure 1.2 The Required Components of the Integrated Farm Assurance Standard

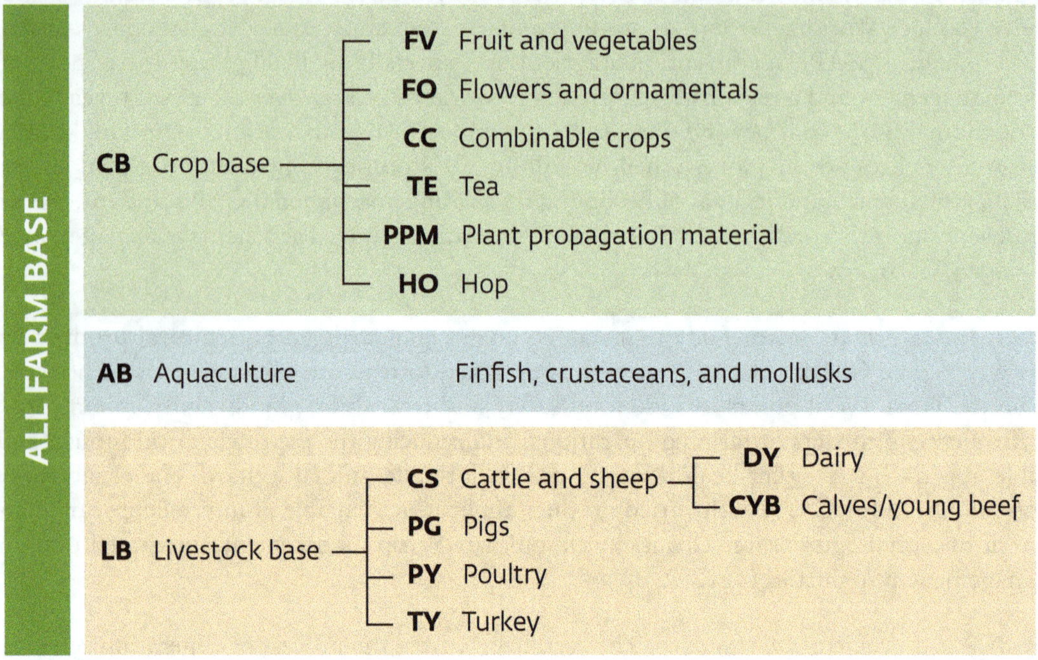

The GLOBALG.A.P. has been GFSI benchmarked (table 1.8).

Table 1.8 The GLOBALG.A.P. Integrated Farm Assurance Standard and the GFSI

Standard	GFSI benchmarked?	GFSI scope
Integrated farm assurance, version 5.2, aquaculture	Yes	AII
Integrated farm assurance, version 5.2, fruits and vegetables	Yes	BI, BIII
Harmonized Produce Safety Standard, version 1.1–2	Yes	BI

Source: GLOBALG.A.P. website, at https://www.globalgap.org/uk_en/.

Note: The letters in the right column refer to GFSI scopes of recognition, as follows: AII = farming of fish and seafood; BI = farming of plants (other than grains and pulses); BIII = primary conversion of plant products.

For more information, see the GLOBALG.A.P. website, at https://www.globalgap.org/uk_en/, and the GFSI website, at https://mygfsi.com/.

The Seafood Processing Standard of the Global Aquaculture Alliance

The Global Aquaculture Alliance is an international, not-for-profit trade association dedicated to advancing environmentally and socially responsible aquaculture. The alliance was established in 1997 and had 59 members in the Americas, Asia, and Europe. It has since grown to 1,100 members in 70 countries, making it the highest profile industrial organization in the global aquaculture business. The scheme is managed by the Global Aquaculture Alliance in the United States.

The alliance develops best aquaculture practice certification standards. The standards cover a full range of aquaculture facilities, from hatchery and feed mills to farm and processing plants, producing shrimp, salmon, tilapia, channel catfish, and the pangasius or basa fish. A specific standard is available for each facility type and category. Additional standards have recently been developed.

The guiding principles underlying the best aquaculture practices aim to assure the environmental, economic, and social sustainability of aquaculture operations for the benefit of local economies and communities by minimizing the environmental effects, promoting the rational use of fresh water, avoiding disease outbreaks, and minimizing risks related to the introduction of exotic species. The standards cover requirements in regulatory management, the quality management system, personnel management, the environment and waste management, food safety management, verification, and traceability.

The Global Aquaculture Alliance standard in seafood processing has been GFSI benchmarked (table 1.9).

Table 1.9 The Global Aquaculture Alliance Seafood Processing Standard and the GFSI

Standard	GFSI benchmarked?	GFSI scope
Seafood processing standard, issue 5.0	Yes	CI

Source: GAA 2019.

Note: The letters in the right column refer to GFSI scopes of recognition, as follows: CI = processing of perishable animal products.

For more information, see the Global Aquaculture Alliance website, at https://www.aquaculturealliance.org/, and the GFSI website, at https://mygfsi.com/.

The Safe Quality Food Institute Standard

The Safe Quality Food Code is a process and product certification standard. It is supported by an HACCP-based food safety and quality management system that relies on the HACCP principles and guidelines of the U.S. National Advisory Committee on Microbiological Criteria for Foods and the Codex Alimentarius Commission (CAC). The scheme is managed by the Safe Quality Food Institute in Arlington, Virginia. The code was developed and pilot programs were implemented in 1994 to ensure applicability to the food industry. The safe quality food 2000 code is relevant in manufacturing, processing, and distribution.

The following Safe Quality Food Institute programs have been established:

The *Fundamentals Program* offers solutions for small and medium food suppliers who do not have a robust food safety management program in place or who want to take an existing program to the next level. Built as a stepwise approach, the Fundamentals Program is designed to help suppliers integrate robust food safety standards into their existing practices, while creating a pathway to achieve globally accepted GFSI certification.

The *Food Safety Program* family of codes are globally accepted GFSI-benchmarked food safety standards. The food safety codes provide sites with an HACCP-based approach to ensure that products meet most regulations. Businesses looking to satisfy the GFSI certification requirements of their retailers and buyers can rely on the Food Safety Program. The Food Safety Program includes safe quality food codes for food safety in primary production; food manufacturing, storage, and distribution; the manufacture of food packaging; and food retail.

The *Quality Program* is aimed at those sites desiring to do more than guarantee food safety. It is designed for monitoring and controlling threats to food quality. It is most suited for sites that have already implemented a successful, robust safe quality food safety plan. The Quality Program can also be implemented in tandem with the Food Safety Program.

The *Ethical Sourcing Program* is an environmental, social, health, and safety management system for the food industry. Developed by professionals with extensive experience in environmental and social compliance, the program assists facilities in documenting and demonstrating commitment to ethical sourcing in daily operations.

The Safe Quality Food Code has been GFSI benchmarked (table 1.10).

Table 1.10 Safe Quality Food Code and the GFSI

Standard	GFSI benchmarked?	GFSI scope
Primary production, edition 8.1	Yes	AI, BI
Manufacturing, edition 8.1	Yes	BIII, C0, CI, CII, CIII, CIV, DI, K
Storage and distribution, edition 8.1	Yes	G
Manufacture of food packaging, edition 8.1	Yes	I
Food retail, edition 8.1	No	
Food service, edition 8.1	No	
Quality code, edition 8.1	No	
Ethical sourcing, edition 2.1	No	
Fundamental factors for social responsibility, edition 1	No	

Sources: SQFI 2017, 2019a, 2019b, 2019c, 2019d, 2019e, 2019f, 2019g, 2019h.

Note: The letters in the right column refer to GFSI scopes of recognition, as follows: AI = farming of animals; BI = farming of plants (other than grains and pulses); BIII = primary conversion of plant products; C0 = animal primary conversion; CI = processing of perishable animal products; CII = processing of perishable plant products; CIII = processing of perishable animal and plant products (mixed products); CIV = processing of ambient stable animal and plant products (mixed products); DI = production of feed; G = provision of storage and distribution services; I = production of food packaging; K = production of (bio)chemicals and biocultures used as food ingredients.

For more information, see the Safe Quality Food Institute website, at https://www.sqfi.com/, and the GFSI website, at https://mygfsi.com/.

The Japan Food Safety Management Association

The Japan Food Safety Management Association, an incorporated foundation formed under Japanese law, was established in 2016. It runs the Japan Food Safety Certification Scheme, a certification program for FSMSs. The scheme is an internationally harmonized certification program in line with ISO–International Electrotechnical Commission 17011:2017, ISO–International Electrotechnical Commission 17021–1:2015, and ISO/TS 22003:2013. The related standards consist of GMPs, the HACCP system, and the FSMS. The scheme is a GFSI-recognized certification program (table 1.11). The Japan food safety standards cover the certification scopes of food processing and the production of (bio)chemicals. The Japan Food Safety Management Association is now working to expand the sectors to cover the entire food supply chain.

Table 1.11 The Japan Food Safety Certification Scheme and the GFSI

Standard	GFSI benchmarked?	GFSI scope
Certification scheme standard, version 2.3	Yes	CIV

Source: JFSM 2018.

Note: The letters in the right column refer to GFSI scopes of recognition, as follows: CIV = processing of ambient stable animal and plant products (mixed products).

In addition to the Food Safety Certification Scheme, the association has launched the Japan Food Safety–A/B Program. The related standards include stepwise processes for FBOs, including small and medium enterprises, to improve FSMSs effectively and efficiently. Making good use of the Japan Food Safety–A/B Program and its standards, FBOs can start with the implementation of a GMP, step up to an HACCP system, and reach an FSMS.

For more information, see the Japan Food Safety Management Association website, at https://www.jfsm.or.jp/eng/, and the GFSI website, at https://mygfsi.com/.

ASIAGAP and Japan GAP

ASIAGAP and Japan GAP represent a set of certification standards created by the Japan GAP Foundation.

Japan GAP includes the control points and compliance criteria needed to create a standard GAP in Japan. It is an agricultural management method for monitoring both food and occupational safety and the environment through control points and compliance criteria. It is the standard GAP in Japan and has already gained the support of many stakeholders. It operates with strict regard for human rights.

ASIAGAP is based on the Japan GAP standards, but encompasses additional requirements, such as HACCP-based risk management, the prevention of food fraud, and so on. ASIAGAP is appropriate as an international GAP standard.

ASIAGAP is a GFSI-recognized certification program (table 1.12).

Table 1.12 The ASIAGAP Standards and the GFSI

ASIAGAP control points and compliance criteria for farms	GFSI benchmarked?	GFSI scope
Subscope fruits and vegetables, version 2	Yes	BI, BIII
Subscope tea, version 2	Yes	BI, BIII
Subscope grains, version 2	Yes	BII, BIII

Source: Japan GAP Foundation website, at https://jgap.asia/en/home-2/.

Note: The letters in the right column refer to GFSI scopes of recognition, as follows: BI = farming of plants (other than grains and pulses); BII = farming of grains and pulses; BIII = primary conversion of plant products.

Agricultural products

In ASIAGAP and Japan GAP, over 120 checkpoints are used to evaluate control criteria, from seeding to harvesting. Cultivation records are kept to document when, where, and how products are grown. Key elements include the following:

- *Soil:* Checking the safety of soil, preventing soil runoff, and making soil sustainable for land use.
- *Water:* Investigating water sources and reservoirs for any harmful industrial waste. The quality and hygiene of washing water reused for agricultural production are also checked.
- *Fertilizer:* Affirming that fertilizer will not harm agricultural products by understanding raw materials, production processes, and inspection results.
- *Pesticide:* Using integrated pest management to consider all available control technologies, including pesticides and herbicides to control pests and weeds. Ensuring the mandatory proper use of pesticides and inspecting for pesticide residues.
- *Sanitation:* Setting rules governing health conditions and clothing among workers and ensuring agricultural products are not damaged at facilities or by machinery or equipment.
- *Radioactive substances:* Identifying and controlling the presence of radioactive substances in the soil, water, fertilizer, compost, and so on used for agricultural cultivation.

Livestock and livestock products

There are 31 items and 113 control criteria in farm management. Key elements include the following:

- *Feed safety:* Investigating livestock feed for harmful components (mold poison, pathogenic microorganisms, and so on). The safety of self-supplied feed is ensured by the inclusion of additional criteria regulating pesticides and fertilizers.
- *Medicine:* Taking measures to prevent contamination from veterinary medicines (antibacterial substances and so on) or injection needles. Antimicrobial substances should be treated carefully.
- *Livestock health:* Working with veterinarians to monitor livestock health and prevent epidemics of infectious disease. This includes complying with animal health control criteria based on the Livestock Infectious Disease Prevention Law.
- *Excrement:* Ensuring farmers meet standards for handling excrement in consideration of the surrounding environment and local residents. Excrement should be used as compost within the community.
- *Animal welfare:* Improving livestock management using checklists based on the international covenants of the World Organisation for Animal Health regarding animal welfare.
- *Radioactive substances:* Confirming the safety of livestock and livestock products against radioactive substances before shipping.

For more information, see the Japan GAP Foundation website, at https://jgap.asia/en/home-2/, and the GFSI website, at https://mygfsi.com/.

Other GFSI-benchmarked standards and certification programs

In addition to the benchmarking and recognition of private certification programs, GFSI has introduced a new category, technical equivalence, which is dedicated to government-owned schemes. To take into account the different structures of these schemes, the new category allows for the acknowledgment of a scheme's equivalence to the relevant technical requirements of GFSI benchmarking. The category is distinguished from GFSI recognition of private certification programs, which also assesses a scheme's governance and operational management components.

The following standards or certification programs have been acknowledged for technical equivalence with the GFSI technical requirements.

China HACCP

The China HACCP is the national certification scheme implemented by the Certification and Accreditation Administration of China. Certification with the China HACCP benefits foreign food producers by bringing them into compliance with Chinese food import regulations and national standards.

The Certification and Accreditation Administration introduced the HACCP system through Announcement 3 in 2002. Since then, the HACCP certification scheme has been updated several times. The latest version of the implementation rules is CNCA-N-008: 2011, "Implementation Rules for Hazard Analysis and Critical Control Point (HACCP) System Certification," and the reference standards are GB / T 27341, "Hazard Analysis and Critical Control Point System: General Requirements for Food Processing Plant" and GB 14881, "General Hygiene Regulation for Food Enterprises Standard."

For more information, see the website of the Certification and Accreditation Administration of China, at http://www.cnca.gov.cn/ (in Chinese), and the GFSI website, at https://mygfsi.com/.

The U.S. Department of Agriculture's Agricultural Marketing Service GAP+

The Harmonized GAP+ Certification Program of the U.S. Department of Agriculture (USDA) Agricultural Marketing Service has achieved mutual technical recognition against version 7.1 of the GFSI benchmarking requirements (GFSI 2017).

GAPs and good handling practices are voluntary audits implemented to verify that fruits and vegetables are produced, packed, handled, and stored as safely as possible to minimize the risk of microbial food safety hazards. The audits verify adherence to the recommendations in the USDA "Guide to Minimize Microbial Food Safety Hazards of Fresh-Cut Fruits and Vegetables" and industry recognized food safety practices (USDA 2008).

For more information, see the USDA Agricultural Marketing Service website, at https://www.ams.usda.gov/, and the GFSI website, at https://mygfsi.com/.

Canadian Grain Commission Standards

The Canadian Grain Commission has achieved technical equivalence for the Canadian Grain Commission HACCP and the Canadian Identity Preserved Recognition System plus HACCP certification programs against version 7.1 of the GFSI benchmarking requirements.

For more information, see the Canadian Grain Commission website, at https://www.grainscanada.gc.ca/, and the GFSI website, at https://mygfsi.com/.

The Global Markets Program

The GFSI Global Markets Program represents a useful resource for small or less highly developed businesses that may frequently face difficulties in accessing market opportunities because they lack the expertise or the wherewithal to meet the food safety requirements of the formal supply chain. The program represents a comprehensive step-by-step tool that guides small or less well developed businesses through a continuous improvement process in their FSMSs. This program is voluntary and helps businesses and other interested parties follow a four-phase approach with the ultimate goal of gaining certification with one of the GFSI-recognized schemes.[6]

Notes

1. No opinion is offered here on which certification program or standard a particular FBO should select.
2. See CGF (Consumer Goods Forum), Levallois-Perret, France, https://www.theconsumergoodsforum.com/who-we-are/overview/.
3. The website of the Danish Agriculture and Food Council is at https://agricultureandfood.dk/.
4. The Canadian Horticultural Council website is at https://www.hortcouncil.ca/en/.
5. See Food Safety Recognition Program, Canadian Food Inspection Agency, Ottawa, https://www.inspection.gc.ca/food-safety-for-industry/archived-food-guidance/safe-food-production-systems/food-safety-enhancement-program/recognition-program/eng/1299860970026/1299861042890.
6. See "Global Markets: A Pathway to Certification," Global Food Safety Initiative, Consumer Goods Forum, Levallois-Perret, France, https://mygfsi.com/how-to-implement/global-markets/.

References

BRCGS (BRC Global Standards). 2015. *Global Standard: Packaging and Packaging Materials*. BRC Global Standards, issue 5 (July). London: BRC Global Standards.

———. 2016a. *Global Standard: Consumer Products, General Merchandise*. BRC Global Standards, issue 4 (November). London: BRC Global Standards.

———. 2016b. *Global Standard: Consumer Products, Personal Care and Household*. BRC Global Standards, issue 4 (November). London: BRC Global Standards.

———. 2016c. *Global Standard: Retail*. BRC Global Standards, issue 1 (September). London: BRC Global Standards.

———. 2016d. *Global Standard: Storage and Distribution*. BRC Global Standards, issue 3 (August). London: BRC Global Standards.

———. 2017. *Global Standard: Agents and Brokers*. BRC Global Standards, issue 2 (August). London: BRC Global Standards.

———. 2018. *Global Standard: Food Safety*. BRC Global Standards, issue 8 (August). London: BRC Global Standards.

———. 2019. *Global Standard: Ethical Trade and Responsible Sourcing*. BRC Global Standards, issue 1 (August). London: BRC Global Standards.

CanadaGAP. 2018a. *CanadaGAP Food Safety Manual for Greenhouse Product*. Version 7.1. Ottawa: CanadaGAP Program.

———. 2018b. *CanadaGAP Food Safety Manual for Fresh Fruits and Vegetables*. Version 7.1. Ottawa: CanadaGAP Program.

GAA (Global Aquaculture Alliance). 2019. "Seafood Processing Standard (SPS)." Issue 5.0 (February 1), GAA, Portsmouth, NH.

GFSI (Global Food Safety Initiative). 2017. "Version 7.1 of GFSI's Benchmarking Requirements Furthering Harmonisation." April 26, GFSI, Consumer Goods Forum, Levallois-Perret, France. https://mygfsi.com/press_releases/version-7-1-of-gfsis-benchmarking-requirements-furthering-harmonisation/.

———. 2018. "GFSI Benchmarking Requirements: Part I, the Benchmarking Process." GFSI Guidance Document Version 7.2, GFSI, Consumer Goods Forum, Levallois-Perret, France.

JFSM (Japan Food Safety Management Association). 2018. "JFS-C Certification Scheme Document." Version 2.3 (April), JFSM, Tokyo.

SQFI (Safe Quality Food Institute). 2017. *SQFI Ethical Sourcing*. Edition 2.1 (April). Arlington, VA: SQFI.

———. 2019a. "Fundamental Factors for Social Responsibility." Edition 1, SQFI, Arlington, VA.

———. 2019b. "SQF Food Safety Code for Manufacture of Food Packaging." Edition 8.1, SQFI, Arlington, VA.

———. 2019c. "SQF Food Safety Code for Manufacturing." Edition 8.1, SQFI, Arlington, VA.

———. 2019d. "SQF Food Safety Code for Primary Production." Edition 8.1, SQFI, Arlington, VA.

———. 2019e. "SQF Food Safety Code for Storage and Distribution." Edition 8.1, SQFI, Arlington, VA.

———. 2019f. "SQF Quality Code." Edition 8.1, SQFI, Arlington, VA.

———. 2019g. "SQF Food Safety Code for Food Retail." Edition 8.1, SQFI, Arlington, VA.

———. 2019h. "SQF Food Safety Code for Foodservice." Edition 8.1, SQFI, Arlington, VA.

USDA (United States Department of Agriculture). 2008. "Guide to Minimize Microbial Food Safety Hazards of Fresh-Cut Fruits and Vegetables." February, Center for Food Safety and Applied Nutrition, USDA, College Park, MD.

CHAPTER 2

Principal Food Safety Regulations

Introduction

The issue of food safety has been addressed by international instruments. One of the most important is the World Trade Organization Agreement on the Application of Sanitary and Phytosanitary Measures. Members of the World Trade Organization are encouraged to base their sanitary and phytosanitary measures on international standards, guidelines, and recommendations. The rules of the World Trade Organization are also applicable to nonmember countries that trade with members.

The food safety standard set forth in the agreement is the Codex Alimentarius, a collection of internationally adopted food standards presented in a uniform manner. The purpose of the codex is to protect the health of consumers, ensure fair practices in the food trade, and promote the harmonization of standards. The Codex Alimentarius Commission (CAC) implements the Joint Food and Agriculture Organization of the United Nations–World Health Organization Food Standards Programme.

This chapter provides an overview of food safety legislation for food business operators (FBOs). Together with various CAC provisions, it addresses relevant regulations of the European Union (EU) and the United States. This has been done because of the importance of these two markets for FBOs throughout the world and the significance of the EU and the United States in developing a regulatory regime for food products that ensures a high level of safety and consumer confidence. The chapter also includes an overview of food safety in relevant regional trade organizations, joint approaches toward food safety legislation, and the production and marketing of food products.

The main food safety regulations

Codex Alimentarius

The Codex Alimentarius is a collection of internationally recognized standards, codes of practice, guidelines, and other recommendations relating to foods, food production, and food safety.[1] Its texts are developed and maintained by the CAC, a body that was established in November 1961 by the Food and Agriculture Organization of the United Nations. The World Health Organization joined in June 1962. Along with standards for various types of food products, the codex contains general standards on animal feed, antimicrobial resistance, contaminants, pesticide residues, nutrition, labeling, and biotechnology. The codex standards and related texts are voluntary. They need to be translated into national legislation or regulations to be enforceable. The codex is set forth in several classes of documents, as follows:

- Product-related *standards* comprising clear definitions of the various food safety issues associated with products
- *Guidelines* that supply general guidance in virtually every aspect of food safety management
- *Codes of practice* that provide general principles of food hygiene practices for a wide range of products and guidelines for the prevention of specific food safety hazards

The Codex Alimentarius is associated with online databases on pesticide residues, veterinary drug residues, and general standards on food additives.[2]

European Union

The EU joined the Codex Alimentarius in 2003 and accepted the obligations established under the codex statutes. The main EU food safety directives and regulations refer to the CAC as the basis for the related requirements.

Among the main EU food regulations is Regulation (EC) No. 178/2002, the General Food Law, which establishes the general principles and requirements of food law and the general concepts of food legislation within the EU and ensures a consistent approach to the development of national food law in EU countries (figure 2.1).[3] It sets out the general principles of EU food law for member states to follow. The main objective is to ensure the free circulation of safe food and feed in the EU for the health and well-being of citizens.

In addition, a package of hygiene regulations was adopted to deliver consistency in the food chain. These include the following:

- Regulation (EC) No. 852/2004 on the hygiene of foodstuffs (general hygiene requirements for food production)
- Regulation (EC) No. 853/2004, which lays down specific hygiene rules for food of animal origin and basic hygiene principles for businesses at all stages of the food chain of animal products

Figure 2.1 The General Food Law of the European Union

Food safety general principles, rules, definitions
Harmonization of EU legislation

	Food hygiene		Control system prior to December 14, 2019		Control system as of December 14, 2019
EU regulation	Regulation (EC) No. 852/2004	Regulation (EC) No. 853/2004	Regulation (EC) No. 882/2004	Regulation (EC) No. 854/2004	Regulation (EC) No. 625/2017
Aims	General rules about food hygiene, FBO responsibility	Specific rules on the hygiene of food of animal origin	Basis for national monitoring and control	Specific rules for the official controls on products of animal origin	Official controls and other official activities performed to ensure the application of food and feed law, rules on animal health and welfare, plant health, and plant protection products
Scope	All stages of production, processing, marketing, and export	Applies to raw and processed foods of animal origin. Does not apply to retailers.	All stages of production, processing, and sale	Only applies in respect of activities and persons to which the Regulation (EC) No. 853/2004 applies	Applies to the entire agrifood chain, with specific details for products of animal origin. Replacing Regulations (EC) No. 882/2004 and 854/2004
General principles	Responsibility of FBOs; transparency of the food chain; flexibility; introduction HACCP	Specific requirements for companies to market products of animal origin; special guarantees for certain types of meat products	Mandatory official control; regularity and proportionality of inspections; inspections as precautions, not punishments	Cooperation with regulatory authorities; risk analysis; specific periods of control depending on the type of product	Mandatory official control; regularity and proportionality of inspections; inspections as precautions, not punishments; cooperation with regulatory authorities; risk analysis

Source: Information in EUR-Lex (database), Publications Office of the European Union, Luxembourg, https://eur-lex.europa.eu/.

- Regulation (EC) No. 625/2017, which sets out specific rules for the organization of official controls on products of animal origin intended for human consumption and on the verification of compliance with feed and food law, animal health, and animal welfare rules, thereby establishing control principles for EU members and third countries

On April 7, 2017, Regulation (EC) No. 625/2017—the Official Controls Regulation—was accepted by the European Parliament and came into force as of April 29, 2017. There are several reasons why Regulation 625/2017 is an important milestone for food safety in the EU and represents the EU response to the U.S. Food Safety Modernization Act (FSMA) of January 4, 2011. First, it simplifies and reduces legal fragmentation. It repeals Regulation No. 854/2004 on official controls on products of animal origin intended for human consumption and Regulation No. 882/2004 on official controls on the verification of compliance with feed and food law, animal health, and animal welfare rules. It also repeals eight more regulations and commission directives and decisions and amends several others. Each segment of the supply chain, such as animal welfare, pesticide residues, controls on products of animal origin, and so on, was previously regulated separately. Now, they are under one legal roof.

Second, the new regulation strengthens the basic principles of previous laws. Although it does not change important principles, such as the transparency of controls and cooperation between member states, it brings more clarity to existing provisions by using more precise language.

Third, it improves the harmonization of procedures and standards. One example is the creation of a single information management system for official controls, which integrates existing systems, such as the Rapid Alert System for Food and Feed, and facilitates the exchange of information among member states.

Fourth, it creates the legal basis for more-sweeping changes in the future. A recurring sentence in the regulation is "the Commission shall adopt delegated acts in accordance with Article 144 to amend this Regulation." The reference is to the article that gives the European Commission (EC) the power to adopt delegated acts. It is an important advance.

Several supportive regulations deal with specific food safety topics, as follows:

- Regulation (EC) No. 2073/2005 on microbiological criteria for foodstuffs
- Regulation (EC) No. 1881/2006 on maximum permitted levels for certain contaminants in foodstuffs
- Regulation (EC) No. 2074/2005 laying down implementing measures for certain products under Regulations
- Regulation (EC) No. 1760/2000 establishing a system for the identification and registration of bovine animals and regarding the labeling of beef and beef products
- Regulation (EU) No. 1169/2011 on the provision of food information to consumers

United States

Because the United States has been a member of CAC since 1963, legislators and responsible agencies there tend to harmonize U.S. food safety laws and regulations with codex requirements. The United States Food Regulatory System consists of numerous statutes, rules, and regulations. This overview focuses on the federal regulation of food safety. However, state regulatory agencies also play an important role, especially in enforcement. In particular, state regulatory agencies are primarily responsible for food sanitation and safe food handling by food retailers, food service providers, and food vending operations.

The main U.S. food safety statutes are listed below.

The FSMA empowers the U.S. Food and Drug Administration (FDA) to implement a science-based system to address food safety hazards and shifts the focus to preventing food contamination rather than only reacting to food contamination.[4] The act covers FDA-regulated foods, including all domestic and imported food products, except meat, poultry, and egg products, which are regulated by the U.S. Department of Agriculture (USDA).

The Federal Food, Drug, and Cosmetic Act of 1938, with amendments, is a set of laws giving authority to the FDA to oversee the safety and efficacy of FDA-regulated food, drugs, and cosmetics.

The Federal Meat Inspection Act of 1906, with amendments, was passed to prevent adulterated or misbranded meat and meat products from being sold as food and to ensure that meat and meat products are slaughtered and processed under sanitary conditions.[5] This bedrock legislation also regulates inspections of imported meat products to ensure that they meet U.S. food safety standards.

The Poultry Products Inspection Act of 1957, as amended, regulates the processing and distribution of poultry products and requires certain sanitary standards and practices, as well as labeling and container standards, to prevent the sale of adulterated or misbranded poultry products. The USDA is responsible for enforcement. It provides inspections on all poultry products sold in interstate commerce and reinspects imported products.

The FDA shares responsibility for egg product safety with the USDA. According to the Egg Products Inspection Act of 1970, as amended, the USDA is responsible for the safety of liquid, frozen, and dried egg products, domestic and imported, and for the safe use or disposition of damaged or dirty eggs.

The Federal Insecticide, Fungicide, and Rodenticide Act of 1947, as amended, provides for federal regulation of pesticide distribution, sale, and use.[6] All pesticides distributed or sold in the United States must be registered (licensed) by the U.S. Environmental Protection Agency.

The Public Health Security and Bioterrorism Preparedness and Response Act of 2002 (Bioterrorism Act) requires the registration of food facilities, the establishment and maintenance of records, and prior notice of the importation of food. Every even year between October and December, each registered facility must reregister with the FDA. The Bioterrorism Act also grants FDA additional enforcement authority. To enforce the statutes related to food safety, regulatory authorities, including the FDA and the USDA, enact rules and regulations that are referred to as administrative law. An example is the Poultry Products Inspection Regulations mentioned above. The Code of Federal Regulations (CFR) is the codification of the general and permanent rules and regulations published in the Federal Register by the executive departments and agencies of the U.S. government.[7]

Additionally, regulatory authorities publish guidance documents and recommendations for both the food industry and consumers. They do not create or confer any rights for or on any person and do not operate to bind the FDA or the public, but they reflect the vision of the FDA on certain issues. For instance, the FDA publishes the Food Code, a model set of guidelines and procedures that assists food control jurisdictions by providing a technical and legal basis for regulating the retail and food service industries, including restaurants and grocery stores.[8]

Case law is another source of the U.S. food safety law system. Precedents are rules established in previous legal cases that are binding on or persuasive among justices in deciding subsequent cases involving similar issues or facts. The National Agricultural Law Center has compiled reported and unreported federal and state court decisions on food safety issued since January 1, 1995.[9]

Australia, China, and Japan

Whereas the main focus of this chapter is food safety regulations affecting food exports to the EU and the United States, the legal regimes in other countries are also informative. A good source of information on exporting food to various countries is the USDA Global Agricultural Information Network.[10] The network compiles useful details on relevant laws and regulations in, for example, Australia, China, and Japan (USDA 2018a, 2018b, 2019).

Requirements for FBOs

General principles, including FBO responsibilities

CODEX ALIMENTARIUS

To protect consumers against unsafe food and ensure that all stages of the food life cycle are safe and do not pose a threat to consumer health, the CAC developed the General Principles of Food Hygiene (CAC 2003). The principles follow the food chain from primary production to final consumption, highlighting the key hygiene controls at each stage and offering recommendations on establishments, personal hygiene, transportation, and the application of an approach based on the hazard analysis critical control point (HACCP) system.

EUROPEAN UNION

General principles of EU food legislation

The basis of EU food legislation is an integrated farm-to-fork approach, combined with risk analysis in relation to food, precautionary principles, the protection of consumer interests, principles of transparency, and the primary legal responsibility of the FBO to ensure food safety.[11]

The *farm-to-fork approach* is the general principle driving European food safety legislation. It aims to cover all potential hazards along the entire food chain, including primary production, processing, transportation and distribution, retail, catering, food service, and home use of food.[12]

The *equivalency principle*: food and feed imported into EU markets must possess food safety characteristics equivalent to food produced in EU member states. In cases in which there may be an agreement between a non-EU country and an EU member state, the food must comply with the provisions in the agreement.

Risk analysis assumes that all measures relating to food safety will be underpinned by strong scientific evidence.

The *precautionary principle* is relevant in those circumstances where health risks are at an unacceptable level, but the supporting data and information are too sparse for comprehensive risk assessment. In such situations, the measures necessary to ensure high standards of health protection, as determined by the EU community, may be adopted pending further scientific research allowing a more comprehensive risk assessment.

According to the *early warning principle*, FBOs must immediately withdraw unsafe food from the market and inform the authorities and consumers about the problem.

The implementation of a *protection of consumer interests principle* requires the maintenance of a status quo whereby consumers are able to make informed choices about the foods they consume.

At all stages of the food production, processing, and distribution involving their businesses, EU FBOs bear the prime responsibility for ensuring that the food under their control satisfies the food law requirements that are relevant to their activities.

General rules for FBOs on hygienic foodstuffs

Principles amplifying the general rules for FBOs on hygienic foodstuffs have been developed in the EU. The main principles are as follows:

- For food that cannot be stored safely at ambient temperatures, particularly frozen food, a cold chain must be maintained.
- The general implementation of procedures based on HACCP principles, together with the application of good hygienic practice, should reinforce FBO responsibility.
- Guides to best practice are a valuable instrument in aiding FBOs at all levels of the food chain to comply with food hygiene rules and in applying HACCP principles.
- Microbiological criteria and temperature-control requirements based on scientific risk assessments must be established.
- Imported foods must meet at least the same hygienic standards as food produced in the EU.

UNITED STATES

The U.S. food safety system is based on strong, flexible, and science-based state and federal laws and the legal responsibility of the food industry to produce safe foods.[13] The system is guided by the following principles:

- Only safe and wholesome foods may be marketed.
- Regulatory decision making in food safety is science based.
- The government has enforcement responsibility.
- Manufacturers, distributors, importers, and others are expected to comply and are liable if they do not.
- The regulatory process is transparent and accessible to the public.

Science and risk analysis are fundamental to U.S. food safety policy making. Regulatory decisions regarding food safety standards and requirements rely on risk analysis performed by competent authorities who are qualified to make scientifically sound decisions.

U.S. food safety statutes, regulations, and policies reflect the *precautionary approaches* that are embedded within them. One example is the premarket approval requirements established for food additives, animal drugs, and pesticides. These products are not allowed on the market unless and until they have been shown by producers to be safe.

HACCP and the traceability requirement

HACCP

All globally recognized food safety management systems (FSMSs) are built on the HACCP risk-based approach, which includes potential hazards analysis and prevention during the production process (see chapter 3). HACCP can be applied throughout the food chain, from primary production to final consumption. Beyond enhancing food safety, HACCP implementation provides other significant benefits. Practice has shown that FSMSs based on HACCP open up new international markets for high–value added food products. Such FSMSs also increase the efficiency of domestic markets. Most private standards, including the International Featured Standards, the BRC Global Standards, and International Organization for Standardization (ISO) 22000 developed and recognized by big retailers, are based on HACCP. Thus, compliance with HACCP principles has become obligatory among FBOs who work or plan to work with large regional or global retailers.

Codex Alimentarius

The recommendation to implement an HACCP-based approach wherever possible in enhancing food safety is fixed in the General Principles of Food Hygiene (CAC 2003). CAC defines HACCP as "a system that identifies, evaluates and controls hazards that are significant for food safety" (WHO and FAO 2009, 6). It also affirms that "food business operators should control food hazards through the use of systems such as HACCP" (WHO and FAO 2009, 12).

The seven HACCP principles, along with additional guidance, are as follows (CAC 2003; FAO and WHO 2006):

- Conduct a hazard analysis: identify all hazards and the possible degree of their severity; consider the control measures that may be applied to confront each hazard.
- Determine the critical control points (CCPs): these are the steps for the application of controls; they are essential in preventing or eliminating a food safety hazard or reducing the hazard to an acceptable minimum.
- Establish critical limits: these are the boundaries of safety for each CCP; they may be set according to specific preventive measures such as temperature, time, physical dimensions, a_w (water activity), pH (acidic or basic water-based solution), and the available chlorine.
- Establish a system to monitor CCP control: monitoring is the measurement or observation of a CCP relative to the CCP's critical limit; this helps detect loss of control at the CCP.
- Establish the corrective action to be taken if monitoring indicates that a particular CCP is not under control: such corrective action must ensure that the CCP is brought under control; this includes the proper disposition of the affected product.
- Establish validation and verification procedures to confirm that the HACCP system is capable of addressing the issue at hand and working effectively: such procedures may include random sampling and analysis, often performed on behalf of a business by external experts.
- Establish documentation on all procedures and recordkeeping relevant to these principles and their application: this includes hazard analysis, CCP determination or a CCP decision tree, the determination of the critical limits of a CCP, CCP monitoring activities, the correction and corrective action to be taken if a CCP deviates, and validation and verification reports.

There are limitations to applying HACCP principles fully in primary production. If HACCP principles cannot be implemented at the farm level, for instance, specific hygienic practices, good agricultural practices, and good veterinary practices should be followed.

Industry-specific codes of practice in line with the peculiarities of the implementation of an HACCP-based approach have been developed and are recommended by the CAC. Examples are the Code of Practice for Fish and Fishery Products (CAC/RCP 52-2003), the Code of Hygienic Practice for Milk and Milk Products (CAC/RCP 57-2004), and the Code of Hygienic Practice for Meat (CAC/RCP 58-2005).[14]

Useful to know: Small or less well developed businesses often face difficulties in developing and implementing an effective HACCP plan because they lack expertise. In such situations, the CAC recommends relying on the guidance of trade and industry associations, independent experts, and regulatory authorities (FAO and WHO 2006). In any case, attention always needs to be paid to the characteristics of the foods and the processes involved.

European Union

EU Regulation No. 852/2004 requires FBOs to establish and maintain a permanent procedure or procedures based on HACCP principles.[15] FBOs must be able to provide the competent authority with evidence of their compliance with the official norms regarding obligatory HACCP implementation.

EU Regulation No. 852/2004 provides the possibility of flexible or simplified HACCP implementation, particularly in the case of small food businesses and, especially, in the management of all required records. This approach enables the application of HACCP in all circumstances regardless of the size and type of activities undertaken by a specific food business.

Useful to know: In another effort to clarify all aspects of the implementation of HACCP principles, the European Commission (2018) has been developing a guidance document on the implementation of certain provisions of Regulation No. 852/2004.

United States

In the United States, HACCP adherence is mandatory for all producers of foodstuffs because of the FSMA. There are specific rules governing HACCP implementation in three classes of manufacturing that are exempt from the more general rules laid down in the FSMA. The classes of manufacturing with separate sets of rules are meat and poultry, seafood, and juice products.[16]

HACCP implementation in meat and poultry is regulated by the USDA, while, in seafood and juice, it is covered by the FDA. These agencies publish guidance documents explaining the HACCP system in specific areas, along with support documents for HACCP implementation and information on HACCP training activities.

Under the FSMA, certain qualifying facilities are exempt from preventive controls and HACCP provisions.[17] Nonetheless, they must still be able to demonstrate that they (1) have identified potential hazards and are implementing preventive controls to address these or (2) are in compliance with local and state food safety laws.

TRACEABILITY

Food traceability is a recordkeeping tool calibrated to enable specific food items to be followed through all processes until they reach consumers. Traceability has become a legal requirement in most parts of the world. Alone, traceability does not enhance food safety, but it contributes considerably to FSMS efficiency if it is combined with food safety measures, such as those implicit in the HACCP-based approach.

Codex Alimentarius

According to the Principles for Traceability/Product Tracing as a Tool within a Food Inspection and Certification System (CAC/GL 60-2006), the traceability or product tracking tool should be able to identify, at any specified stage of the food chain (from production to distribution), the provenance of the food (one step back) and its destination (one step forward), as appropriate to the objectives of the food inspection and certification system.[18]

The CAC considers traceability a requirement in the case of some food businesses, for example: (1) the Code of Practice for the Prevention and Reduction of Aflatoxin Contamination in Tree Nuts (CAC/RCP 59-2005), (2) the Code of Practice for the Prevention and Reduction of Aflatoxin Contamination in Peanuts (CAC/RCP 55-2004), and (3) Principles and Guidelines for the Conduct of Microbiological Risk Management (CAC/GL 63-2007).

European Union

According to the EU General Food Law, Regulation (EC) No. 178/2002, traceability is the ability to track a food, feed, food-producing animal, or substance intended to be, or expected to be incorporated into a food or feed, through all stages of production, processing, and distribution. Thus, a traceability system should be constructed to ensure an ability to identify any person supplying FBOs or other businesses with a food or feed product. It follows that labeling and identifying products through relevant documentation is an integral component of a traceability system.

In addition to EU food law, specific traceability requirements have been established in EU legislation or regulations on certain categories of food, such as beef, fish, and genetically modified organisms, as in the following examples:

- Regulation (EC) No. 1760/2000, which establishes a system for the identification and registration of bovine animals and for labeling beef and beef products
- Regulation (EC) No. 1420/2013 and Regulation (EU) No. 1379/2013 on the organization of the markets in fishery and aquaculture products
- Regulation (EC) No. 1830/2003 on the traceability and labeling of genetically modified organisms and the traceability of food and feed products produced from genetically modified organisms

Useful to know: The European Commission (2007) has published a factsheet that provides details on the scope and implementation of the traceability requirement.

United States

In the United States, many producers, manufacturers, and retailers must have product tracing systems in place as part of the Public Health Security and Bioterrorism Preparedness and Response Act of 2002. The act requires all producers and manufacturers to be able to trace one step forward and one step back in the supply chain. Retailers need to be able to trace one step back only.

The FSMA directs the FDA to build a system that enhances the FDA's ability to track and trace both domestic and imported foods that are adulterated. In particular, the FDA, along with the USDA and state agencies, is directed to establish pilot projects to explore and evaluate methods to identify recipients of food as a means of preventing or controlling outbreaks of foodborne illnesses. As an aid in tracing products, the FSMA also requires the FDA to establish recordkeeping requirements on high-risk foods delivered to FBOs.

According to "Traceability for Livestock Moving Interstate," a USDA rule published on January 9, 2013, livestock moved interstate, unless specifically exempted, must be officially identified and accompanied by an interstate certificate of veterinary inspection or other relevant documentation. Covered livestock include cattle and bison, horses and other equine species, poultry, sheep and goats, swine, and captive cervids. Additional guidance is available on animal disease traceability.

Food labeling

Food labeling is the primary means of communication linking the producer and the seller of food with the purchaser and consumer of the food. The most important rule of labeling is that the consumer should not be misled.

CODEX ALIMENTARIUS

The CAC has developed various standards and guidelines on the labeling of prepackaged food, food additives, and food for special dietary uses as well as guidelines on related claims about benefits, nutrition labeling, and so on (FAO and WHO 2007). Codex Alimentarius standards and guidelines enable the wide use of food labeling that can be readily understood by government agents, regulatory authorities, food industry representatives, retailers, and consumers.

The core standard is the General Standard for the Labeling of Prepackaged Food, which applies to the labeling of all prepackaged foods sold or catered to consumers and covers certain required features of the label. Thus, it establishes the sort of information that must appear on the label of prepackaged food, such as the name of the food, a list of ingredients, the net contents, the drained weight, the name and address of the final producer or packager, the country of origin, lot identification, date marking, storage instructions, and instructions for use. There may be additional requirements for quantitative ingredient declarations and irradiated food.

The CAC has also issued more specific standards for the labeling of food additives sold as such, the labeling of prepackaged foods for special dietary uses and claims about the associated benefits, the labeling of foods for special medical purposes and claims about the associated benefits, the labeling of organically produced foods, and the labeling of genetically modified foods.

EUROPEAN UNION

Regulation (EU) No. 1169/2011 requires that the following appear on the label: the name under which the product is sold, a list of ingredients, the quantity of certain specified ingredients, allergens, nutritional values (including guideline daily amounts on a voluntary basis), the net quantity, the date by which the integrity of the product can no longer be assured, any claims about benefits, any special storage instructions or conditions of use, the name or business name and address of the manufacturer, packager, or seller within the EU, and the place of origin of the foodstuff if the absence of the information might mislead the consumer to a material degree (FoodDrinkEurope and EuroCommerce 2013). The regulation also stipulates that a minimum font size of 1.2 millimeters must be used on the label information to ensure legibility.

Besides these general labeling requirements, there are special requirements for some products. For instance, specific legislation has been passed on the labeling of beef products. Regulation (EC) No. 1760/2000 requires the beef label to contain the following:

- A reference number or reference code ensuring the link between the meat and the animal or animals
- The approval number of the slaughterhouse at which the animal or group of animals was slaughtered, and the member state or nonmember country in which the slaughterhouse is located
- The approval number of the cutting hall that performed the cutting operation on the carcass or group of carcasses and the member state or other country in which the hall is located

Regulation (EC) No. 1830/2003 sets out specific labeling requirements for foods that are to be delivered as foods to final consumers or mass caterers and that contain or consist of genetically modified organisms or are produced from or contain ingredients produced from genetically modified organisms.

UNITED STATES

Under the Federal Food, Drug, and Cosmetic Act, food labeling is required for most prepared foods. The act specifies that food labels must include six types of information: the name of the food, the name and place of business of the manufacturer, a statement of the ingredients, the net quantity of the contents,[19] the nutrient content, and benefit claims.

Regulations require retailers to notify customers about the source of (1) muscle cuts and ground meats consisting of lamb, goat, or chickens; (2) wild and farm-raised fish and shellfish; (3) fresh and frozen fruits and vegetables; (4) peanuts, pecans, and macadamia nuts; and (5) ginseng.

The Food Allergen Labeling and Consumer Protection Act of 2004 requires that food labels indicate the presence of any of eight major food allergens, such as milk, eggs, fish (for example, bass, flounder, or cod), crustacean shellfish (for instance, crab, lobster, or shrimp), tree nuts (for example, almonds, pecans, or walnuts), wheat, peanuts, and soybeans.

To assist the food industry, the FDA (2013, 2019) has developed guidance on the labeling of general food products and foods produced using genetically modified plants. The guidance documents also contain nonbinding recommendations on labeling.

Useful to know: nutrition labeling of raw produce (fruits and vegetables) and fish is voluntary.

Withdrawal and recall

The withdrawal or recall of unsafe food is one of the core responsibilities of FBOs aiming to protect customers from unsafe food. In cases of withdrawal or recall, FBOs are also responsible for cooperating with the relevant regulatory authorities.

CODEX ALIMENTARIUS

According to provisions of the Recommended International Code of Practice General Principles of Food Hygiene, effective measures are required to ensure the rapid and complete recall of any lot of unsafe food from the market. In addition, if a product has been withdrawn because of an immediate health hazard, other products produced under similar conditions and which may therefore present a similar hazard to public health should be evaluated for safety and may ultimately need to be withdrawn as well. The principles also include the requirement to notify the public about existing hazards. Recalled products are to be under supervision until they are destroyed, used for purposes other than human consumption, determined to be safe for human consumption, or reprocessed in a manner to ensure safety.

EUROPEAN UNION

The withdrawal and recall provisions in EU food safety legislation are set out in the EU General Food Law. The obligation to withdraw a food from the market applies if (1) the food is considered by the FBO not to be in compliance with food safety requirements and (2) the food is on the market, but is no longer under the immediate control of the initial FBO.

Withdrawal is the removal of a food from the market if the food has been taken to market, but has not yet reached the public. A recall is implemented if the product has reached customers and other measures have not been sufficient to ensure a high level of health protection.

Withdrawal and recall procedures must include steps to inform and collaborate with the relevant regulatory authorities even if the FBO only suspects that the food is not fit for consumption.

UNITED STATES

Before the FSMA, all FDA-regulated food recalls were voluntary for the industry except in the case of infant formula. Since the introduction of the FSMA, the FDA is authorized to issue mandatory recalls of any FDA-regulated food, including all domestic and imported food products except meat, poultry, and egg products. Nonetheless, apart from the case of infant formula, the FDA must follow a three-step process prior to ordering the recall, as follows:

- Determine that there is evidence of a threat that meets a certain standard of proof
- Offer the company the opportunity to recall the product voluntarily before the mandatory recall is ordered
- Provide the company with the opportunity to challenge a recall decision

In 2003, the FDA issued recall guidance to companies that addresses both voluntary and mandatory recalls.[20] The recall procedure consists of the recall submission to the FDA, public notification, and evaluation of the recall. Meat and poultry recalls are voluntary, and they are initiated by the manufacturer or distributor, sometimes at the request of the Food Safety and Inspection Service (FSIS).[21] If a company refuses to recall its products, however, FSIS has the legal authority to detain and seize any of the products that are on the market.

If FSIS learns through inspections, sampling programs, or other activity that a potentially unsafe or mislabeled meat or poultry product is being marketed, it investigates the need for a recall. In case of an actual recall, FSIS notifies the public. The recall information is issued to media outlets in the areas where the product has been distributed, and this information is likewise posted on the FSIS website.[22]

Useful to know: A market withdrawal may occur if a product displays a minor violation that would not be subject to FDA or FSIS legal action. For example, a product that shows no evidence of manufacturing or distribution problems may be withdrawn from the market because of tampering.

Microbiological criteria for food and residues control

MICROBIOLOGICAL CRITERIA

Microbiological criteria play an important role in the validation and verification of HACCP procedures and other hygiene control measures. Thus, appropriate microbiological criteria must be set to determine limits of acceptability, along with food safety microbiological criteria to establish the limits above which a foodstuff should be considered unacceptably contaminated by the microorganisms that are the subject of the criteria.

Codex Alimentarius

The Codex Alimentarius sets microbiological criteria for food that determine the acceptability of a product or a food lot based on the absence, presence, or number of microorganisms, including parasites, or the quantity of toxins or metabolites per unit of mass, volume, area, or lot.

In general, regulatory authorities or FBOs may use microbiological criteria to distinguish between acceptable and unacceptable raw materials, ingredients, products, or lots. The Codex Alimentarius also emphasizes the importance of microbiological criteria in the verification or validation of the efficacy of HACCP plans.

European Union

Commission Regulation (EC) No. 2073/2005 on the microbiological criteria for foodstuffs establishes the food safety criteria for certain important foodborne bacteria, including their toxins and metabolites.

These include *Salmonella*, *Listeria monocytogenes*, *Cronobacter sakazakii*, Staphylococcal enterotoxin, and Histamine in specific foodstuffs. These microbiological criteria have been developed in accordance with Codex Alimentarius. FBOs are required to ensure that foodstuffs comply with the relevant microbiological criteria set out in the regulation.

The regulation requires FBOs to perform tests as appropriate against the microbiological criteria to validate or verify that the procedures are functioning correctly based on HACCP principles and best hygienic practice.

United States

There is no uniform microbiological standard in the U.S. food safety system. A standard has not been adopted because of the wide variation in products and processing procedures, which are constantly changing. Instead, FDA and FSIS simply provide guidance documents that include microbiological criteria for certain foods. A good starting point is the FDA's (2012) *Bad Bug Book*, which contains a wealth of information on foodborne illness-causing microorganisms. FSIS has developed guidance documentation on *Escherichia coli*, *Listeria monocytogenes*, *Salmonella*, and *Trichinella*.[23]

The FDA is in the process of updating its guidance documentation to establish a more harmonized framework for addressing biological hazards in food. The FDA is also aiming to provide guidance on the implementation aspects of hazard analysis and risk-based preventive controls for human food (FDA 2018).

The FDA has developed compliance policy guides for product categories that describe its policies on compliance and set out specific criteria that must be met by producers.[24] The contaminants covered by these guides include foodborne pathogens, bacterial toxins, mycotoxins, and bacterial indicators, for example, *Escherichia coli*. Some states also have their own microbiological standards for foods.

RESIDUES CONTROL

Residues control aims to protect public safety by setting maximum residue levels (MRLs) in accordance with generally recognized principles of safety assessment, taking into account any other scientific assessment of the safety of the substances concerned that may have been undertaken by international organizations, particularly the CAC.

Codex Alimentarius

The CAC has addressed the residues control issue through its Committee on Pesticides Residues and its Committee on Residues of Veterinary Drugs in Food. The former is responsible for establishing MRLs of pesticides in specific food items or in groups of food. The latter committee determines the priorities in considering the residues of veterinary drugs in foods and recommends MRLs on veterinary drugs. The MRLs on pesticides and veterinary drugs are constantly being developed and updated.

European Union

Regulation (EC) No. 396/2005 on the MRLs of pesticides in or on food and feed of plant and animal origin sets out the regulatory framework for the MRLs on pesticides.[25] All MRLs also apply to products after processing, although they may be adjusted to take account of dilution or concentration as a result of the processing. EU legislation stipulates that, in the absence of scientific evidence, the MRL of any substance is 0.01 parts per million.

The use of hormones in animals is forbidden in the EU. Commission Regulation (EU) No. 37/2010 sets out the regulatory framework for the MRLs on veterinary drugs.[26]

United States

The U.S. government regulates pesticides under a broad authority granted in two major statutes, the Federal Insecticide, Fungicide, and Rodenticide Act and the Federal Food, Drug, and Cosmetic Act. These laws have been amended by the Food Quality Protection Act of 1996 and the Pesticide Registration Improvement Act of 2003.

The FDA maintains and updates a list of all allowed pesticides and the associated MRLs. It uses this list for enforcement: any substance not on the list must not be present in food (zero tolerance). This is a notable difference with the EU.

The Center for Veterinary Medicine is tasked with ensuring the safety of animal-derived foods and the availability of safe and effective animal drugs.[27] It monitors the adherence of the industry to legal, administrative, and regulatory programs and policies.

In recent years, the FDA has outsourced the MRL database. However, a free login is available to search current MRLs on pesticides, veterinary drugs, contaminants, and food additives.[28]

Import and export

Laws and regulations must be followed by domestic and foreign FBO importers and exporters in each country. Moreover, the world food community has established shared principles on the import and export trade that provide the possibility of developing an efficient system of state control, while avoiding deceptive marketing practices.

Codex Alimentarius

Recognizing that quality and safety become more secure through the application of well-designed food control systems for exports and imports, principles on food import and export inspection and certification have been developed by the Codex Committee on Food Import and Export Inspection and Certification Systems (FAO and WHO 2012).

European Union

The main rule for food imports to the EU is that the products should meet the same hygiene and safety standards as food produced in the EU. However, the requirements differ for imports of food of animal origin and food of nonanimal origin or food containing ingredients of animal and plant origin.[29]

LIVE ANIMALS AND FOOD OF ANIMAL ORIGIN

Live animals and food of animal origin can be imported into the EU only from non-EU countries included in a list compiled by the community and only from establishments approved by the community.[30] Such food products are also subject to compulsory controls at border control points, at which official veterinarians are responsible for carrying out required health checks on incoming consignments. Official border controls are conducted on a fee basis. On December 14, 2019, the fees were updated based on Regulation (EC) No. 625/2017. The fees may vary in each EU member state. Prior notification of consignment arrival is required, and a consignment must be accompanied by the required documents, including an appropriate certificate issued by the competent authority in the third country. Special import conditions may be imposed on the consignment if the imported products are named on the List of Special Import Conditions.[31] This list indicates the products from each country that are to be checked as well as any control actions that may be taken. Only after these checks prove successful and all necessary information cited in the common entry veterinary document has been received is the consignment allowed to enter the EU. Consignments that are found not to be compliant with EU legislation are destroyed or, under certain conditions, redispatched within 60 days. If any of the checks indicates that a consignment of animals or animal products is likely to constitute a danger to animal or human health, the consignment is immediately seized and destroyed by the competent authorities.

CHAPTER 2

FOOD OF NONANIMAL ORIGIN

Food of nonanimal origin may be imported into the EU from any third country. No special approval of the country or of the exporting establishment in the third country is required. Import controls on food of nonanimal origin take place in accordance with national law in the different member states. This may be at the point of entry, the point of release for free circulation, the importer's premises, or retail outlets. Certain food of nonanimal origin are imported into the EU through designated points of entry. With certain exceptions, such food is not required to undergo a prenotification procedure. It may also be allowed to enter the EU without certification by competent authorities in the third country of dispatch. Only certain plants and plant products must be accompanied by a phytosanitary certificate issued by the national plant protection organization of the exporting country. According to Regulation (EU) No. 2016/2031, upon entry into the community, the phytosanitary certificate may be replaced by a plant passport.

United States

Food imported to the United States must meet the same legal standards as food produced domestically. The FSMA contains significant requirements for importers. In particular, importers must verify the safety of the food offered for import through the Foreign Supplier Verification Program. This program requires importers to conduct risk-based verification activities to ensure that imported food is not adulterated or misbranded and is produced in compliance with FDA preventive controls requirements and produce safety standards. Verification activities may include monitoring records on shipments, lot-by-lot certification compliance, annual on-site inspections of the hazard analysis and risk-based preventive control plans of the foreign suppliers, and periodic testing and sampling of shipments.[32]

The verification program is mandatory, unlike the Voluntary Qualified Importer Program, which is entirely voluntary and gives importers a green light to import foods from trusted, certified suppliers.[33] Noncompliance with the verification program represents a basis for rejecting an imported article. The FSMA authorizes the FDA to require that imported foods considered high risk because of potential health consequences be accompanied by a credible third-party certification or other assurance of compliance as a condition of entry into the United States.

Before products may be imported to the United States, the FDA also requires both domestic and foreign food facilities to register with the FDA.[34] The FDA is also to be provided with advance notice on shipments of imported food, including the product code of the food to be imported.[35]

A foreign facility that manufactures, processes, packs, or stores food is required to register with the FDA unless food from the facility undergoes additional processing, including packaging, at another foreign facility before the food is exported to the United States. Food facilities may be registered and prior notice may be submitted online. Food facilities are required to renew the registration between October 1 and December 31 of every even year.

Imported food products are subject to FDA inspection if they are offered for import at U.S. ports of entry. The FDA may detain shipments of products offered for import if the shipments are found not to be in compliance with U.S. requirements.

Unlike the FDA, for which inspection requirements are company-specific, the FSIS coordinates with the government of the exporting country before accepting meat, poultry, or egg products for sale in the United States. In particular, to import meat, poultry, or eggs into the United States, these products must originate from certified countries and establishments within these countries that are eligible to export to the United States.[36]

Regulatory authority

The issue of food safety regulation is one of the most important in terms of ensuring both customer health and effective FBO operations. Indeed, the ability to produce safe food and to be trusted by potential customers is crucial for food producers aiming to integrate their businesses into the international food trade. This means that food safety systems are a key issue in the private sector. At the same time, however, food safety regulations can also impose a heavy administrative burden on businesses.

Codex Alimentarius

The Codex Alimentarius international food standards, guidelines, and codes of practice contribute to the safety, quality, and fairness of international food trade. Consumers can trust the safety and quality of the food products they buy, and importers can trust that the food they order will meet their specifications. While the standards are only recommendations for national authorities and are voluntary, they serve in many cases as a basis for national legislation.

The reference to codex food safety standards in the World Trade Organization's Agreement on the Application of Sanitary and Phytosanitary Measures means that the codex has far-reaching implications for the resolution of trade disputes. Members of the World Trade Organization that wish to apply stricter food safety measures than those set out in the codex may be required to justify these measures scientifically.

Since its foundation in 1963, the codex system has evolved in an open, transparent, and inclusive manner to meet emerging challenges. The codex system contains four levels of documentation, as follows:

- *Guidelines:* these are intended as a regulatory structure for authorities and are less relevant for FBOs.
- *Codes of practice:* these are broad lists of documents covering hygienic practices across various processes associated with the food industry and with preventive measures for contaminants; these documents are a good starting point for FBOs.
- *Standards:* these give a comprehensive description of foodstuffs, including specifications, levels of defects, and product acceptability criteria; the documents can serve as a starting point among FBOs in establishing raw material and product specifications.[37]
- *MRLs and related databases:* the MRL document provides guidelines for the management by authorities of residues of veterinary drugs in food, whereas the related databases supply information on pesticide residues, veterinary drug residues, and general standards for food additives; these are thus a good source of information for FBOs.

European Union

The European Food Safety Authority is an independent European agency funded by the EU budget that operates separately from the European Commission, European Parliament, and EU member states. The agency's role involves assessing and communicating risks associated with the food chain. Through its risk communication activities, it seeks to raise awareness and explain the implications of its scientific work. It aims to provide appropriate, consistent, accurate, and timely communications on food safety issues to all stakeholders and the public at large based on its risk assessments and scientific expertise.

In the EU, the regulatory authority in each member state is responsible for coordinating the enforcement of national food safety legislation, and the food and veterinary office is tasked with supervising the performance of the regulatory authority.[38]

The Rapid Alert System for Food and Feed has been established to provide authorities with an effective tool to exchange information about measures taken in response to serious risks detected in relation to food or feed.[39] The information exchange helps authorities act more rapidly and in a coordinated manner in response to a health threat rooted in food or feed. If network participants have any information about serious health risks linked to food or feed, they must immediately notify the European Commission using the system. According to Regulation (EC) No. 625/2017, the system is eventually to be replaced by an information management system for official controls.

United States

Two primary federal agencies are responsible for the U.S. food system, namely, the FDA and the USDA. The USDA oversees the regulation of meat, poultry, and processed egg products. Within the USDA, the FSIS inspects and regulates meat, poultry, and processed egg products that are produced in federally inspected plants. The FSIS ensures that these products are safe, wholesome, and correctly labeled and packaged.

The FDA regulates virtually all other foods. In particular, the Center for Food Safety and Applied Nutrition (CFSAN) ensures that the food supply is safe, sanitary, wholesome, and honestly labeled.

Among other agencies responsible for food system–related issues are the following:

- The Department of Homeland Security coordinates food security activities, including at U.S. borders.
- The National Marine Fisheries Service in the U.S. Department of Commerce conducts voluntary fee-for-service inspections of seafood safety and quality.
- The Environmental Protection Agency monitors pesticide use and the MRLs in food commodities and animal feed.
- The Centers for Disease Control and Prevention, within the U.S. Department of Health and Human Services, investigates outbreaks of foodborne illness and tracks individual cases.

In the states, food safety regulatory functions may be carried out by departments of health, agriculture, or environment or some combination of these. State agencies perform a wide range of food safety functions, including outbreak response and recalls, laboratory testing, and retail, food service, processing, and farm inspections.[40] Local public health departments normally carry out restaurant inspections and other community food safety activities.

Other relevant regulations

European Union

There are several reasons why Regulation (EC) No. 625/2017, enacted on December 14, 2019, is an important milestone for food safety in the EU. First, it reduces legal fragmentation by repealing 10 regulations and commission directives and decisions, and amends others. Previously, animal welfare, pesticide residues, products of animal origin, and other segments of the supply chain were regulated separately. They are now under one legal roof. Second, it strengthens the basic principles of previous laws. Although the new regulation does not change important principles, such as the transparency of controls and cooperation among member states, it clarifies existing provisions through greater precision. Third, it improves the harmonization of procedures and standards. For example, it creates a single information management system for official controls, by integrating systems, such as the Rapid Alert System for Food and Feed, and facilitates the exchange of information among member states. Fourth, it repeatedly invokes Article 144, which gives the European Commission the power to adopt delegated acts. It is an important step because it establishes the legal basis for more sweeping future changes.

It is anticipated that the implementation of Regulation (EC) No. 625/2017 will lead the U.S. government to classify the EU member states among the countries with FSMSs equivalent to the U.S. system. Relevant future exports from the EU to the United States will thus be facilitated in light of the favorable provisions of the FSMA.

United States

The passage of the FSMA in 2011 represented the first major reform of the FDA food safety authority in over 70 years. The law requires the enhanced regulation of produce from farm to sale and of other FDA-regulated foods from processing to sale. It also introduces food defense requirements. The FSMA alters the role of the FDA in food safety through five key changes: (1) a shift in focus from reaction to prevention, including the prevention of intentional contamination; (2) an increase in the authority to inspect and to ensure compliance with inspection frequencies based on risk; (3) the recognition of a new authority to make recalls mandatory; (4) controls on imports are strengthened to ensure that U.S. food safety standards are met; and (5) partnerships with other government agencies and private entities are reinforced.

PREVENTIVE CONTROLS: HUMAN FOOD

The FSMA requires FBOs to produce written food safety plans that include a hazard analysis and establish preventive controls. The first step in hazard analysis is hazard identification, which involves diligence in considering known or reasonably foreseeable biological, chemical, and physical hazards. These hazards may occur naturally or be intentionally introduced, including for economic gain.

Preventive controls are the measures implemented to ensure that hazards are minimized or prevented. They include process controls, food allergen controls, sanitation controls, supply chain controls, and a recall plan. The oversight and management of preventive controls involves monitoring, corrective action, and verification.

- Monitoring is the set of procedures undertaken as needed to ensure that preventive controls are consistently applied.
- Corrective actions are steps taken to identify and correct a minor, isolated problem that occurs during food production. They include the application of preventive controls to reduce the likelihood that the problem will recur, evaluate affected food for safety, and remove any unsafe food from the market. Corrective actions are always documented.
- Verification is the set of activities required to ensure that preventive controls are consistently applied and effective. It involves producing scientific evidence that a preventive control is able to identify and eliminate a hazard, undertaking calibration or accuracy checks of process monitoring and instruments such as thermometers, and reviewing records to verify that monitoring and necessary corrective actions are being conducted.

Operations defined as farms in the FSMA are not subject to the preventive controls rule. The supply chain is made more flexible by instituting separate compliance dates. The rule requires manufacturing or processing facilities to establish risk-based supply chain control programs for raw materials and other inputs that have been identified as hazard risks. However, manufacturing or processing facilities that use preventive controls on hazards or that follow regulations allowing reliance on customers to control hazards do not need to have a supply chain program for those risk hazards. Relevant food facilities are responsible for ensuring that they receive foods only from approved suppliers or, on a temporary basis, from unapproved suppliers whose materials are subject to verification activities before being accepted for use.

The FSMA updated and clarified good manufacturing practices (GMPs) as follows:

- FBO management is required to ensure that all employees who manufacture, process, pack, or store food are qualified to perform their assigned duties.
- FBO employees must have the education, training, or experience necessary to manufacture, process, pack, or store clean and safe food.
- FBO employees must receive training in the principles of food hygiene and food safety, including the principles of employee health and hygiene.

Foreign supplier verification programs

According to the FSMA, importers of food for humans and animals must be subject to foreign supplier verification programs. The rule requires that importers perform certain risk-based activities to verify that food imported into the United States has been produced in a manner that meets applicable U.S. safety standards. Importers covered by the rule must have a system in place to verify that their foreign suppliers are producing food in a manner that provides the same level of public health protection as the preventive controls or apply safety regulations as appropriate to ensure that the food is not adulterated and is not misbranded with respect to allergen labeling. Importers are responsible for the following:

- Determining known or reasonably foreseeable hazards with each food
- Evaluating the risk posed by an imported food based on hazard analysis or food safety performance indicators on foreign suppliers, such as complaints, withdrawals, or recalls
- Using the evaluation of the risk associated with an imported food and a supplier's performance to approve suppliers and determine appropriate supplier verification activities
- Conducting supplier verification activities
- Conducting corrective actions

Third-party certification

The third-party certification rule establishes a voluntary program for the accreditation of third-party certification bodies, also known as auditors, to conduct food safety audits and issue certifications for foreign facilities and the foods they produce for humans and animals. These requirements cover legal authority, competency, capacity, safeguards against conflicts of interest, quality assurance, and documentation procedures. The certificates may be used by importers to establish eligibility for participation in the Voluntary Qualified Importer Program, which offers the expedited review and entry of imported food. To prevent potentially harmful food from reaching U.S. consumers, the FDA can also require in specific circumstances that a food offered for import be accompanied by certification from an accredited third-party certification body.

Produce safety

The produce safety rule establishes, for the first time, science-based minimum standards for the safe growing, harvesting, packing, and storage of fruits and vegetables grown for human consumption. According to the rule, for certain uses, no detectable generic *Escherichia coli* can be present in agricultural water if there is a reasonable likelihood that potentially dangerous microbes may be transferred to produce through direct or indirect contact. Agricultural water that is directly applied to growing produce other than sprouts is subject to a second set of criteria based on two values: the geometric mean and the statistical threshold. The geometric mean of samples is 126 or fewer colony-forming units of generic *Escherichia coli* in 100 milliliters of water, and the statistical threshold of samples is 410 or fewer colony-forming units of generic *Escherichia coli* in 100 milliliters of water.

Testing is required for untreated water used for certain purposes based on the testing frequency for the type of water source, that is, surface or groundwater.

A biological soil amendment is a material, including manure, that is intentionally added to the soil to improve its chemical or physical condition for growing plants or to improve its capacity to hold water. Untreated biological soil amendments of animal origin, such as raw manure, must be applied so that they do not come into contact with covered produce during application, and the potential for contact with covered produce after application must also be minimized. The FDA does not object if farmers comply with USDA National Organic Program standards, which call for a 120-day interval between the application of raw manure to crops in contact with the soil and a 90-day interval for crops not in contact with the soil. Microbial standards that impose limits on the detectable amounts of bacteria, including *Listeria monocytogenes*, *Salmonella* species pluralis, fecal coliforms, and *Escherichia coli*, have been established for processes used to treat biological soil amendments, including manure.

The FSMA imposes new requirements to help prevent the contamination of sprouts, which have been frequently associated with outbreaks of foodborne illness. Sprouts are especially vulnerable to dangerous microbes because of the warm, moist, and nutrient-rich conditions needed for their growth.

Food defense

Food defense is the effort to protect the food supply against intentional contamination arising because of sabotage, terrorism, counterfeiting, or other illegal, intentionally harmful means. Potential contaminants include biological, chemical, and radiological hazards that are generally not found in foods or their production environment. The FDA intentional adulteration rule requires domestic and foreign facilities to address vulnerable processes in their operations to prevent acts on the food supply intended to cause large-scale public harm. While all large FBOs were required to comply with this rule by July 27, 2019, small FBOs and very small FBOs are required to comply only by July 27, 2020, and July 26, 2021, respectively.

THE SANITARY TRANSPORTATION OF HUMAN AND ANIMAL FOOD

The sanitary transportation rule establishes requirements for vehicles and transportation equipment, transportation operations, training, and recordkeeping. Operators of motor vehicles, railcars, and other equipment used in food transportation are required to set out written procedures, subject to recordkeeping requirements, for cleaning their vehicles and transportation equipment.

ADMINISTRATIVE DETENTION

The FSMA enhances the FDA administrative detention authority by authorizing the FDA to administratively detain articles of food that the FDA has reason to believe may be adulterated or misbranded.

REGULATIONS ON FRUITS AND VEGETABLES

All imported fruits and vegetables must be free of plant litter, debris, or any parts of plants that are specifically prohibited in the regulations. Whether commercial or noncommercial consignments, they must also be imported under permits issued by the Animal and Plant Health Inspection Service.

Port of entry

Fruits and vegetables must be imported into specific ports if so required or they may be imported into any port listed in 19 CFR 101.3(b)(1). Fruits and vegetables that are to be cold treated at ports in the United States may only be imported into specific ports.

Inspection, treatment, and other requirements

All imported fruits or vegetables are subject to inspection and disinfection at the port of first arrival as may be required by a border control post inspector and are subject to reinspection at other locations at the option of an inspector. If an inspector finds plants or portions of plants, or a plant pest or noxious weed, or evidence of a plant pest or noxious weed on or in any fruit or vegetable or its container, or finds that the fruit or vegetable may have been associated with other articles infested with plant pests or noxious weeds, the owner or agent of the owner of the fruit or vegetable must clean or treat the fruit or vegetable and its container as required by the inspector, and the fruit or vegetable is also subject to reinspection, cleaning, and treatment at the option of the inspector at any place and time until all applicable requirements have been accomplished.

Any person importing fruits and vegetables into the United States must offer those agricultural products for inspection and entry at the port of first arrival. The owner or agent must assemble the fruits and vegetables for inspection at the port of first arrival, or at any other place designated by an inspector and in a manner designated by the inspector. All fruits and vegetables must be accurately disclosed and made available to an inspector for examination. The owner or the agent must provide an inspector with the name and address of the consignee and must make full disclosure of the type, quantity, and country and locality of origin of all fruits and vegetables in the consignment either orally for noncommercial consignments or on an invoice or similar document for commercial consignments.

If an inspector finds that an imported fruit or vegetable is prohibited, or is not accompanied by required documentation, or is so infested with a plant pest or noxious weed that, in the judgment of the inspector, it cannot be cleaned or treated, or contains soil or other prohibited contaminants, the entire lot or consignment may be refused entry into the United States.

No person may move a fruit or vegetable from the port of first arrival unless an inspector has (1) released it; (2) ordered treatment at the port of first arrival and, after treatment, released the fruit or vegetable; (3) authorized movement of the fruit or vegetable to another location for treatment, further inspection, or destruction; or (4) ordered the fruit or vegetable to be reexported.

If an inspector orders any disinfection, cleaning, treatment, reexportation, recall, destruction, or other action with regard to imported fruits or vegetables while the consignment is in foreign commerce, the inspector will issue an emergency action notification to the owner of the fruits or vegetables or to the owner's agent. The owner must, within the time and in the manner specified in the emergency action notification, destroy the fruits and vegetables, ship them to a point outside the United States, move them to an authorized site, or apply treatments or other safeguards to the fruits and vegetables as prescribed to prevent the introduction of plant pests or noxious weeds into the United States.

The Animal and Plant Health Inspection Service is responsible only for the costs of providing the services of an inspector during regularly assigned hours of duty and at the usual places of duty. The owner of the imported fruits or vegetables is responsible for all additional costs of inspection, treatment, movement, storage, destruction, or other measures ordered by an inspector, including any labor, chemicals, packing materials, or other supplies. The inspection service will not be responsible for any costs or charges other than those identified in this section.

Other jurisdictions: The Eurasian Economic Union

This section provides information on the production and marketing of food products in the Eurasian Economic Union (EAEU). It is based on a report of the World Bank (2015).

The EAEU, an international organization for regional economic integration, was established by the Treaty on the Eurasian Economic Union of May 29, 2014. The member states are Armenia, Belarus, Kazakhstan, the Kyrgyz Republic, and the Russian Federation. The EAEU ensures the free circulation of goods, services, capital, and labor as well as coordinated, coherent, or unified economic policy. It has been established to promote comprehensive modernization, cooperation, competitiveness, the improvement of the national economies, and the creation of an environment for sustainable development to raise the living standards of the citizens of the member states.

Legal instruments

The EAEU system of normative regulation consists of general and product-specific technical regulations (TRs) that provide a framework for food controls (figure 2.2). The common framework is supported by the laws, regulations, and standards of member states. The main legal instruments are TRs. There are also standards, which are voluntary, though products must comply with laws on TRs to promote regional standardization. Standardization is fostered through national standards and TRs that are applied by manufacturers, who demonstrate compliance by using appropriate product labeling for the benefit of end users or for transportation and following proper procedures and requirements. The standards and TRs are directly applicable in member states. With some exceptions, however, they do not address implementation mechanisms, which are covered only in the national laws and regulations of EAEU member states. While directly applicable, the TRs focus mainly on the technical aspects of products. They establish specifications, not policies. Among food categories for which product-specific TRs have not yet been developed, the national laws of EAEU member states apply.

To be compliant within the EAEU, stakeholders must take into consideration compliance with the EAEU TRs as well as the laws and standards of the member states. Furthermore, FBOs must be aware that the EAEU system does not address enforcement, fines, penalties, incident management, recalls and withdrawals, authorization and approval of new substances (pesticides or veterinary medicines), and so on. All these issues are tackled within the framework of the national legislation of member states. This increases the complexity of the regulatory environment and of compliance among actors wishing to export to the EAEU and among governments wishing to model or harmonize their approaches relative to the EAEU.

Organizational arrangements

The legal framework in the EAEU combines horizontal and vertical regulations. Several TRs on general food safety, labeling, packaging, and food additives and flavorings cover issues horizontally across all food products and, in the case of the TR on packaging, also nonfood items. A horizontal TR is also currently being drafted on materials that come into contact with food. The vertical TRs are specific to product groups, particularly grains, oils and fats, fruit and vegetable juices, meat and meat products, and milk and dairy products. Additional vertical TRs are being drafted on alcohol products, poultry and poultry products, fish and fish products, and bottled potable water, including mineral water.

The EAEU TRs include requirements that relate to market circulation and distribution. The most important requirement is that food must pass through compliance assessment procedures and bear a special EAEU logo as proof of compliance (box 2.1). Furthermore, because the EAEU is based on compliance assessments, many food products have to meet compositional standards, as well as requirements on chemical and physical properties, nutritional properties, organoleptic properties (appearance, taste, touch, smell), and, in some cases, size.

Figure 2.2 Technical Regulations, Food Safety, Eurasian Economic Union

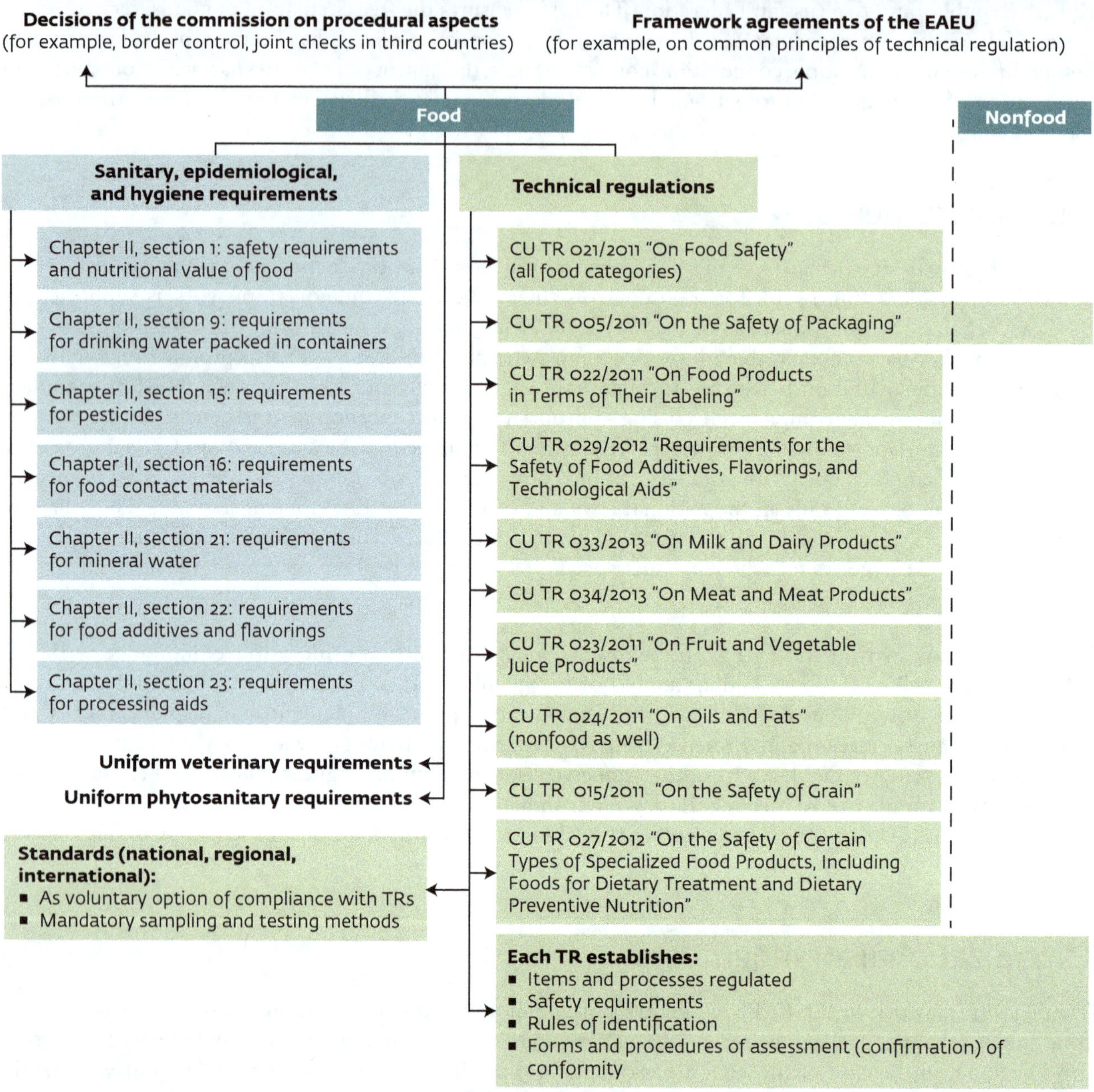

Note: The list of TRs is not exclusive. Thus, CU TR 040/2016 "On the Safety of Fish and Fish Products," CU TR 044/2017 "On the Safety of Bottled Water, Including Natural Mineral Water," and CU TR 047/2018 "On the Safety of Alcohol Products" are not shown. Relevant sections of the uniform sanitary, epidemiology, and hygiene requirements are no longer adequate in the case of several TRs that were issued subsequently.

Box 2.1 The EAEU Conformity Mark

The Eurasian compliance logo—Cyrillic: Евразийское соответствие (EAC, Eurasian Compliant)—is a certification mark to indicate that a product marked with the logo conforms to all requirements of the corresponding TRs and has passed all compliance assessment procedures stipulated by EAEU TRs that apply to the product. The mark was introduced in August 2013.

The food control system

In the EAEU, the food control system incorporates two levels: (1) food control through all-union compliance assessment and (2) individual member state controls (supervision) of sanitary, veterinary, and phytosanitary features (figure 2.3).

Enforcement is carried out by national bodies designated as competent authorities for specific areas of state control (supervision) and competent authorities in the area of TRs. Compliance assessment is carried out by authorized certification bodies that are listed in a single EAEU table. Testing required for the purposes of enforcement is carried out by authorized testing laboratories, and there is a separate EAEU table of such laboratories.

Specially designated bodies are responsible for registration and certification for several groups of products for which registration is required as one of the means of compliance assessment, such as specialized products, genetically modified organisms, and so on.

Within the EAEU, foodstuffs are subject to compliance assessment. The evaluation of food compliance is conducted according to the following steps: (1) confirmation (declaration) of the compliance of the food products, (2) state registration of the specialized food products, (3) state registration of new types of food products, and (4) veterinary-sanitary expertise assessment.

Figure 2.3 Food Control System, Eurasian Economic Union

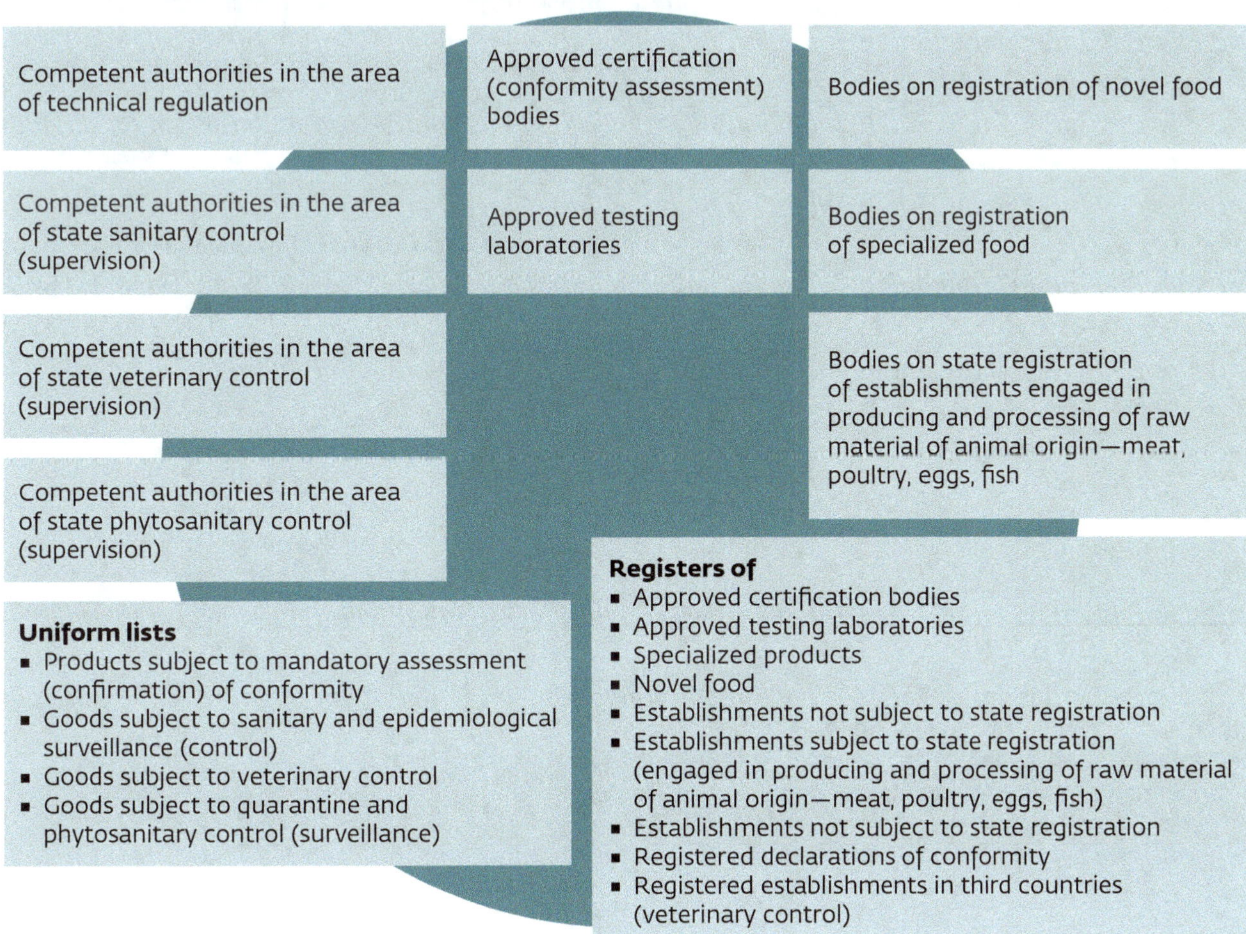

Food products are also subject to state controls (supervision) (table 2.1).[41] This combines border controls (people, vehicles, and goods) and internal controls in the member states. Food products are divided into three groups that are subject, respectively, to sanitary controls (epidemiological and hygiene), veterinary controls, and phytosanitary controls (supervision). The purpose of, for instance, the state sanitary controls (epidemiological and hygiene) is to eliminate or prevent the introduction or spread of infections and toxins that are hazardous to human health. The goal of state controls is the prevention of outbreaks, other emergencies, and

Table 2.1 General Framework of Food Safety Controls, Eurasian Economic Union

	Sanitary	Veterinary	Phytosanitary
Key legal act	EAEU agreement on sanitary measures	Agreement of the EAEU on veterinary and sanitary measures	EAEU agreement on plant quarantine
Competent authority	Competent authorities in the area of state sanitary control (supervision) in member states		
Scope	Common list of goods subject to sanitary and epidemiological control (supervision) • Foodstuffs (products in natural or processed form used for human food) including those derived from genetically engineered or modified (transgenic) organisms • Materials, products, and equipment contacting with foodstuffs • Pesticides and agrochemicals	Common list of goods subject to veterinary control • Live animals • All food of animal origin, fresh and processed • Food that has ingredients of animal origin • Yeasts, enzymes, starter cultures • Grains and other plant origin items when they are intended for manufacture of feed	List of goods subject to quarantine and phytosanitary control (supervision) • Vegetables, fresh or chilled • Dried leguminous vegetables • Fruits, fresh, dried • Nuts, fresh or dried, whether or not shelled or peeled • Coffee, not roasted, whether or not decaffeinated • Cocoa beans • Grains • Cereal flours • Seeds, whether or not broken
Point of control	At the border and within the customs territory of the customs union		
Documents that establish compliance criteria	Uniform sanitary, epidemiological, and hygiene requirements for goods subject to veterinary control (supervision)	Uniform veterinary requirements for goods subject to veterinary control (supervision)	–

continued

Table 2.1 (Continued)

	Sanitary	Veterinary	Phytosanitary
Procedural documents	Procedure for state sanitary and epidemiological control (supervision) over persons crossing the EAEU customs border and goods subject to control that are being moved through the customs border and customs territory of the EAEU Common templates of product (goods) safety documentation	Procedure for carrying out veterinary control at the customs border and on the customs territory of the EAEU Procedure for carrying out joint inspections and sampling of goods (products) subject to veterinary control (supervision) on the territory of the EAEU member states and third countries Consolidated list of highly dangerous and quarantine diseases of animals Common templates of veterinary certificates (movement, import)	List of Quarantine Products subject to quarantine and phytosanitary control (supervision) while being imported to the common customs territory of the EAEU Procedure for carrying out the quarantine and phytosanitary control (supervision) at the external border of the EAEU Procedure for carrying out the quarantine and phytosanitary control in respect of quarantine products that are moved within the common customs territory of the EAEU
Registers	Common register of state registration certificates for certain products	Register of food production objects (facilities) that are subject to state registration	–

acts of terrorism through the use of biological agents, chemicals, or radioactive substances.[42] All food with ingredients of animal origin are subject to veterinary checks.

When they are first imported or produced in the EAEU, certain specific products are subject to state registration. This includes mineral, therapeutic, and bottled water; beverages, such as tonics and beer; food for special purposes, including food for babies and older children and food for pregnant and nursing women; food additives; foodstuffs derived from genetically engineered or modified (transgenic) organisms; and some foods that come into contact with other materials. Whether such products have been registered is verified during the implementation of state controls (supervision).

Certain production or processing facilities must be registered. This requirement extends to facilities engaged in the production and processing of meat and meat products, milk and dairy products, poultry and poultry products, and fish and fishery products. The state registration of production or processing facilities is conducted by the agencies authorized for this purpose by the EAEU member states. This procedure begins with the registration application of the processor. An inspection of the facility follows to determine conformity with the requirements governing the processes established by the relevant TRs for production, processing, storage, transportation, sale, and disposal. The details of the procedure are established by legislation in the EAEU member states.

Upon satisfactory completion of the inspection and the review of the findings, the agency designated with the relevant responsibility assigns an identification (record) number to the facility and adds the production facility in the register of food facilities subject to state registration. The state registration of a production or processing facility has no fixed expiration date; however, the registration may be suspended or cancelled in the case of a serious breach of the TR requirements.

The regulation of food quality issues

In the EAEU, food product quality is usually defined in product definitions that include minimum requirements on composition. These requirements may be found in the section in vertical (product-specific) TRs that is usually labeled "Safety requirements for" followed by the name of the product group. For example, the label might read "Safety requirements for fruit and/or vegetable juice products" as in customs union (CU) TR 023/2011 "On Fruit and Vegetable Juice Products." This might also be the formulation in annexes on the microbiological, physical, and chemical properties and organoleptic characteristics of a product. The intent in the EAEU TRs is to ensure that products entering the marketplace conform to the specific TRs in all attributes. The quality characteristics outlined in the TRs are aimed at ensuring the uniformity of the food products offered to consumers, satisfy the needs of vulnerable groups of consumers, and, for the purposes of product identification, to establish whether the products are subject to conformity assessment under the relevant TRs.

According to a general rule that applies to compliance with the EAEU TRs, a manufacturer may choose whether to comply with the TR itself or with regional standards that are listed in support of each TR. Compliance with these standards is voluntary, but meets the requirements for compliance with the TR. Furthermore, if norms are absent in the EAEU TRs, the national norms of the member states apply.

Food labeling

In the EAEU, labeling requirements focus on consumer packaging and transport packaging. A packing list envelope must be attached, affixed, or enclosed with the product packaging. The EAEU has established that packaged food product labeling may include additional information that is otherwise not required.

In the EAEU, labeling is one of the requirements in the specifications for mandatory compliance assessment. In this case, the label represents a declaration of compliance. Noncompliance may result in exclusion from the EAEU market.

Food and food-related articles and materials requiring special authorization

The EAEU has established that certain types of food, classes of substances, and materials that are added to food or come into contact with food must meet special requirements to ensure food safety and require special authorization to be placed on the market (World Bank 2015). These products include novel food, food supplements, food additives, food packaging, and other articles and materials that are in contact with food. These are broad groups of substances, materials, and articles. Each group is the subject of special laws and TRs, definitions, and authorization procedures.

In the EAEU, only packing and bottling materials must be regulated as materials and items coming into contact with food (CU TR 005/2011). Work is under way to draft a TR on the safety of materials coming into contact with food.

The EAEU approach to compliance assessment based on testing to define the safety of packaging and bottling materials is efficient. However, this approach assumes that, before releasing a product onto the market, the technical specifications (TSs) of the product will have been established. This means that the EAEU regulatory framework in this case is based mainly on the TSs set out for existing and approved packaging and bottling materials.

The microbiological criteria for food safety

The EAEU relies on both vertical and horizontal regulations to establish microbiological requirements for foodstuffs. This legal foundation combines the general requirements for all foodstuffs in the horizontal CU TR 021/2011 "On Food Safety" with additional requirements established in vertical product-specific TRs for certain types of food. In the combined form, the requirements can be found in the uniform sanitary, epidemiological, and hygiene requirements for products subject to state control (supervision).[43] At the same time, as a general rule, if a product-specific TR is adopted, the relevant sections of the uniform sanitary, epidemiological, and hygiene requirements lose their validity for the products covered by the new TR.

The microbiological requirements of the EAEU focus on a combination of pathogens as well as indicative microorganisms and microorganisms associated with spoilage in finished products. This is because the intention behind the EAEU regulatory framework is based on the compliance assessment of finished products as a mechanism to control food safety and quality as well as to identify foods.

Approaches to laboratory control, sampling, and testing

Testing, sampling, and laboratory work are a part of the overall EAEU compliance assessment process, which aims at ensuring food safety through documents establishing conformity with the TRs, that is, the safety regulations. Sampling is performed to ensure that products meet the requisite TRs. The samples are used to validate a variety of parameters, including pathogens, residues of pesticides, veterinary medicines, heavy metals, radionuclides, and mycotoxins. Testing methods and the specific requirements for testing are codified in the approved lists that support each TR and are established in national standards that are approved regionally within the EAEU. This establishes a certain degree of uniformity.

Notes

1. See "Codex Alimentarius: International Food Standards," Joint Food and Agriculture Organization of the United Nations–World Health Organization Food Standards Programme, Rome, http://www.fao.org/fao-who-codexalimentarius/en/.
2. See Codex Online Databases, Codex Alimentarius, International Food Standards, Codex Alimentarius Commission Secretariat, Food and Agriculture Organization of the United Nations, Rome, http://www.fao.org/fao-who-codexalimentarius/codex-texts/dbs/pt/.
3. For the latest full consolidated texts of all EU regulations and legislation mentioned in this chapter, see EUR-Lex (database), Publications Office of the European Union, Luxembourg, https://eur-lex.europa.eu/.
4. For guidance and other information on U.S. laws, statutes, and regulations under the statutory authority of the FDA that are mentioned in this chapter, see Guidance Documents and Regulatory Information by Topic (Food and Dietary Supplements) (database), U.S. Food and Drug Administration, Silver Spring, MD, https://www.fda.gov/food/guidance-regulation-food-and-dietary-supplements/guidance-documents-regulatory-information-topic-food-and-dietary-supplements.
5. For the meat, poultry products, and egg products inspection acts mentioned here, see "Inspection Acts, Related Laws, and Guidance," U.S. Department of Agriculture, Washington, DC, http://www.fsis.usda.gov/rulemaking.
6. See "Federal Insecticide, Fungicide, and Rodenticide Act (FIFRA) and Federal Facilities," U.S. Environmental Protection Agency, Washington, DC, https://www.epa.gov/enforcement/federal-insecticide-fungicide-and-rodenticide-act-fifra-and-federal-facilities.
7. See Federal Register (database), Office of the Federal Register, National Archives and Records Administration, College Park, MD, https://www.federalregister.gov/.
8. See FDA Food Code (database), U.S. Food and Drug Administration, Silver Spring, MD, https://www.fda.gov/food/retail-food-protection/fda-food-code.

9. See Case Law Indexes (database), National Agricultural Law Center, University of Arkansas, Fayetteville, AR, https://nationalaglawcenter.org/aglaw-reporter/case-law-index/.
10. See GAIN (Global Agricultural Information Network) (database), U.S. Department of Agriculture, Washington, DC, https://gain.fas.usda.gov/Pages/Default.aspx.
11. General Principles of European food legislation came into force in 2002 with the adoption of the EU Food Law, Regulation (EC) No. 178/2002, of January 28, 2002, which set out the general principles and requirements of food law, established the European Food Safety Authority, and defined procedures to ensure food safety.
12. For details about the approach, see European Commission (2004).
13. For a description of the U.S. food safety system, see FDA and USDA (2000).
14. For any Codex Alimentarius codes of practice mentioned in this chapter, see Codes of Practice (database), Codex Alimentarius, International Food Standards, Codex Alimentarius Commission Secretariat, Food and Agriculture Organization of the United Nations, Rome, http://www.fao.org/fao-who-codexalimentarius/codex-texts/codes-of-practice/en/.
15. This requirement does not apply to primary production, however. For the primary sector, a separate piece of legislation has been established since 1962 that provides for the organization of the market in agricultural products (EU Regulation No. 1308/2013).
16. Any U.S. federal rules or regulations mentioned in this chapter are accessible on a single searchable website. See e-CFR (Electronic Code of Federal Regulations) (database), Office of the Federal Register, National Archives and Records Administration, Government Publishing Office, Washington, DC, https://www.ecfr.gov/cgi-bin/ECFR?page=browse.
17. These are (1) small businesses as defined by FDA rules and (2) facilities or businesses with average annual food sales of less than $500,000 during the previous three years, but only so long as the majority of the food was sold directly to consumers, restaurants, or grocery stores, rather than third-party food brokers, and the sales activity occurred in the same state in which the facility is located or within 275 miles of the facility.
18. For any Codex Alimentarius guidelines mentioned in this chapter, see Guidelines (database), Codex Alimentarius, International Food Standards, Codex Alimentarius Commission Secretariat, Food and Agriculture Organization of the United Nations, Rome, http://www.fao.org/fao-who-codexalimentarius/codex-texts/guidelines/en/.
19. Weight is expressed in pounds and ounces or common or decimal fractions of the pound; in the case of liquids, the measure is the largest whole unit in quarts, quarts and pints, or pints, as appropriate.
20. "Guidance for Industry: Product Recalls, Including Removals and Corrections," U.S. Food and Drug Administration, Silver Spring, MD, http://www.fda.gov/Safety/Recalls/IndustryGuidance/ucm129259.htm.
21. See the FSIS website, at https://www.fsis.usda.gov/wps/portal/fsis/home.
22. See "Current Recalls and Alerts," Food Safety and Inspection Service, U.S. Food and Drug Administration, Washington, DC, https://www.fsis.usda.gov/wps/portal/fsis/topics/recalls-and-public-health-alerts/current-recalls-and-alerts.
23. See "Biological Hazard Guidance," Food Safety and Inspection Service, U.S. Department of Agriculture, Washington, DC, https://www.fsis.usda.gov/wps/portal/fsis/topics/regulatory-compliance/compliance-guides-index/bacteria-guidance.
24. For any FDA policy guides mentioned in this chapter, see Manual of Compliance Policy Guides (database), U.S. Food and Drug Administration, College Park, MD, https://www.fda.gov/inspections-compliance-enforcement-and-criminal-investigations/compliance-manuals/manual-compliance-policy-guides.
25. For information on the MRLs on pesticides, see EU Pesticides Database, Directorate-General for Health and Food Safety, European Commission, Brussels, http://ec.europa.eu/food/plant/pesticides/eu-pesticides-database/public/?event=homepage&language=EN.
26. For these MRLs, see the veterinary category at Medicines (database), European Medicines Agency, Amsterdam, https://www.ema.europa.eu/en/medicines/field_ema_web_categories%253Aname_field/Veterinary.
27. See "Compliance and Enforcement," Center for Veterinary Medicine, Rockville, MD, https://www.fda.gov/animal-veterinary/compliance-enforcement.
28. See the BCGlobal databases, Bryant Christie Inc., Sacramento, CA, http://www.bryantchristie.com/BCGlobal-Subscriptions.
29. To facilitate the export or import of goods between the EU and other countries, the European Commission has established the Market Access Database, a tool that provides users with information on import and export formalities

and tariffs based on the imported or exported product and the country of export or import. See MADB (Market Access Database), Directorate-General for Trade, European Commission, Brussels, https://madb.europa.eu/madb/indexPubli.htm.

30. See Non-EU Country Establishments Database, Directorate-General for Trade, European Commission, https://ec.europa.eu/food/safety/international_affairs/trade/non-eu-countries_en.

31. See "Special Import Conditions," Directorate-General for Trade, European Commission, https://ec.europa.eu/food/animals/vet-border-control/special-import-conditions_en.

32. However, importers are not required to conduct verification on products that are subject to low-acid canned food regulations or regulations on seafood or juice because these products undergo other processes.

33. See "Voluntary Qualified Importer Program (VQIP)," U.S. Food and Drug Administration, Silver Spring, MD, https://www.fda.gov/food/importing-food-products-united-states/voluntary-qualified-importer-program-vqip.

34. See "Online Registration of Food Facilities," U.S. Food and Drug Administration, Silver Spring, MD, https://www.fda.gov/food/registration-food-facilities-and-other-submissions/online-registration-food-facilities.

35. See "Product Codes and Product Code Builder," U.S. Food and Drug Administration, Silver Spring, MD, https://www.fda.gov/industry/import-program-resources/product-codes-and-product-code-builder.

36. For a list of certified countries and establishments, see "Eligible Foreign Establishments," Food Safety and Inspection Service, U.S. Department of Agriculture, Washington, DC, https://www.fsis.usda.gov/wps/portal/fsis/topics/international-affairs/importing-products/eligible-countries-products-foreign-establishments/eligible-foreign-establishments.

37. For any Codex Alimentarius standards mentioned in this chapter, see Standards (database), Codex Alimentarius, International Food Standards, Codex Alimentarius Commission Secretariat, Food and Agriculture Organization of the United Nations, Rome, http://www.fao.org/fao-who-codexalimentarius/codex-texts/list-standards/en/.

38. Appendix A, table A.1, lists the web addresses of various relevant national, regional, and international regulatory authorities.

39. See "RASFF: Food and Feed Safety Alerts," Rapid Alert System for Food and Feed, European Commission, Brussels, https://ec.europa.eu/food/safety/rasff_en.

40. For an overview of state departments of public health and state departments of agriculture and other information about food safety initiatives in the United States, see "Your Gateway to Food Safety Information," U.S. Department of Health and Human Services, Washington, DC, https://www.foodsafety.gov/.

41. State sanitary controls (supervision) are implemented based on the product and process requirements set forth in the documentation on uniform sanitary, epidemiological, and hygiene requirements for products subject to state control (supervision). State veterinary controls are carried out according to documentation on uniform veterinary (veterinary and sanitary) requirements for goods subject to veterinary inspection (supervision).

42. Food products and other items covered by certain TRs are exempt from compliance with uniform sanitary, epidemiological, and hygiene requirements. These include, for example, materials and articles produced of polymer and other materials intended for contact with food and food media, labeling requirements, food additives and flavorings, and technological aids, as well as meat, meat products, milk, and dairy products.

43. CU TR 021/2011 "On Food Safety"; CU TR 005/2011 "On the Safety of Packaging"; CU TR 023/2011 "On Fruit and Vegetable Juice Products"; CU TR 027/2012 "On the Safety of Certain Types of Specialized Food Products, Including Foods for Dietary Treatment and Dietary Preventive Nutrition"; CU TR 033/2013 "On Milk and Dairy Products"; CU TR 034/2013 "On Meat and Meat Products." As set out in the explanatory note of CU TR 021/2011, the requirements, including on microbiological safety, are based on national laws of the EAEU member states and on international requirements.

References

CAC (Codex Alimentarius Commission). 2003. "General Principles of Food Hygiene." CAC/RCP 1-1969, Rev. 4-2003, CAC Secretariat, Food and Agriculture Organization of the United Nations, Rome.

European Commission. 2004. "From Farm to Fork: Safe Food for Europe's Consumers." Office for Official Publications of the European Communities, Luxembourg. https://www.paceadesso.it/ew/ew_sitepage/3/from%20farm%20to%20fork%20-An%20European%20Programme.pdf.

CHAPTER 2

European Commission. 2007. "Food Traceability." Factsheet (June), Directorate-General for Health and Consumer Protection, European Commission, Brussels. https://ec.europa.eu/food/sites/food/files/safety/docs/gfl_req_factsheet_traceability_2007_en.pdf.

———. 2018. "Guidance Document on the Implementation of Certain Provisions of Regulation (EC) No. 852/2004: On the Hygiene of Foodstuffs." Directorate-General for Health and Food Safety, European Commission, Brussels. https://ec.europa.eu/food/sites/food/files/safety/docs/biosafety_fh_legis_guidance_reg-2004-852_en.pdf.

FAO (Food and Agriculture Organization of the United Nations) and WHO (World Health Organization). 2006. "FAO/WHO Guidance to Governments on the Application of HACCP in Small and/or Less-Developed Food Businesses." FAO Food and Nutrition Paper 86 (November), FAO, Rome.

———. 2007. "Food Labelling." 5th ed., Codex Alimentarius Commission Secretariat, FAO, Rome.

———. 2012. *Food Import and Export Inspection and Certification Systems.* 5th ed. Rome: Codex Alimentarius Commission Secretariat, FAO, Rome.

FDA (U.S. Food and Drug Administration). 2012. *Bad Bug Book: Handbook of Foodborne Pathogenic Microorganisms and Natural Toxins.* College Park, MD: FDA. https://www.fda.gov/files/food/published/Bad-Bug-Book-2nd-Edition-%28PDF%29.pdf.

———. 2013. *A Food Labeling Guide: Guidance for Industry.* January. College Park, MD: Office of Nutrition, Labeling, and Dietary Supplements, Center for Food Safety and Applied Nutrition, FDA.

———. 2018. *Hazard Analysis and Risk-Based Preventive Controls for Human Food: Guidance for Industry.* Draft guidance (January). College Park, MD: Center for Food Safety and Applied Nutrition, FDA. https://www.fda.gov/media/100002/download.

———. 2019. "Voluntary Labeling Indicating Whether Foods Have or Have Not Been Derived from Genetically Engineered Plants: Guidance for Industry." March, Office of Nutrition and Food Labeling, Center for Food Safety and Applied Nutrition, FDA, College Park, MD.

FDA (U.S. Food and Drug Administration) and USDA (U.S. Department of Agriculture). 2000. "A Description of the U.S. Food Safety System." March 3, U.S. Codex Office, Food Safety and Inspection Service, USDA, Washington, DC.

FoodDrinkEurope and EuroCommerce. 2013. *Guidance on the Provision of Food Information to Consumers: Regulation (EU) No. 1169/2011.* September. Brussels: FoodDrinkEurope and EuroCommerce. https://www.fooddrinkeurope.eu/uploads/publications_documents/FDE_Guidance_WEB.pdf.

USDA (U.S. Department of Agriculture). 2018a. "Australia: Food and Agricultural Import Regulations and Standards Report." FAIRS Annual Country Report, GAIN Report AS1832 (November 16), Global Agricultural Information Network, Foreign Agricultural Service, USDA, Canberra, Australia.

———. 2018b. "Japan: Food and Agricultural Import Regulations and Standards Report." FAIRS Annual Country Report, GAIN Report JA8113 (December 21), Global Agricultural Information Network, Foreign Agricultural Service, USDA, Tokyo.

———. 2019. "China, Peoples Republic of: Food and Agricultural Import Regulations and Standards Report." FAIRS Export Certificate Report, GAIN Report CH 18087 (February 22), Global Agricultural Information Network, Foreign Agricultural Service, USDA, Beijing.

WHO (World Health Organization) and FAO (Food and Agriculture Organization of the United Nations). 2009. *Codex Alimentarius: Food Hygiene, Basic Texts.* 4th ed. Rome: Codex Alimentarius Commission Secretariat, FAO.

World Bank. 2015. *Comparative Analysis of Certain Requirements of Food Legislation in the European Union and the Customs Union of Russia, Belarus, and Kazakhstan.* Washington, DC: World Bank.

CHAPTER 3

Food Safety Tools and Techniques

An overview of prerequisite programs

The World Health Organization (WHO 1999, 4) defines a prerequisite program (PRP) as the "practices and conditions needed prior to and during the implementation of HACCP [hazard analysis critical control point (HACCP)] and which are essential for food safety." PRPs provide a foundation for effective HACCP systems. They are often facility-wide programs rather than process or product specific. They aim to prevent or reduce the likelihood of food safety hazards. PRPs are outside the hazard control plan, but still within the HACCP system (figure 3.1).

International Organization for Standardization (ISO) 22000:2018 defines a PRP as the "basic conditions and activities that are necessary within the organization . . . and throughout the food chain . . . to maintain food safety."[1]

Food business operators (FBOs) can meet their food safety responsibilities by implementing food safety management systems (FSMSs) along the food production chain. PRPs are the initial controls established by an FBO. The PRPs needed by an FBO depend on the segment of the food production chain in which the operator is active and the type of the food business. Examples of PRPs include good agricultural practice (GAP), good distribution practice, good hygiene practice, good manufacturing practice (GMP), good production practice, good trading practice, good veterinary practice, and good warehouse practice.

The ISO, the largest source of international standards, has issued numerous PRP standards. The PRP food safety standards and related guidelines—ISO/technical specification (TS) 22002—are as follows: part 1: food manufacturing (2009), ISO/TS 22002-1; part 2: catering (2013), ISO/TS 22002-2; part 3: farming (2011), ISO/TS 22002-3; part 4: food packaging manufacturing (2013), ISO/TS 22002-4; part 5: transport and storage (2019), ISO/TS 22002-5; and part 6: feed and animal food production (2016), ISO/TS 22002-6.

Figure 3.1 Position of PRPs in an HACCP System

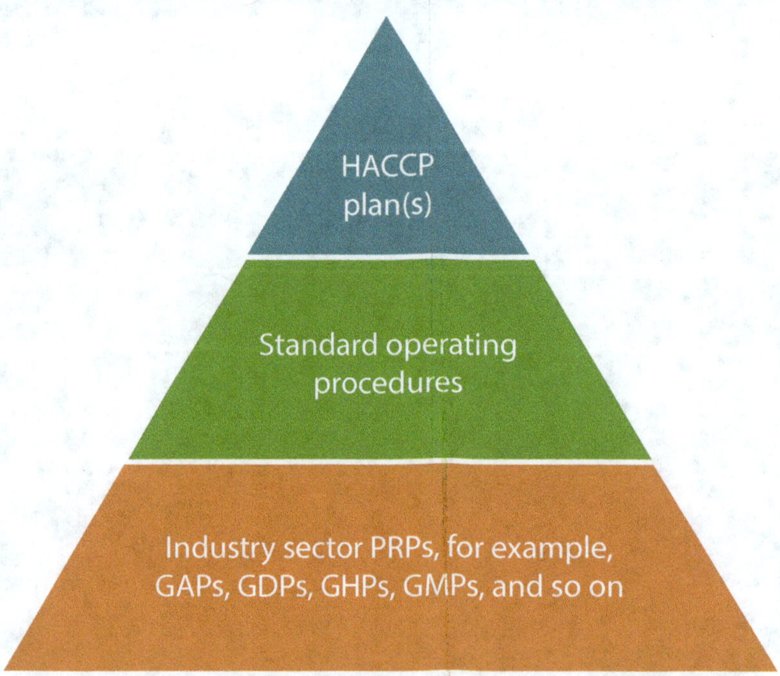

Note: GAP = good agricultural practice; GDP = good distribution practice; GHP = good hygiene practice; GMP = good manufacturing practice.

PRPs thus represent the foundation of food safety. Without well-developed PRPs that are properly documented, implemented, and maintained, an FBO risks serious problems. Most foodborne outbreaks are caused not by a breakdown or failure at critical control points (CCPs), but by a failure in one or more PRPs. The word "maintained" is used for a reason.

Many businesses may face challenges, but small-scale producers and traders in developing countries especially need support in planning and implementing food safety management programs in line with international requirements and the guidelines and recommendations of the Codex Alimentarius Commission (CAC).[2]

Because ISO/TS 22002–specific guideline standards are aligned with the Codex Alimentarius, this chapter provides a high-level overview of PRPs and PRP requirements. A particular focus is ISO/TS 22002-1, the Food Manufacturing Specification Standard.[3]

ISO/TS 22002-1:2009 specifies the requirements for establishing, implementing, and maintaining PRPs to assist in controlling food safety hazards.

ISO/TS 22002-1:2009 is applicable to all organizations, regardless of size or complexity, that are involved in manufacturing along the food chain and that are seeking to implement PRPs to address the requirements specified in ISO 22000:2018, clause 8.2.

ISO/TS 22002-1:2009 is not designed or intended for use in other parts of the food supply chain.

Food manufacturing operations are diverse, and the requirements specified in ISO/TS 22002-1:2009 may not apply to all individual establishments or processes. However, exclusions and alternative measures need to be justified and documented through a hazard analysis, as described in ISO 22000:2018, 8.2. Any exclusions or alternative measures adopted should not affect the ability of an FBO to comply with the requirements. Examples of such exclusions include the other aspects considered relevant to manufacturing operations that are listed below under (1)–(5), beginning with (1) rework and ending with (5) food defense, biovigilance, and bioterrorism.

ISO/TS 22002-1:2009 specifies the detailed requirements to be specifically considered in relation to ISO 22000:2018, 8.2.4, as follows: (1) the construction and layout of buildings and associated utilities; (2) the layout of premises, including zoning, workspace, and employee facilities; (3) the supply of air, water, energy, and other utilities; (4) pest control, waste and sewage disposal, and support services; (5) the suitability of equipment and the accessibility of equipment for cleaning and maintenance; (6) supplier approval and assurance processes (raw materials, ingredients, chemicals, and packaging); (7) the reception of incoming materials, storage, dispatch, transport, and product handling; (8) measures for the prevention of cross-contamination; (9) cleaning and disinfecting; (10) personnel hygiene; (11) product information and consumer awareness; and (12) other.

ISO/TS 22002-1:2009 also adds other aspects considered relevant to manufacturing operations, as follows: (1) rework, (2) product recall procedures, (3) warehousing, (4) product information and consumer awareness, and (5) food defense, biovigilance, and bioterrorism.

General information on PRPs

PRPs support the HACCP plan

PRPs deal with the good housekeeping concerns of individual establishments, whereas an HACCP manages specific process hazards.

FBOs must provide all documentation, including written programs, records, and results of all PRPs that support an HACCP system. For example, an establishment may conclude that *Escherichia coli* O157:H7 is a hazard that is not reasonably likely to occur during the establishment's processing because the establishment has a PRP with purchase specifications addressing *Escherichia coli* O157:H7.

PRP supporting documentation must be maintained. Without this documentation, the auditor of the Global Food Safety Initiative (GFSI) would question the adequacy of the establishment's HACCP system and hazard analysis. GFSI auditors expect the PRP supporting documentation to include the program's procedures and operational controls in written form. In addition, GFSI auditors expect the documentation to include records that demonstrate that the program is effective and that *Escherichia coli* O157:H7 is not reasonably likely to occur. Generally, an FBO's own food safety inspectors are required to review testing and PRP records at least once a week.

Differences between CCPs in establishment hazard control plans and in PRPs

PRPs are outside the hazard control plan, but still within the HACCP system. FBO auditors cannot apply the same criteria as they would in verifying the regulatory requirements of the hazard control plan. Inspection program personnel should evaluate PRPs and determine if they continue to support the decision in the hazard analysis. So what is the difference between a CCP in an establishment's hazard control plan and a PRP? A CCP is designed to control a food safety hazard that has been determined to be reasonably likely to occur. A PRP may prevent a food safety hazard from occurring.

PRPs set the stage for a hazard control plan and provide ongoing support for an FBO's FSMS. They keep potential hazards from becoming sufficiently serious to affect adversely the safety of the foods produced. Thus, if an establishment fails to follow its PRP addressing the occurrence of *Escherichia coli* O157:H7, there is a significant food safety concern.

The role of PRPs

FBOs should revise their PRPs, as necessary, to ensure their effectiveness, and they should take appropriate corrective actions if they determine that their PRPs may have failed to prevent the contamination or adulteration of a food product. Suppose an establishment addresses *Escherichia coli* O157:H7 in a PRP, but not in a hazard control plan. If the establishment produces an *Escherichia coli* O157:H7–positive product, this would be considered a deviation not covered by a specific corrective action or an unforeseen hazard. The establishment would therefore be required to take the corrective action, including reassessment. The PRP was not effective in reducing the likely risk in the processing environment.

The review of records generated by PRPs

PRP implementation must be associated with supporting documentation, such as records verifying implementation if this is referenced in the hazard analysis, hazard control plan, or sanitation standard operating procedure (SOP). Records on monitoring and testing may include instances of less than perfect control without resulting in a threat to food or product safety. However, records generated by PRPs must support the decisions made in the establishment's hazard analysis. When GFSI auditors review PRP records, they should review the records, results, and supporting documentation of the FBO's hazard control plan. Hence, if the FBO is reviewing the results and records on a weekly basis, it may identify trends, missing records, and so on indicating that a PRP may no longer support the decisions made in the hazard analysis, which would represent noncompliance.

Planning, developing, and managing PRPs

During the identification and development of PRPs, it is essential to consider information on statutory and regulatory requirements; industry standards and codes of practice; CAC principles and codes of practice; and international food safety standards, for example, Food Safety System Certification (FSSC) 22000, BRC Global Standards, Safe Quality Food Programs, GLOBALG.A.P., and so on. Customer requirements include historic data, such as audit reports and customer complaints.

All PRPs should be documented, regularly audited, reviewed periodically, and modified whenever necessary. As a general rule, PRPs and hazard control plans are managed separately. However, certain parts of PRPs may sometimes be integrated into a hazard control plan.

There are three challenges in PRP development: (1) developing and implementing effective PRPs, (2) maintaining the PRPs once they have been implemented, and (3) ensuring that the programs will stand up to auditor scrutiny.

Establishing an effective PRP is a good start, but FSMSs that are proscriptive may be too restrictive to be effective. Proper PRP maintenance is often overlooked. In the field, PRPs may appear beautifully designed and written, but they are simply not being followed in FBO operations. FBO operations must match the documented procedures.

In building PRPs, FBOs should seek to realize the following elements: responsibility, development, documentation, implementation, training, monitoring and recording, verifying and auditing, and reviewing and updating.

PRP workbook: Instructions and examples

This section offers guidance on the methodology for developing a PRP. The methodology may be applied to any food product, but the examples focus on particular dairy sector PRPs and the related FSMS documentation based on ISO/TS 22002-1.

The examples cover the documentation and other steps needed to establish six relevant dairy sector PRPs. The six PRPs are PRP 6: utilities—supply of air, water, and energy; PRP 9: management of purchased materials; PRP 11: cleaning and sanitizing; PRP 12: pest control; PRP 13: personnel hygiene and employee facilities; and PRP 14: rework.

The process for each PRP involves the use of up to six work sheets. Although the procedures are broadly the same, instructions on how to fill out the work sheets are supplied for all six PRPs. Complete sample work sheets are also provided in tables for all six PRPs.

The section is organized as follows. Each of the six PRPs is the subject of a separate subsection. The subsections consist mainly of the relevant sample work sheets and instructions on how to complete them. Editable work sheets and templates can be found at http://www.ifc.org/foodsafety/handbook/templates.

PRP 6: Utilities—supply of air, water, and energy

WORK SHEET 1: PRP SCOPE

Work sheet 1 defines the scope of an FBO PRP. The information on the work sheet needs to be clear, especially in detailing the product or products, including the production lines, that are the subject of the PRP study. The work sheet should also provide information about the individuals making up the study team, along with any revision history of the PRP. The PRP scope work sheet contains five sections (see table 3.1). Instructions for completing these sections are outlined in the box below.

Instructions for Completing PRP Work Sheet 1: PRP Scope	
A. PRP study scope	Provide the PRP title from the standard or scheme (for example, Utilities). Provide the standard PRP number (for example, in ISO/TS 22002-1, 6 Utilities—supply of air, water, and energy). Provide the facility name, product category, processes, product, PRP start date, status of the PRP (for example, draft, approved), and end date.
B. PRP review history	In this section, record information about the history of the PRP revision, with an explanation of the reason why this update has been done: "according to plan" or "unscheduled." For an unscheduled revision, why has this revision been undertaken? (What reason?)
C. PRP team members	For every PRP study, the organization needs to establish an HACCP team with specific responsibilities and roles. Names within the company, department name, and responsibilities should be detailed. The competence of each team member should also be documented.
D. Specialist input	To establish PRP studies, companies may need advice from an outsourced expert (consultant/subject matter expert). The expert's role should be explained: input/specialist advice.
E. Authorization	Team members must indicate their approval of the document by providing their names, positions, responsibilities held, and signature. The authorized team member should provide his/her signature and the date signed.

Table 3.1 PRP 6, Work Sheet 1: PRP Scope

PRP	6 Utilities—supply of air, water, and energy		
A. PRP study scope			
Facility	Joe Bloggs Dairy Plant	Start date	February 17, 2019
Product category	Grade A Interstate Milk Shippers registered whole milk	Status	Draft
Processes	High-temperature/short-time pasteurizer, aseptic filling, retort	End date	Ongoing
Products	Grade A aseptically processed and packaged milk		
B. PRP review history	Check as appropriate	Notes/reason for unscheduled review	Dates of last three reviews
New PRP study	✓	Current PRPs underwent a comprehensive review for compliance with ISO/TS 22002-1 and ISO 22000:2018 starting on February 15, 2019, and completed on February 17, 2019. These management sheets describe each PRP in place at the dairy plant facility.	
Scheduled review	December 20, 2019		
Unscheduled review			
C. PRP team members			
Name	Position	Department	Responsibility/role
G Moran	Food safety manager	Food safety	Food safety/quality assurance
O Brown	Hygienist/microbiologist	Food safety	Hygienist/microbiologist
M Rodrigues	Milk processing manager	Milk processing	Milk processing
B Murphy	Laboratory manager	Quality assurance	Laboratory
D Small	Warehouse manager	Warehousing	Warehousing
O Murphy	Engineering manager	Engineering	Engineering
C Flack	Factory manager	Management	Management
D. Specialist input			
Name	Location/job title	Input/specialist advice	
Angela Yard	Consultant	PRP team facilitator	
E. Authorization			
Food safety team leader/quality assurance manager		Signature: *G Moran*	Date: *February 17, 2019*
Management team member		Signature: *C Flack*	Date: *February 17, 2019*

CHAPTER 3

WORK SHEET 2: PRP MANAGEMENT

The purpose of this work sheet is to identify and document hazards and to cite the measures needed to control the hazards through relevant PRPs. The work sheet identifies the corrective actions to be taken should hazard levels rise above acceptable limits. It cites the records that need to be kept by FBOs, and the verification procedures required for each PRP. The work sheet consists of 11 columns (see table 3.2). The instructions for completing these sections are outlined in the box directly below and continuing on page 63.

Instructions for Completing PRP Work Sheet 2: PRP Management	
Column A	Describe the ISO/TS 22002-1 requirements.
Column B	Describe the hazard agent, for example, biological (B), chemical (C), physical (P), or a combination.
Column C	Describe how the hazard is manifest as a threat, including presence, increase, or survival.
Column D	Describe the cause, origin, condition, source, or vector of a hazard.
Column E	Describe the control measures the FBO has in place to control relevant hazards.
Column F	Describe the hazard measurement parameters and the monitoring frequency of the measurement parameters.

Table 3.2 PRP 6, Work Sheet 2: PRP Management

A. PRP (go step-by-step through ISO/TS 22002-1) 6 Utilities—supply of air, water, and energy	Hazards					
	B. Agent(s)	**C.** Presence, growth, survival, increase, contamination	**D.** Origin, cause, source, vector, condition	**E.** Control measures	**F.** What is monitored and when	
6.1 General requirements	B, C, P (see below)	Contamination	Contamination by pathogens	Utilities specifications, for example, air, water, gas Hygienic design of the dairy plant Pathogen monitoring procedure Supplier management procedure Product inspection procedure Cleaning/sanitizing awareness training Audits/inspections	Audits/inspection, hygiene, cleaning, sanitization, segregation/physical breaks between circuits containing cleaning solutions, temperature and pathogen monitoring program (each batch, daily, weekly)	
6.2 Water supply	B	Contamination	Contamination by pathogens may be introduced from the supplier of water (ground, surface)	Water supply specification Supplier certificate of analysis Supplier management program Incoming, in process, and wastewater treatment laboratory testing	Audits/inspections, temperature and pathogen monitoring each batch	

WORK SHEET 2: PRP MANAGEMENT

The purpose of this work sheet is to identify and document hazards and to cite the measures needed to control the hazards through relevant PRPs. The work sheet identifies the corrective actions to be taken should hazard levels rise above acceptable limits. It cites the records that need to be kept by FBOs, and the verification procedures required for each PRP. The work sheet consists of 11 columns (see table 3.2). The instructions for completing these sections are outlined in the box directly below and continuing on page 63.

Instructions for Completing PRP Work Sheet 2: PRP Management	
Column A	Describe the ISO/TS 22002-1 requirements.
Column B	Describe the hazard agent, for example, biological (B), chemical (C), physical (P), or a combination.
Column C	Describe how the hazard is manifest as a threat, including presence, increase, or survival.
Column D	Describe the cause, origin, condition, source, or vector of a hazard.
Column E	Describe the control measures the FBO has in place to control relevant hazards.
Column F	Describe the hazard measurement parameters and the monitoring frequency of the measurement parameters.

Table 3.2 PRP 6, Work Sheet 2: PRP Management

A. PRP (go step-by-step through ISO/TS 22002-1) 6 Utilities—supply of air, water, and energy	Hazards					
	B. Agent(s)	**C.** Presence, growth, survival, increase, contamination	**D.** Origin, cause, source, vector, condition	**E.** Control measures	**F.** What is monitored and when	
6.1 General requirements	B, C, P (see below)	Contamination	Contamination by pathogens	Utilities specifications, for example, air, water, gas Hygienic design of the dairy plant Pathogen monitoring procedure Supplier management procedure Product inspection procedure Cleaning/sanitizing awareness training Audits/inspections	Audits/inspection, hygiene, cleaning, sanitization, segregation/physical breaks between circuits containing cleaning solutions, temperature and pathogen monitoring program (each batch, daily, weekly)	
6.2 Water supply	B	Contamination	Contamination by pathogens may be introduced from the supplier of water (ground, surface)	Water supply specification Supplier certificate of analysis Supplier management program Incoming, in process, and wastewater treatment laboratory testing	Audits/inspections, temperature and pathogen monitoring each batch	

Table 3.1 PRP 6, Work Sheet 1: PRP Scope

PRP	6 Utilities—supply of air, water, and energy		
A. PRP study scope			
Facility	Joe Bloggs Dairy Plant	Start date	February 17, 2019
Product category	Grade A Interstate Milk Shippers registered whole milk	Status	Draft
Processes	High-temperature/short-time pasteurizer, aseptic filling, retort	End date	Ongoing
Products	Grade A aseptically processed and packaged milk		
B. PRP review history	Check as appropriate	Notes/reason for unscheduled review	Dates of last three reviews
New PRP study	✓	Current PRPs underwent a comprehensive review for compliance with ISO/TS 22002-1 and ISO 22000:2018 starting on February 15, 2019, and completed on February 17, 2019. These management sheets describe each PRP in place at the dairy plant facility.	
Scheduled review	December 20, 2019		
Unscheduled review			
C. PRP team members			
Name	Position	Department	Responsibility/role
G Moran	Food safety manager	Food safety	Food safety/quality assurance
O Brown	Hygienist/microbiologist	Food safety	Hygienist/microbiologist
M Rodrigues	Milk processing manager	Milk processing	Milk processing
B Murphy	Laboratory manager	Quality assurance	Laboratory
D Small	Warehouse manager	Warehousing	Warehousing
O Murphy	Engineering manager	Engineering	Engineering
C Flack	Factory manager	Management	Management
D. Specialist input			
Name	Location/job title	Input/specialist advice	
Angela Yard	Consultant	PRP team facilitator	
E. Authorization			
Food safety team leader/quality assurance manager		Signature: *G Moran*	Date: *February 17, 2019*
Management team member		Signature: *C Flack*	Date: *February 17, 2019*

Instructions for Completing PRP Work Sheet 2: PRP Management (*continued*)	
Column G	Describe the job role or title of the department/function within the FBO responsible for monitoring the relevant hazard measurement parameters.
Column H	Describe the correction and corrective action aimed at preventing a reoccurrence of a rise above the allowable or permitted hazard measurement parameters.
Column I	Indicate the monitoring and hazard measurement parameter records to be maintained.
Column J	Describe the verification activities necessary to confirm the accuracy of the monitoring and hazard measurement parameters.
Column K	Describe the FBO documents and relevant external documents, for example, statutory and regulatory requirements.

	G. Who is responsible	**H.** Correction/ corrective action	**I.** Records	**J.** Verification activities	**K.** Reference documents
	Dairy plant quality assurance/ laboratory Dairy plant engineering Dairy plant maintenance Dairy plant hygienist/ hygiene team Cleaning/sanitization operatives	Awareness/training 100% product inspection Product disposal, where relevant	Product inspection Audits Good hygiene practices inspections Pathogen monitoring Awareness/training Wastewater treatment Product spoilage/ disposal	Product (water supply) inspection Supplier management program Pathogen monitoring Chemical residue Product spoilage/ disposal	Dairy Plant Layout of Premise and Workspace PRP Dairy Plant Waste Disposal PRP Product Inspection Procedure Dairy Plant Audit Procedure Dairy Plant Good Hygiene Practices Inspection Procedure Dairy Plant Awareness/Training Procedure Dairy Plant Wastewater Treatment Monitoring Procedure Dairy Plant Pathogen Monitoring Procedure
	Dairy plant quality assurance/laboratory	Resterilization of piping, equipment, and containers	Audits Good hygiene practices inspections Awareness/training Pathogen monitoring	Water supply inspection Product disposal	Dairy Plant Cleaning and Sanitizing PRP Dairy Plant Product Inspection Procedure Dairy Plant Pathogen Monitoring Procedure

continued

CHAPTER 3

Table 3.2 PRP 6, Work Sheet 2: PRP Management *(Continued)*

A. PRP (go step-by-step through ISO/TS 22002-1) 6 Utilities—supply of air, water, and energy	Hazards			**E.** Control measures	**F.** What is monitored and when
	B. Agent(s)	**C.** Presence, growth, survival, increase, contamination	**D.** Origin, cause, source, vector, condition		
6.2 Water supply *(continued)*	C	Contamination	Cleaning and sanitizing solution residues, that is, without proper separation between cleaning and sanitizing solutions and product; there could be product contamination	Maintain proper separation or physical break between circuits containing cleaning solutions and containers and pipelines used to contain product. Particular attention is needed to assure that the required separation remains in place during partial/short/interwashes completed during an operating day	Audits/inspection Segregation or physical break between circuits containing cleaning solutions and containers and pipelines used to contain product Solution temperature, concentration, duration of application, cleaning sequence, flow rates, and so on (daily)
	P	None	None	Not applicable	Not applicable
6.3 Boiler chemicals	B	None	None	Not applicable	Not applicable
	C	Contamination	Boiler additives. Some boiler water compounds used in the production of steam to be used in contact with food or food contact surfaces may contain toxic substances	Boiler additives specification Supplier management program	Boiler water additives (daily/weekly)
	P	None	None	Not applicable	Not applicable
6.4 Air quality and ventilation	B	Contamination	Contamination by pathogens may be introduced into the air supply and may come in contact with the product or food contact surface if negative air pressure in the dairy plant is allowed to occur	Hygienic dairy plant design incorporating heating, ventilation, and air-conditioning system (creation of positive air pressure zones), air ducts, air filtration, exhaust stacks, intake ducts Cleaning of air ducts Air filtration Environmental pathogen monitoring Air testing (past the filtration)	Environment pathogen monitoring Air filtration Air quality Air turns Cleaning of air ducts (daily/weekly)
	C	None	Not applicable	Not applicable	Not applicable
	P	None	Not applicable	Not applicable	Not applicable
6.5 Compressed air and gases	B	Contamination	Contamination by pathogens may be introduced into the air supply and may come in contact with the product or food contact surface	Specification for the supply of compressed air Air is drawn from a clean area, is filtered at the intake as needed, and is provided to the point of use oil free and free of excess moisture. A final filter is provided as near as possible to the point of use to verify these aspects	Environment pathogen monitoring (daily/weekly)

	G. Who is responsible	**H.** Correction/ corrective action	**I.** Records	**J.** Verification activities	**K.** Reference documents
	Dairy plant hygienist/ hygiene team Cleaning/sanitization operatives	Awareness/training, product disposal, where relevant	Audits Good hygiene practices inspections Awareness/training Product spoilage/ disposal	Product inspection Product disposal	Dairy Plant Cleaning and Sanitizing PRP Dairy Plant Waste Disposal PRP Dairy Plant Awareness/Training Procedure
	Not applicable	Not applicable	Not applicable	Not applicable	Not applicable
	Not applicable	Not applicable	Not applicable	Not applicable	Not applicable
	Quality assurance	Return to supplier: products that are not to specification	Incoming product	Incoming product supplier management program	Management of Purchased Materials PRP Dairy Plant Product Inspection Procedure Dairy Plant Supplier Management Procedure
	Not applicable	Not applicable	Not applicable	Not applicable	Not applicable
	Dairy plant quality assurance/laboratory (environment pathogen monitoring/air testing) Dairy plant engineering (dairy plant hygienic design heating, ventilation, and air-conditioning system) Dairy plant maintenance (preventive maintenance of filters/ cleaning, or air ducts, and so on)	Product hold/ withdrawal/ recall Testing of all production lots Implementation of intensive cleaning/ sanitization Review/revisions of process controls	Environment pathogen monitoring Heating, ventilation, and air-conditioning system design/ drawings Air testing Preventive maintenance (filter/ cleaning)	Environment pathogen monitoring	Dairy Plant Layout of Premises and Workspace PRP Dairy Plant Cleaning and Sanitizing PRP Dairy Plant Environment Pathogen Monitoring Program Dairy Plant Product Inspection Procedure
	Not applicable	Not applicable	Not applicable	Not applicable	Not applicable
	Not applicable	Not applicable	Not applicable	Not applicable	Not applicable
	Dairy quality assurance laboratory	Replace compressed air/filter	Environment pathogen monitoring Preventive maintenance (filter)	Environment pathogen monitoring	Dairy Plant Environment Suitability, Cleaning, and Maintenance PRP Dairy Plant Pathogen Monitoring Program

continued

Table 3.2 PRP 6, Work Sheet 2: PRP Management *(Continued)*

A. PRP (go step-by-step through ISO/TS 22002-1) 6 Utilities—supply of air, water, and energy	Hazards				
	B. Agent(s)	**C.** Presence, growth, survival, increase, contamination	**D.** Origin, cause, source, vector, condition	**E.** Control measures	**F.** What is monitored and when
6.5 Compressed air and gases *(continued)*	C	Contamination	Toxic substances, that is, air compressor lubricants, may be carried over the air and may be toxic	Specification for the supply of compressor lubricants (food grade) Air is drawn from a clean area, is filtered at the intake as needed, and is provided to the point of use oil free and free of excess moisture. A final filter is provided as near as possible to the point of use to verify these aspects	Environment pathogen monitoring (daily/weekly)
	P	None	Not applicable	Not applicable	Not applicable
6.6 Lighting	B	Contamination	Poor or inadequate lighting (intensity) may contribute to personnel applying poor hygiene standards and, as a result, material, products, or equipment may become contaminated	Hygienic design of the dairy plant Throughout the dairy plant, storage, preparation, processing areas are provided with natural or artificial lighting (or both). A minimum light intensity of 200 lux is recommended. Reference the relevant national lighting standard for recommended lighting standards. All lights are fitted with light diffusers/covers or shatterproof tubes to facilitate cleaning and to prevent contamination of food	Hygienic design, light intensity, dirt, spills, pest (daily/weekly)
	C	None	Not applicable	Not applicable	Not applicable
	P	Contamination	Poor or inadequate lighting (intensity) may contribute to personnel applying poor hygiene standards and, as a result, material, product, or equipment may become contaminated, for example, breakages and/or dirt	Hygienic design of the dairy plant; for example, all lights are fitted with light diffusers/covers or shatterproof tubes to facilitate cleaning and to prevent contamination of food and the premises should breakage occur Hygiene inspections to detect breakages or dirt	Hygienic design, breakages, and dirt (daily/weekly)

	G. Who is responsible	**H.** Correction/ corrective action	**I.** Records	**J.** Verification activities	**K.** Reference documents
	Dairy quality assurance laboratory	Replace compressed air filter	Environment pathogen monitoring Preventive maintenance (filter)	Environment pathogen monitoring	Dairy Plant Environment Suitability, Cleaning, and Maintenance PRP Dairy Plant Environment Pathogen Monitoring Program Dairy Plant Preventive Maintenance Procedure
	Not applicable	Not applicable	Not applicable	Not applicable	Not applicable
	Dairy plant engineering/food safety (dairy plant hygienic design) Dairy plant maintenance (lighting maintenance) Cleaning/sanitization program, including spills Dairy plant hygienist and hygiene team	Capital expenditure projects (hygiene related) Preventive maintenance Cleaning/sanitization program	Capital expenditure projects Preventive maintenance Cleaning Good hygiene practices inspection	Cleaning/sanitizing Good hygiene practices inspections, audits	Dairy Plant Design and Construction of Buildings PRP Dairy Plant Site Location and Standards PRP Dairy Plant Layout of Premises and Workspace PRP Dairy Plant Internal Structure PRP Dairy Plant Environment Suitability, Cleaning, and Maintenance PRP Preventive Maintenance Procedure Hygiene Procedures Cleaning/Sanitization Procedures
	Not applicable	Not applicable	Not applicable	Not applicable	Not applicable
	Dairy plant engineering/food safety (dairy plant hygienic design) Dairy plant maintenance (lighting maintenance) Dairy plant hygienist and hygiene team	Capital expenditure projects (hygiene related) Preventive maintenance Cleaning/sanitization program	Capital expenditure projects Preventive maintenance Cleaning/sanitizing Good hygiene practices inspection	Cleaning/sanitizing Good hygiene practices inspection Audits	Dairy Plant Design and Construction of Buildings PRP Dairy Plant Site Location and Standards PRP Dairy Plant Layout of Premises and Workspace PRP Dairy Plant Environment Suitability, Cleaning, and Maintenance PRP Dairy Plant Internal Structure PRP Preventive Maintenance Procedure Hygiene Procedures Cleaning/Sanitization Procedures

WORK SHEET 3: PRP VERIFICATION ACTION PLAN

Verification is a confirmation, replete with objective evidence, specifying that requirements have been fulfilled. Original PRP verification is carried out after a PRP has been developed and implemented. Additional planned verifications, at least once a year, and unscheduled verifications are required in the case of alterations in the PRP. An FBO should draw up a verification plan. However, verification may only be carried out by an authorized individual. The FBO must document all verification activities for each PRP. This work sheet (see table 3.3) aids in planning for PRP verification. Instructions for completing the work sheet are outlined in the box directly below.

Instructions for Completing PRP Work Sheet 3: PRP Verification Action Plan	
Column A	The establishment should provide details on the number and title of the PRP. It is recommended that the number should match the number of the relevant part in the appropriate FSMS scheme standard, for example, in ISO/TS 22002-1, 6 Utilities—supply of air, water, and energy.
Column B	The establishment should provide details on the verification actions associated with the PRP and on the individual or entity responsible for reviewing these verification actions.

Table 3.3 PRP 6, Work Sheet 3: PRP Verification Action Plan

A. PRP	**B.** Verification action
6 Utilities—supply of air, water, and energy	Reviewed by regulated utilities price plan team
	Review of referenced documents, for example, PRP-related procedures and utility specifications Review of pathogen monitoring records Review of product inspection records Review of records on cleaning/sanitizing Review of preventive maintenance records Review of product spoilage/disposal records Review of rework records Review of awareness/training records Review of consumer complaints FSMS audits Internal good manufacturing practice audits/good hygiene practice inspections Frequency and criticality review

CHAPTER 3

WORK SHEET 4: PRP MEETING SUMMARY

This work sheet is focused on assisting in maintaining records on PRP meetings and any meeting decisions. It consists of seven columns (see table 3.4). The instructions for completing the work sheet are summarized in the box directly below.

Instructions for Completing PRP Work Sheet 4: PRP Meeting Summary	
Column A	List meeting dates.
Column B	List attendees among the team and other invitees.
Column C	Provide the reason for the meeting.
Column D	Record decisions and next steps.
Column E	Identify the individuals or entities responsible for executing any decisions.
Column F	Record deadlines.
Column G	Indicate the dates of relevant actions.

Table 3.4 PRP 6, Work Sheet 4: PRP Meeting Summary

A. Date	**B.** Participants	**C.** Purpose	**D.** Outcome (decisions/actions)	**E.** Responsibility	**F.** Deadline	**G.** Deadline reached
April 20, 2018	G Moran, O Brown, M Rodrigues, B White, D Collins, O Murphy, C Flack	Initial PRP review	Update PRP management work sheet Review related PRPs	G Moran to complete verification sheet	May 15, 2018	May 15, 2018
April 28, 2018	G Moran, O Brown, M Rodrigues, B White, D Collins, O Murphy, C Flack	Complete GAP sheet Review PRP management work sheet	Completed and approved Reviewed and approved	G Moran to update PRP work sheets	May 15, 2018	May 15, 2018
February 17, 2019	G Moran, O Brown, M Rodrigues, B White, D Collins, O Murphy, C Flack	Review and update of utility specifications	Complete the update of the water supply specification	PRP team to complete	February 17, 2019	February 17, 2019
February 17, 2019	G Moran, O Brown, M Rodrigues, B White, D Collins, O Murphy, C Flack	Review and update based on changes to ISO 22000:2018	The current PRPs underwent comprehensive reviews for compliance with ISO/TS 22002-1 and ISO 22000:2018 starting on February 17, 2019, and completed on February 20, 2019	PRP team to complete	February 20, 2019	February 20, 2019

CHAPTER 3

WORK SHEET 5: PRP GAP REGISTRATION AND RESOLUTION

This work sheet defines gaps between the PRP requirements according to a certain standard or standards, for example, ISO/TS 22002-1, and other requirements with which an FSMS imposes compliance. The work sheet assists in eliminating these gaps. In completing the work sheet, an FBO may rely on different standards and documents to determine PRP requirements, for instance, ISO/TS 22002-1. The standards and documents should correspond with the FSMS requirements with which the particular FBO must engage. The work sheet consists of eight columns (see table 3.5). The instructions for completing the work sheet are summarized in the box directly below.

Instructions for Completing PRP Work Sheet 5: PRP Gap Registration and Resolution	
Column A	Provide a description of the FSMS scheme requirement.
Column B	Provide a description of the requirement arising from the FSMS scheme where the gap exists.
Column C	Provide a short description of the specific requirement where the gap exists within the FBO.
Column D	Detail the relevant FSMS policy.
Column E	Describe the gap.
Column F	Provide the action to be taken to address the requirement identified as not having been fulfilled.
Column G	Provide details of the actions taken to address the gap and the date of the completion of the actions.
Column H	Add any additional relevant comments as required.

Table 3.5 PRP 6, Work Sheet 5: PRP Gap Registration and Resolution

Fill out this work sheet only if gaps have been identified.

A. ISO/TS 22002-1, 6 Utilities—supply of air, water, and energy	**B.** Description (of the requirement of the standard)	**C.** Specific requirement	**D.** Associated dairy policy	**E.** Gap	**F.** Action plan (including time frame for completion)	**G.** Gap resolution (actions completed, with date)	**H.** Comments
6.3 Boiler chemicals	The provision and distribution routes for utilities to and around processing and storage areas shall be designed to minimize the risk of product contamination. Utility quality shall be monitored to minimize the risk of product contamination	Boiler chemicals, if used, shall be approved food additives that meet relevant additive specifications	Dairy plant food safety policy	PRP management work sheet incomplete, related PRPs and procedures to be reviewed and updated	All documents to be reviewed and updated prior to next PRP team meeting May 15, 2018	All documented, reviewed, and updated; see PRP team meeting	None
6.4 Air quality and ventilation	The organization shall establish requirements for the filtration, humidity (% relative humidity), and microbiology of the air used as an ingredient or for direct product contact	Specification for pressurized air	Dairy plant food safety policy	Utility specifications to be set	Create pressurized air specification February 17, 2019	Air specification completed; see PRP team meeting	None

WORK SHEET 6: HAZARD AGENT

The purpose of this work sheet (see table 3.6) is to define a standard classification system for recording hazardous agents. The hazardous agents classification system is based on the food and beverage industry hazardous agent classification system. The work sheet is supplied for reference and guidance only. The instructions for completing the work sheet are summarized in the box directly below.

Instructions for Completing PRP Work Sheet 6: Hazard Agent	
Column A	Classify food safety hazard agents, for example, biological, chemical, or physical hazard agents.
Column B	Indicate the food safety hazard agent code, for example, allergen = A, biological = B, chemical = C, physical = P.

Table 3.6 PRP 6, Work Sheet 6: Hazard Agent

A. Hazardous agents	**B.** Hazard class
Biological (for example, vegetative or spores, depending on circumstances)	B
Chemical (for example, cleaning chemicals, nonfood-grade lubricants, oils and greases, and chemical residues)	C
Physical (for example, various types of foreign material, including metal, wood, plastic, or other foreign bodies)	P
Allergens (for example, milk, soy, wheat, eggs, fish, shellfish, tree nuts, peanuts)	A

PRP 9: Management of purchased materials

Sample completed work sheets for PRP 9 follow (tables 3.7–3.12). For instructions on filling out each PRP work sheet, see the boxes that precede each sample completed work sheet.

Instructions for Completing PRP Work Sheet 1: PRP Scope	
A. PRP study scope	Provide the PRP title from the standard or scheme (for example, Management of purchased materials). Provide the standard PRP number (for example, in ISO/TS 22002-1, 9—Management of purchased materials). Provide the facility name, product category, processes, product, PRP start date, status of the PRP (for example, draft, approved), and end date.
B. PRP review history	In this section, record information about the history of the PRP revision, with an explanation of the reason why this update has been done: "according to plan" or "unscheduled." For an unscheduled revision, why has this revision been undertaken? (What reason?)
C. PRP team members	For every PRP study, the organization needs to establish an HACCP team with specific responsibilities and roles. Names within the company, department name, and responsibilities should be detailed. The competence of each team member should also be documented.
D. Specialist input	To establish PRP studies, companies may need advice from an outsourced expert (consultant/subject matter expert). The expert's role should be explained: input/specialist advice.
E. Authorization	Team members must indicate their approval of the document by providing their names, positions, responsibilities held, and signature. The authorized team member should provide his/her signature and the date signed.

CHAPTER 3

Table 3.7 PRP 9, Work Sheet 1: PRP Scope			
PRP	9 Management of purchased materials		
A. PRP study scope			
Facility	Joe Bloggs Dairy Plant	Start date	February 17, 2019
Product category	Grade A Interstate Milk Shippers registered whole milk	Status	Draft
Processes	High-temperature/short-time pasteurizer, aseptic filling, retort	End date	Ongoing
Products	Grade A aseptically processed and packaged milk		
B. PRP review history	Check as appropriate	Notes/reason for unscheduled review	Dates of last three reviews
New PRP study	✓	Current PRPs underwent a comprehensive review for compliance with ISO/TS 22002-1 and ISO 22000:2018 starting on February 15, 2019, and completed on February 17, 2019. These management sheets describe each PRP in place at the dairy plant facility.	
Scheduled review	December 20, 2019		
Unscheduled review			
C. PRP team members			
Name	Position	Department	Responsibility/role
G Moran	Food safety manager	Food safety	Food safety/quality assurance
O Brown	Hygienist/microbiologist	Food safety	Hygienist/ microbiologist
M Rodrigues	Milk processing manager	Milk processing	Milk processing
B Murphy	Laboratory manager	Quality assurance	Laboratory
D Small	Warehouse manager	Warehousing	Warehousing
O Murphy	Engineering manager	Engineering	Engineering
C Flack	Factory manager	Management	Management
D. Specialist input			
Name		Location/job title	Input/specialist advice
Angela Yard		Consultant	PRP team facilitator
E. Authorization			
Food safety team leader/quality assurance manager		Signature: *G Moran*	Date: *February 17, 2019*
Management team member		Signature: *C Flack*	Date: *February 17, 2019*

CHAPTER 3

Instructions for Completing PRP Work Sheet 2: PRP Management	
Column A	Describe the ISO/TS 22002-1 requirements.
Column B	Describe the hazard agent, for example, biological (B), chemical (C), physical (P), or a combination.
Column C	Describe how the hazard is manifest as a threat, including presence, increase, or survival.
Column D	Describe the cause, origin, condition, source, or vector of a hazard.
Column E	Describe the control measures the FBO has in place to control relevant hazards.
Column F	Describe the hazard measurement parameters and the monitoring frequency of the measurement parameters.

Table 3.8 PRP 9, Work Sheet 2: PRP Management

A. PRP (go step-by-step through ISO/TS 22002-1) 9 Management of purchased materials	Hazards			**E.** Control measures	**F.** What is monitored and when
	B. Agent(s)	**C.** Presence, growth, survival, increase, contamination	**D.** Origin, cause, source, vector, condition		
9.1 General requirements	B, C, P (see below)	Presence, contamination	Supplier management, hygiene, cleaning, sanitization, and incoming material inspection in place as well as pathogen, environmental, and extraneous material monitoring	Supplier management program/procedure, audits/inspection, hygiene, cleaning, sanitization, and raw material monitored Pathogen, mycotoxin, and extraneous material monitoring program in place	Audits/inspection, hygiene, cleaning, sanitization, and raw material monitored Pathogen, mycotoxin, and extraneous material monitoring program in place

Instructions for Completing PRP Work Sheet 2: PRP Management (*continued*)	
Column G	Describe the job role or title of the department/function within the FBO responsible for monitoring the relevant hazard measurement parameters.
Column H	Describe the correction and corrective action aimed at preventing a reoccurrence of a rise above the allowable or permitted hazard measurement parameters.
Column I	Indicate the monitoring and hazard measurement parameter records to be maintained.
Column J	Describe the verification activities necessary to confirm the accuracy of the monitoring and hazard measurement parameters.
Column K	Describe the FBO documents and relevant external documents, for example, statutory and regulatory requirements.

	G. Who is responsible	**H.** Correction/ corrective action	**I.** Records	**J.** Verification activities	**K.** Reference documents
	Dairy plant (see below for details)	Awareness/ training Cleaning of area where deviation was found Raw material is sent back to supplier or discarded if not compliant	Various (see below for details)	Supplier management program Tank truck cleaning and sanitization records Raw milk temperature records Raw milk intake records	Dairy Plant Supplier Management Procedure Dairy Plant Audit Procedure Dairy Plant Awareness/Training Procedure Dairy Plant Good Hygiene Practices Inspection Procedure Dairy Plant Mycotoxin Analysis Testing Dairy Plant Raw Material Handling Procedure Dairy Plant Tank Truck Cleaning and Sanitizing Procedure Dairy Plant Record Control Procedure (manifest) Dairy Plant Product Inspection Procedure

continued

Table 3.8 PRP 9, Work Sheet 2: PRP Management *(Continued)*

A. PRP (go step-by-step through ISO/TS 22002-1) 9 Management of purchased materials	Hazards			**E.** Control measures	**F.** What is monitored and when	
	B. Agent(s)	**C.** Presence, growth, survival, increase, contamination	**D.** Origin, cause, source, vector, condition			
9.2 Selection and management of suppliers	B, C, P (see below)	Presence, contamination	Supplier management, hygiene, cleaning, sanitization, and incoming material inspection in place as well as pathogen, environmental, and extraneous material monitoring	Supplier management program/procedure, audits/inspection, hygiene, cleaning, sanitization, and raw material monitored Pathogen, mycotoxin, and extraneous material monitoring program in place	Audits/inspection, hygiene, cleaning, sanitization, and raw material monitored Pathogen, mycotoxin, and extraneous material monitoring program in place	
9.3 Incoming material requirements	B	Presence	Based on scientific studies, vegetative pathogens (*Brucella abortus*, *Campylobacter jejuni*, *Campylobacter coli*, *Coxiella burnetii*, pathogenic *Escherichia coli* 0157:H7, *Listeria monocytogenes*, *Mycobacterium tuberculosis*, *Mycobacterium bovis*, *Salmonella enterica* serotypes, *Streptococcus pyogenes*, and *Yersinia enterocolitica*) may be present in raw milk	Supplier management program Minimize the incoming bacterial load by purchasing Grade A listed raw milk and testing incoming product Verify that tank trucks were cleaned and sanitized prior to picking up the milk being unloaded Milk temperature records from the dairy farm to the dairy plant	Incoming product Tank truck cleaning and sanitizing records Milk temperature records (each batch)	

	G. Who is responsible	**H.** Correction/ corrective action	**I.** Records	**J.** Verification activities	**K.** Reference documents
	Dairy plant (see below for details)	Awareness/ training Cleaning of area where deviation was found Raw material is sent back to supplier or discarded if not compliant	Supplier inspections/audits Certificate of analysis requirements On-site (dairy farm) incoming product specification	Supplier management program Tank truck cleaning and sanitization records Raw milk temperature records Raw milk intake records	Dairy Plant Supplier Management Procedure Dairy Plant Audit Procedure Dairy Plant Awareness/Training Procedure Dairy Plant Good Hygiene Practices Inspection Procedure Dairy Plant Mycotoxin Analysis Testing Dairy Plant Raw Material Handling Procedure Dairy Plant Tank Truck Cleaning and Sanitizing Procedure Dairy Plant Record Control Procedure (manifest) Dairy Plant Product Inspection Procedure
	Dairy plant quality assurance/ laboratory Dairy plant truck driver (cleaning/ sanitization/milk temperature)	Pasteurization/ sterilization Investigation	Wash tags Plant cleaning Manifest Quality assurance/laboratory incoming product	Supplier management program Tank truck cleaning and sanitization records Raw milk temperature records Raw milk intake records	Dairy Plant Environment Suitability, Cleaning, and Maintenance PRP Dairy Plant Supplier Management Procedure Dairy Plant Raw Material Handling Procedure Dairy Plant Tank Truck Cleaning and Sanitizing Procedure Dairy Plant Record Control Procedure (manifest) Dairy Plant Product Inspection Procedure Dairy Farm Hygiene Inspection/ Audit Procedure

continued

Table 3.8 PRP 9, Work Sheet 2: PRP Management *(Continued)*

A. PRP (go step-by-step through ISO/TS 22002-1) 9 Management of purchased materials	Hazards				
	B. Agent(s)	**C.** Presence, growth, survival, increase, contamination	**D.** Origin, cause, source, vector, condition	**E.** Control measures	**F.** What is monitored and when
9.3 Incoming material requirements *(continued)*	C	Presence	Presence of therapeutic drugs	Supplier management program Screen all tankers for animal drug residues The dairy plant should also screen for other residues	Therapeutic drugs/ (antibiotics) and other residues (each batch)
	C	Presence of mycotoxins	Based on historical data, mold growth in animal feed can contaminate milk with aflatoxin M1. This is dependent on geographic location, growing season conditions, and so on.	Supplier management program Supplier supplied certificates of analysis Periodic quality assurance/laboratory testing by the dairy plant (ELISA [enzyme-linked immunosorbent assay] screening)	AFM1 (aflatoxin hydroxymetabolites), daily analysis
	C	Presence	Milk protein is considered an allergen	Labeling verification procedure	Statutory and regulatory requirements regarding labeling, as changes occur
	P	Contamination	If dairy cattle are not kept clean or if milk is drawn in an unclean environment and is not properly protected, physical objects from the dairy farm environment may become incorporated in the raw milk	Dairy farm hygiene practices Supplier management program Dairy farm inspection during milk collection	Dairy farm hygiene practices, as per supplier management program Dairy farm inspections (daily)

	G. Who is responsible	**H.** Correction/ corrective action	**I.** Records	**J.** Verification activities	**K.** Reference documents
	Dairy plant quality assurance/ laboratory Dairy plant truck driver (raw milk samples at the dairy farm)	Awareness/ training Return raw milk to dairy farm or environmental disposal/inves- tigation at dairy farm	Delvo test Quality assur- ance/laboratory incoming product	Milk samples at the dairy farm Laboratory incoming product records	Dairy Plant Supplier Management Procedure Dairy Plant Awareness/Training Procedure Dairy Plant Raw Milk Sample Procedure Dairy Plant Raw Material Handling Procedure Dairy Plant Record Control Procedure (manifest) Dairy Plant Product Inspection Procedure
	Dairy plant quality assurance/ laboratory	Awareness/ training Product with- drawal by dairy farm/suspend delivery of raw milk from dairy farm	ELISA/HPLC (high- performance liquid chromatography) screening	Screening records	Dairy Plant Mycotoxin Analysis Testing Dairy Plant Product Inspection Procedure
	Dairy plant marketing Quality assurance Food safety	Product hold/ withdrawal Product rework Investigation Consumer alert	Evaluation of compliance Labeling qual- ity assurance verification	Document/record review	Dairy Plant Evaluation of Compli- ance Procedure Dairy Plant Labeling Verification Procedure
	Dairy farm Dairy plant quality assurance/ food safety	Consumer awareness Refusal to accept product at source Supplier manage- ment program	Manifest Quality assurance/ laboratory incom- ing product Supplier hygiene inspection/audit	Document/record review	Dairy Farm Hygiene Inspection/ Audit Procedure Dairy Plant Supplier Management Procedure Dairy Plant Raw Material Handling Procedure

continued

Table 3.8 PRP 9, Work Sheet 2: PRP Management (*Continued*)

A. PRP (go step-by-step through ISO/TS 22002-1) 9 Management of purchased materials	Hazards			**E.** Control measures	**F.** What is monitored and when	
	B. Agent(s)	**C.** Presence, growth, survival, increase, contamination	**D.** Origin, cause, source, vector, condition			
9.3 Incoming material requirements ***(continued)***	B	Presence	Based on scientific studies, vegetative pathogens may be present in ingredients	Supplier management program, for example, supplier certificates of analysis and dairy plant periodic quality assurance/laboratory testing	Incoming product (each batch)	
	C	Contamination	Based on historical data, adulteration with toxic or carcinogenic chemicals may contaminate raw milk	Approved packaging suppliers Supplier certificates of analysis Supplier management program	Product packaging specification conformity Supplier certificates of analysis Period quality assurance/laboratory Packaging testing (each batch)	
	P	Contamination	Based on historical data, foreign materials may constitute food safety hazards	Approved packaging suppliers Supplier certificates of analysis Supplier management program	Product packaging specification conformity Supplier certificates of analysis Period quality assurance/laboratory Packaging testing (each batch)	

CHAPTER 3

	G. Who is responsible	**H.** Correction/ corrective action	**I.** Records	**J.** Verification activities	**K.** Reference documents
	Dairy farm audit Plant quality assurance/ food safety	Refusal to accept product at source Supplier management program	Manifest Quality assurance/laboratory incoming product Supplier good hygiene practices inspections Audit reports	Document/record review	Dairy Plant Good Hygiene Practices Inspection Procedure Dairy Plant Audit Procedure Dairy Plant Supplier Management Procedure
	Dairy plant quality assurance/ food safety	Awareness/ training Product hold/ return material to supplier Supplier management program	Quality assurance/ laboratory incoming product Supplier certificate of analysis Supplier inspection/audit	Document/record review	Dairy Plant Product Inspection Procedure Dairy Plant Awareness/Training Procedure Dairy Plant Supplier Management Procedure
	Dairy plant quality assurance/ food safety	Awareness/ training Product hold/ return material to supplier Supplier management program	Quality assurance/laboratory incoming product Dairy farm certificates of analysis Dairy farm supplier audit	Document/record review	Dairy Plant Product Specifications Dairy Plant Product Inspection Procedure Dairy Plant Supplier Management Procedure

Instructions for Completing PRP Work Sheet 3: PRP Verification Action Plan	
Column A	The establishment should provide details on the number and title of the PRP. It is recommended that the number should match the number of the relevant part in the appropriate FSMS scheme standard, for example, in ISO/TS 22002-1, 9—Management of purchased materials.
Column B	The establishment should provide details on the verification actions associated with the PRP and on the individual or entity responsible for reviewing these verification actions.

Table 3.9 PRP 9, Work Sheet 3: PRP Verification Action Plan

A. PRP	**B.** Verification action
9 Management of purchased materials	Review by management of purchased materials PRP team
	Review of tank truck cleaning and sanitizing records Review of raw milk temperature records Review of manifest records Review of ELISA/HPLC (enzyme-linked immunosorbent assay screening/high-performance liquid chromatography) records Review of labeling verification records Review of product inspection records Review of supplier performance records Review of awareness/training records Review of consumer complaints Food safety management system audits Internal GMP audits/good hygiene practices inspections Frequency and criticality review

Instructions for Completing PRP Work Sheet 4: PRP Meeting Summary	
Column A	List meeting dates.
Column B	List attendees among the team and other invitees.
Column C	Provide the reason for the meeting.
Column D	Record decisions and next steps.
Column E	Identify the individuals or entities responsible for executing any decisions.
Column F	Record deadlines.
Column G	Indicate the dates of relevant actions.

Table 3.10 PRP 9, Work Sheet 4: PRP Meeting Summary

A. Date	**B.** Participants	**C.** Purpose	**D.** Outcome (decisions/actions)	**E.** Respon-sibility	**F.** Deadline	**G.** Deadline reached
February 17, 2018	G Moran, O Brown, M Rodrigues, B White, D Collins, O Murphy, C Flack	Initial review of PRP	Update PRP management work sheet Review related PRPs	G Moran to complete verification sheet	May 15, 2018	May 15, 2018
March 20, 2018	G Moran, O Brown, M Rodrigues, B White, D Collins, O Murphy, C Flack	Complete GAP sheet Review PRP management work sheet	Completed and approved Reviewed and approved	G Moran to update PRP work sheets	May 15, 2018	May 15, 2018
February 15, 2019	G Moran, O Brown, M Rodrigues, B White, D Collins, O Murphy, C Flack	Review and update verification of labeling procedure and introduction of the management of inputs (periodic testing of raw material/ ingredients/packaging)	Complete update of labeling verification procedure Introduce management of inputs based on risk of raw material/ ingredients/ packaging	PRP team to complete	February 17, 2019	February 17, 2019
February 17, 2019	G Moran, O Brown, M Rodrigues, B White, D Collins, O Murphy, C Flack	Review and update based on changes to ISO 22000:2018	Current PRPs underwent a comprehensive review for compliance with ISO/TS 22002-1 and ISO 22000:2018 starting on February 17, 2019, and completed on February 20, 2019	PRP team to complete	February 20, 2019	February 20, 2019

CHAPTER 3

Instructions for Completing PRP Work Sheet 5: PRP Gap Registration and Resolution	
Column A	Provide a description of the FSMS scheme requirement.
Column B	Provide a description of the requirement arising from the FSMS scheme where the gap exists.
Column C	Provide a short description of the specific requirement where the gap exists within the FBO.
Column D	Detail the relevant FSMS policy.
Column E	Describe the gap.
Column F	Provide the action to be taken to address the requirement identified as not having been fulfilled.
Column G	Provide details of the actions taken to address the gap and the date of the completion of the actions.
Column H	Add any additional relevant comments as required.

Table 3.11 PRP 9, Work Sheet 5: PRP Gap Registration and Resolution

Fill out this work sheet only if gaps have been identified.

A. Reference guide: ISO/TS 22002-1, 9 Management of purchased materials	**B.** Description (of the requirement of the standard)	**C.** Specific requirement	**D.** Associated dairy policy	**E.** Gap	**F.** Action plan (including time frame for completion)	**G.** Gap resolution (actions completed, with date)	**H.** Comments
9.3 Incoming material requirements	Materials shall be inspected, tested, or covered by certificates of analysis to verify conformity with specified requirements prior to acceptance or use. The method of verification shall be documented.	Management of inputs/verification of raw materials, ingredients, and packaging	Food safety policy	Reliance 100% on supplier certificates of analysis	Introduce management of inputs by fourth quarter 2018	Critical raw materials, ingredients, and packaging verified as conforming to dairy plant product specifications by fourth quarter 2018	Closed

Instructions for Completing PRP Work Sheet 6: Hazard Agent	
Column A	Classify food safety hazard agents, for example, biological, chemical, or physical hazard agents.
Column B	Indicate the food safety hazard agent code, for example, allergen = A, biological = B, chemical = C, physical = P.

Table 3.12 PRP 9, Work Sheet 6: Hazard Agent

A. Hazardous agents	**B.** Hazard class
Biological (for example, vegetative or spores, depending on circumstances)	B
Chemical (for example, cleaning chemicals, nonfood-grade lubricants, oils and greases, and chemical residues)	C
Physical (for example, various types of foreign material, including metal, wood, plastic, or other foreign bodies)	P
Allergens (for example, milk, soy, wheat, eggs, fish, shellfish, tree nuts, peanuts)	A

PRP 11: Cleaning and sanitizing

Sample completed work sheets for PRP 11 follow (tables 3.13–3.18). For instructions on filling out each PRP work sheet, see the boxes that precede each sample completed work sheet.

Instructions for Completing PRP Work Sheet 1: PRP Scope	
A. PRP study scope	Provide the PRP title from the standard or scheme (for example, Cleaning and sanitizing). Provide the standard PRP number (for example, in ISO/TS 22002-1, 11—Cleaning and sanitizing). Provide the facility name, product category, processes, product, PRP start date, status of the PRP (for example, draft, approved), and end date.
B. PRP review history	In this section, record information about the history of the PRP revision, with an explanation of the reason why this update has been done: "according to plan" or "unscheduled." For an unscheduled revision, why has this revision been undertaken? (What reason?)
C. PRP team members	For every PRP study, the organization needs to establish an HACCP team with specific responsibilities and roles. Names within the company, department name, and responsibilities should be detailed. The competence of each team member should also be documented.
D. Specialist input	To establish PRP studies, companies may need advice from an outsourced expert (consultant/subject matter expert). The expert's role should be explained: input/specialist advice.
E. Authorization	Team members must indicate their approval of the document by providing their names, positions, responsibilities held, and signature. The authorized team member should provide his/her signature and the date signed.

Table 3.13 PRP 11, Work Sheet 1: PRP Scope			
PRP	11 Cleaning and sanitizing		
A. PRP study scope			
Facility	Joe Bloggs Dairy Plant	Start date	February 17, 2019
Product category	Grade A Interstate Milk Shippers registered whole milk	Status	Draft
Processes	High-temperature/short-time pasteurizer, aseptic filling, retort	End date	Ongoing
Products	Grade A aseptically processed and packaged milk		
B. PRP review history	Check as appropriate	Notes/reason for unscheduled review	Dates of last three reviews
New PRP study	✓	Current PRPs underwent a comprehensive review for compliance with ISO/TS 22002-1 and ISO 22000:2018 starting on February 15, 2019, and completed on February 17, 2019. These management sheets describe each PRP in place at the dairy plant facility.	
Scheduled review	December 20, 2019		
Unscheduled review			
C. PRP team members			
Name	Position	Department	Responsibility/role
G Moran	Food safety manager	Food safety	Food safety/quality assurance
O Brown	Hygienist/microbiologist	Food safety	Hygienist/microbiologist
M Rodrigues	Milk processing manager	Milk processing	Milk processing
B Murphy	Laboratory manager	Quality assurance	Laboratory
D Small	Warehouse manager	Warehousing	Warehousing
O Murphy	Engineering manager	Engineering	Engineering
C Flack	Factory manager	Management	Management
D. Specialist input			
Name	Location/job title	Input/specialist advice	
Angela Yard	Consultant	PRP team facilitator	
E. Authorization			
Food safety team leader/quality assurance manager	Signature: *G Moran*	Date: *February 17, 2019*	
Management team member	Signature: *C Flack*	Date: *February 17, 2019*	

Instructions for Completing PRP Work Sheet 2: PRP Management	
Column A	Describe the ISO/TS 22002-1 requirements.
Column B	Describe the hazard agent, for example, biological (B), chemical (C), physical (P), or a combination.
Column C	Describe how the hazard is manifest as a threat, including presence, increase, or survival.
Column D	Describe the cause, origin, condition, source, or vector of a hazard.
Column E	Describe the control measures the FBO has in place to control relevant hazards.
Column F	Describe the hazard measurement parameters and the monitoring frequency of the measurement parameters.

Table 3.14 PRP 11, Work Sheet 2: PRP Management

A. PRP (go step-by-step through ISO/TS 22002-1) 11 Cleaning and sanitizing	Hazards			**E.** Control measures	**F.** What is monitored and when	
	B. Agent(s)	**C.** Presence, growth, survival, increase, contamination	**D.** Origin, cause, source, vector, condition			
11.1 General requirements	B, C, P (see below)	Presence, contamination	Contamination by pathogens Cleaning/sanitizing solution residues	Hygiene, cleaning, sanitization Separation between cleaning and sanitizing solution Master cleaning/sanitizing schedule Temperature	Pathogen monitoring (daily) Separation (weekly) Temperature (daily/weekly [7 day])	

Instructions for Completing PRP Work Sheet 2: PRP Management (*continued*)	
Column G	Describe the job role or title of the department/function within the FBO responsible for monitoring the relevant hazard measurement parameters.
Column H	Describe the correction and corrective action aimed at preventing a reoccurrence of a rise above the allowable or permitted hazard measurement parameters.
Column I	Indicate the monitoring and hazard measurement parameter records to be maintained.
Column J	Describe the verification activities necessary to confirm the accuracy of the monitoring and hazard measurement parameters.
Column K	Describe the FBO documents and relevant external documents, for example, statutory and regulatory requirements.

	G. Who is responsible	**H.** Correction/ corrective action	**I.** Records	**J.** Verification activities	**K.** Reference documents
	Quality assurance/ laboratory Cleaning/sanitizing operators	Reclean/sanitize Review/update master cleaning/sanitizing schedule or program Revalidate the effectiveness of the cleaning/sanitizing schedule/program	Good hygiene practices inspections Audit Master cleaning/ sanitizing Temperature	Record review Inspection Audit	Dairy Plant Environment Suitability, Cleaning, and Maintenance PRP Dairy Plant Management of Purchased Materials PRP Dairy Plant Personnel Hygiene and Employee Facilities PRP Utilities PRP Dairy Plant Master Cleaning/ Sanitizing Program/Schedule Dairy Plant Cleaning/Sanitizing Procedures Dairy Plant Awareness and Training Procedure Dairy Plant Product Traceability Procedure Dairy Plant Environmental and Pathogen Monitoring Procedure

continued

Table 3.14 PRP 11, Work Sheet 2: PRP Management *(Continued)*

A. PRP (go step-by-step through ISO/TS 22002-1) 11 Cleaning and sanitizing	Hazards			**E.** Control measures	**F.** What is monitored and when	
	B. Agent(s)	**C.** Presence, growth, survival, increase, contamination	**D.** Origin, cause, source, vector, condition			
11.2 Cleaning and sanitizing agents and tools	B	Presence, contamination	Contamination by vegetative pathogens	Clean water Restricted use of condensing water from milk evaporators and water reclaimed from milk or milk products Training of cleaning/sanitizing operators Hygienic design/suitability of tools, for example, brushes used for manual washing are nonabsorbent, nylon, or plastic bristled type and designed not to retain soil, quick to dry Utensils manually cleaned using a two-compartment wash and rinse sink Color-coding of tools 5S program (sort, set in order, shine, standardize, sustain), including protecting tools once cleaned, for example, stored off the contact floor, protected from splashes following cleaning, and so on	Pathogen monitoring (daily)	
	C	Presence, contamination	Without proper separation between cleaning and sanitizing solutions and products, there could be product contamination	Material safety data sheets for (chemicals) chlorine/acids used Approved chemicals Chemical storage Maintain proper separation or physical break between circuits containing cleaning solution and vessels and lines used to contain product Manual sanitizing with chemicals to be accomplished using a third treatment vat, unless heat is used for sanitizing	Toxic residues Alkaline detergents/acid cleaner not mixed Daily/each batch	
	P	None				

CHAPTER 3

	G. Who is responsible	**H.** Correction/ corrective action	**I.** Records	**J.** Verification activities	**K.** Reference documents
	Quality assurance/ laboratory Cleaning/sanitizing operators	Replacement tools Retraining, if required Reclean/resanitize	Good hygiene practices inspections Audits Milk tank truck wash tags or logbook Manual cleaning logbook	Good hygiene practices inspections Audit Document/ record review	Dairy Plant Environment Suitability, Cleaning, and Maintenance PRP Management of Purchased Materials PRP Utilities PRP Master Cleaning/Sanitizing Program/Schedule Cleaning/Sanitizing Procedures Dairy Plant Awareness and Training Procedure Dairy Plant Product Traceability Procedure Dairy Plant Environmental and Pathogen Monitoring Procedure
	Quality assurance/ laboratory Cleaning/sanitizing operators	Monitoring frequency review Retraining, if required Reclean/resanitize	Good hygiene practices inspections Audits Cleaning/ sanitizing	Good hygiene practices inspections Audit Document/ record review	Dairy Plant Environment Suitability, Cleaning, and Maintenance PRP Dairy Plant Management of Purchased Materials PRP Dairy Plant Utilities PRP Master Cleaning/Sanitizing Program/Schedule Dairy Plant Cleaning/Sanitizing Procedures Dairy Plant Awareness and Training Procedure Dairy Plant Product Traceability Procedure Dairy Plant Environmental and Pathogen Monitoring Procedure

continued

CHAPTER 3

Table 3.14 PRP 11, Work Sheet 2: PRP Management *(Continued)*

A. PRP (go step-by-step through ISO/TS 22002-1) 11 Cleaning and sanitizing	Hazards					
	B. Agent(s)	**C.** Presence, growth, survival, increase, contamination	**D.** Origin, cause, source, vector, condition	**E.** Control measures	**F.** What is monitored and when	
11.3 Cleaning and sanitizing programs	B	Presence, contamination	Contamination by vegetative pathogens	Master Cleaning/ Sanitizing Program Master Cleaning/ Sanitizing Schedule Validated Cleaning/ Sanitizing Program/ Schedule (including revalidation)	Pathogen monitoring (daily) Temperature (daily/weekly [7 day]) for milk storage tanks	
	C	Presence, contamination	Without proper separation between cleaning and sanitizing solutions and products, there could be product contamination	Master Cleaning/ Sanitizing Program Master Cleaning/ Sanitizing Schedule Validated Cleaning/ Sanitizing Program/ Schedule (including revalidation)	Toxic residues Alkaline detergents/acid cleaner not mixed Daily/each batch	
	P	None				
11.4 Cleaning-in-place systems	B	Presence, contamination	Contamination by pathogens	Clean-in-place parameters, for example, temperature Clean-in-place venting door device associated with larger tanks and silos Water characteristics with water hardness exceeding 100 parts per million hardness	Temperature Pathogen monitoring	

CHAPTER 3

	G. Who is responsible	**H.** Correction/ corrective action	**I.** Records	**J.** Verification activities	**K.** Reference documents
	Hygienist Cleaning/ sanitization	Review/update Master Cleaning/ Sanitizing Schedule or Program Revalidate the effectiveness of the Cleaning/Sanitizing Schedule/Program	Good hygiene practices inspections Audits Master Cleaning/ Sanitizing Validation/ Revalidation Study	Good hygiene practices inspections Audit Document/ record review	Dairy Plant Environment Suitability, Cleaning, and Maintenance PRP Dairy Plant Master Cleaning/ Sanitizing Program/Schedule Dairy Plant Cleaning/Sanitizing Procedures Dairy Plant Awareness and Training Procedure Dairy Plant Product Traceability Procedure Dairy Plant Environmental and Pathogen Monitoring Procedure
	Hygienist Cleaning/ sanitization	Review/update Master Cleaning/ Sanitizing Schedule or Program Revalidate the effectiveness of the Cleaning/Sanitizing Schedule/Program	Good hygiene practices inspections Audits Master Cleaning/ Sanitizing Validation/ Revalidation Study	Good hygiene practices inspections Audit Document/ record review	Dairy Plant Environment Suitability, Cleaning, and Maintenance PRP Dairy Plant Master Cleaning/ Sanitizing Program/Schedule Dairy Plant Cleaning/Sanitizing Procedures Dairy Plant Awareness and Training Procedure Dairy Plant Product Traceability Procedure Dairy Plant Environmental and Pathogen Monitoring Procedure
	Quality assurance laboratory Cleaning operator	Reclean	Clean-in-place charts for all dairy plant processing equipment	Good hygiene practices inspections Audit Document/ record review	Dairy Plant Environment Suitability, Cleaning, and Maintenance PRP Dairy Plant Master Cleaning/ Sanitizing Program/Schedule Dairy Plant Cleaning/Sanitizing Procedures Dairy Plant Awareness and Training Procedure Dairy Plant Product Traceability Procedure Dairy Plant Environmental and Pathogen Monitoring Procedure

continued

CHAPTER 3

Table 3.14 PRP 11, Work Sheet 2: PRP Management *(Continued)*

A. PRP (go step-by-step through ISO/TS 22002-1) 11 Cleaning and sanitizing	Hazards			**E.** Control measures	**F.** What is monitored and when	
	B. Agent(s)	**C.** Presence, growth, survival, increase, contamination	**D.** Origin, cause, source, vector, condition			
11.4 Cleaning-in-place systems *(continued)*	C	Presence, contamination	Without proper separation between cleaning and sanitizing solutions and products, there could be product contamination	Clean-in-place parameters, for example, temperature, type, concentration, concentration time, and so on Clean-in-place venting door device associated with larger tanks and silos Water characteristics with water hardness exceeding 100 parts per million hardness	Chemical type, concentration Contact time and temperature	
	P	None				
11.5 Monitoring sanitation effectiveness	B	Presence, contamination	Contamination by pathogens	Master Cleaning/Sanitizing Schedule Good hygiene practices inspection Audit Pathogen monitoring	Pathogen monitoring frequency daily/weekly	
	C	Presence, contamination	Without proper separation between cleaning and sanitizing solutions and products, there could be product contamination	Master Cleaning/Sanitizing Schedule Good hygiene practices inspection Audit Pathogen monitoring	Chemical type, concentration Contact time and temperature	
	P	None				

CHAPTER 3

	G. Who is responsible	**H.** Correction/ corrective action	**I.** Records	**J.** Verification activities	**K.** Reference documents
	Quality assurance laboratory Cleaning operator	Reclean	Clean-in-place charts for all dairy plant processing equipment	Good hygiene practices inspections Audit Document/ record review	Dairy Plant Environment Suitability, Cleaning, and Maintenance PRP Dairy Plant Master Cleaning/ Sanitizing Program/Schedule Dairy Plant Cleaning/Sanitizing Procedures Dairy Plant Awareness and Training Procedure Dairy Plant Product Traceability Procedure Dairy Plant Environmental and Pathogen Monitoring Procedure
	Cleaning/sanitizing supervisor Quality assurance/ laboratory	Review/update Master Cleaning/ Sanitizing Schedule or Program Revalidate the effectiveness of the Cleaning/Sanitizing Schedule/Program	Cleaning/ sanitizing Good hygiene practices inspections Audits Cleaning/ sanitizing validation/ revalidation	Good hygiene practices inspections Audit Document/ record review	Dairy Plant Environment Suitability, Cleaning, and Maintenance PRP Dairy Plant Master Cleaning/ Sanitizing Program/Schedule Dairy Plant Cleaning/Sanitizing Procedures Dairy Plant Awareness and Training Procedure Dairy Plant Product Traceability Procedure Dairy Plant Environmental and Pathogen Monitoring Procedure
	Cleaning/sanitizing supervisor Quality assurance/ laboratory	Review/update Master Cleaning/ Sanitizing Schedule or Program Revalidate the effectiveness of the Cleaning/Sanitizing Schedule/Program	Cleaning/ sanitizing Good hygiene practices inspections Audits Cleaning/ sanitizing validation/ revalidation	Good hygiene practices inspections Audit Document/ record review	Dairy Plant Environment Suitability, Cleaning, and Maintenance PRP Dairy Plant Master Cleaning/ Sanitizing Program/Schedule Dairy Plant Cleaning/Sanitizing Procedures Dairy Plant Awareness and Training Procedure Dairy Plant Product Traceability Procedure Dairy Plant Environmental and Pathogen Monitoring Procedure

Instructions for Completing PRP Work Sheet 3: PRP Verification Action Plan	
Column A	The establishment should provide details on the number and title of the PRP. It is recommended that the number should match the number of the relevant part in the appropriate FSMS scheme standard, for example, in ISO/TS 22002-1, 11—Cleaning and sanitizing.
Column B	The establishment should provide details on the verification actions associated with the PRP and on the individual or entity responsible for reviewing these verification actions.

Table 3.15 PRP 11, Work Sheet 3: PRP Verification Action Plan

A. PRP	**B.** Verification action
11 Cleaning and sanitizing	Reviewed by hygienist and cleaning and sanitizing PRP team
	Review of environment, pathogen, and foreign objects monitoring Review of good hygiene practices inspections Food safety management system audits Internal GMP/hygiene audits Review of chemicals/material safety data sheets and chemical storage Review of cleaning/sanitizing validation/revalidation study Review of traceability Review of training Frequency and criticality review

Instructions for Completing PRP Work Sheet 4: PRP Meeting Summary	
Column A	List meeting dates.
Column B	List attendees among the team and other invitees.
Column C	Provide the reason for the meeting.
Column D	Record decisions and next steps.
Column E	Identify the individuals or entities responsible for executing any decisions.
Column F	Record deadlines.
Column G	Indicate the dates of relevant actions.

Table 3.16 PRP 11, Work Sheet 4: PRP Meeting Summary

A. Date	**B.** Participants	**C.** Purpose	**D.** Outcome (decisions/actions)	**E.** Responsibility	**F.** Deadline	**G.** Deadline reached
April 20, 2018	G Moran, O Brown, M Rodrigues, B White, D Collins, O Murphy, C Flack	Initial review of PRP	Update PRP management work sheet Review related PRPs	G Moran to complete verification sheet	May 15, 2018	May 15, 2018
April 28, 2018	G Moran, O Brown, M Rodrigues, B White, D Collins, O Murphy, C Flack	Complete GAP sheet Review PRP management work sheet Review cleaning/sanitization revalidation study	Completed and approved Reviewed and approved Appointed designated person	G Moran to update PRP work sheets	May 15, 2018	May 15, 2018
February 17, 2019	G Moran, O Brown, M Rodrigues, B White, D Collins, O Murphy, C Flack	Review of cleaning tool program awareness, for example, 5S program (sort, set in order, shine, standardize, sustain), storage, replacement	Reviewed/updated training and improvements shown following improved coaching and supervising by supervisors	PRP team to complete	February 17, 2019	February 17, 2019
February 17, 2019	G Moran, O Brown, M Rodrigues, B White, D Collins, O Murphy, C Flack	Review and update based on changes in ISO 22000:2018	Current PRPs underwent a comprehensive review for compliance with ISO/TS 22002-1 and ISO 22000:2018 starting on February 15, 2019, and completed on February 20, 2019	PRP team to complete	February 20, 2019	February 20, 2019

CHAPTER 3

Instructions for Completing PRP Work Sheet 5: PRP Gap Registration and Resolution	
Column A	Provide a description of the FSMS scheme requirement.
Column B	Provide a description of the requirement arising from the FSMS scheme where the gap exists.
Column C	Provide a short description of the specific requirement where the gap exists within the FBO.
Column D	Detail the relevant FSMS policy.
Column E	Describe the gap.
Column F	Provide the action to be taken to address the requirement identified as not having been fulfilled.
Column G	Provide details of the actions taken to address the gap and the date of the completion of the actions.
Column H	Add any additional relevant comments as required.

Table 3.17 PRP 11, Work Sheet 5: PRP Gap Registration and Resolution

Fill out this work sheet only if gaps have been identified.

A. ISO/TS 22002-1, 11 Cleaning and sanitizing	**B.** Description (of the requirement of the standard)	**C.** Specific requirement	**D.** Associated dairy policy	**E.** Gap	**F.** Action plan (including time frame for completion)	**G.** Gap resolution (actions completed, with date)	**H.** Comments
11.2 Cleaning and sanitizing agents and tools	Tools and equipment must be of hygienic design and maintained in a condition that does not present a potential source of extraneous matter	Review effectiveness of awareness of the ISO/TS 22002-1 requirement	Food safety policy	Enhance awareness of 5S (sort, set in order, shine, standardize, sustain), storage, tool protection procedures	Update awareness/training and monitoring effectiveness through greater supervision of FBO supervisors	Reviewed/updated awareness/training effectiveness; see PRP team meeting February 17, 2019	Need to continue to monitor for next six months to sustain improvements to date
11.3 Cleaning and sanitizing programs	Cleaning and sanitizing programs should be established and validated by the organization	Revalidate cleaning/sanitizing validation study	Food safety policy	Previous validation study incomplete/inadequate	Revalidation study review/approved	Review/approved revalidation study; see PRP team meeting February 17, 2019	Need to continue to monitor for next 12 months

CHAPTER 3

Instructions for Completing PRP Work Sheet 6: Hazard Agent	
Column A	Classify food safety hazard agents, for example, biological, chemical, or physical hazard agents.
Column B	Indicate the food safety hazard agent code, for example, allergen = A, biological = B, chemical = C, physical = P.

Table 3.18 PRP 11, Work Sheet 6: Hazard Agent

A. Hazardous agents	**B.** Hazard class
Biological (for example, vegetative or spores, depending on circumstances)	B
Chemical (for example, cleaning chemicals, nonfood-grade lubricants, oils and greases, and chemical residues)	C
Physical (for example, various types of foreign material, including metal, wood, plastic, or other foreign bodies)	P
Allergens (for example, milk, soy, wheat, eggs, fish, shellfish, tree nuts, peanuts)	A

CHAPTER 3

PRP 12: Pest control

Sample completed work sheets for PRP 12 follow (tables 3.19–3.24). For instructions on filling out each PRP work sheet, see the boxes that precede each sample completed work sheet.

Instructions for Completing PRP Work Sheet 1: PRP Scope	
A. PRP study scope	Provide the PRP title from the standard or scheme (for example, Pest control). Provide the standard PRP number (for example, in ISO/TS 22002-1, 12—Pest control). Provide the facility name, product category, processes, product, PRP start date, status of the PRP (for example, draft, approved), and end date.
B. PRP review history	In this section, record information about the history of the PRP revision, with an explanation of the reason why this update has been done: "according to plan" or "unscheduled." For an unscheduled revision, why has this revision been undertaken? (What reason?)
C. PRP team members	For every PRP study, the organization needs to establish an HACCP team with specific responsibilities and roles. Names within the company, department name, and responsibilities should be detailed. The competence of each team member should also be documented.
D. Specialist input	To establish PRP studies, companies may need advice from an outsourced expert (consultant/subject matter expert). The expert's role should be explained: input/specialist advice.
E. Authorization	Team members must indicate their approval of the document by providing their names, positions, responsibilities held, and signature. The authorized team member should provide his/her signature and the date signed.

Table 3.19 PRP 12, Work Sheet 1: PRP Scope

PRP	12 Pest control		
A. PRP study scope			
Facility	Joe Bloggs Dairy Plant	Start date	February 17, 2019
Product category	Grade A Interstate Milk Shippers registered whole milk	Status	Draft
Processes	High-temperature/short-time pasteurizer, aseptic filling, retort	End date	Ongoing
Products	Grade A aseptically processed and packaged milk		
B. PRP review history	Check as appropriate	Notes/reason for unscheduled review	Dates of last three reviews
New PRP study	✓	Current PRPs underwent a comprehensive review for compliance with ISO/TS 22002-1 and ISO 22000:2018 starting on February 15, 2019, and completed on February 17, 2019. These management sheets describe each PRP in place at the dairy plant facility.	
Scheduled review	December 20, 2019		
Unscheduled review			
C. PRP team members			
Name	Position	Department	Responsibility/role
G Moran	Food safety manager	Food safety	Food safety/quality assurance
O Brown	Hygienist/microbiologist	Food safety	Hygienist/microbiologist
M Rodrigues	Milk processing manager	Milk processing	Milk processing
B Murphy	Laboratory manager	Quality assurance	Laboratory
D Small	Warehouse manager	Warehousing	Warehousing
O Murphy	Engineering manager	Engineering	Engineering
C Flack	Factory manager	Management	Management

D. Specialist input		
Name	Location/job title	Input/specialist advice
Angela Yard	Consultant	PRP team facilitator
E. Authorization		
Food safety team leader/quality assurance manager	Signature: *G Moran*	Date: *February 17, 2019*
Management team member	Signature: *C Flack*	Date: *February 17, 2019*

Instructions for Completing PRP Work Sheet 2: PRP Management	
Column A	Describe the ISO/TS 22002-1 requirements.
Column B	Describe the hazard agent, for example, biological (B), chemical (C), physical (P), or a combination.
Column C	Describe how the hazard is manifest as a threat, including presence, increase, or survival.
Column D	Describe the cause, origin, condition, source, or vector of a hazard.
Column E	Describe the control measures the FBO has in place to control relevant hazards.
Column F	Describe the hazard measurement parameters and the monitoring frequency of the measurement parameters.

Table 3.20 PRP 12, Work Sheet 2: PRP Management

	Hazards					
A. PRP (go step-by-step through ISO/TS 22002-1) 12 Pest control	**B.** Agent(s)	**C.** Presence, growth, survival, increase, contamination	**D.** Origin, cause, source, vector, condition	**E.** Control measures	**F.** What is monitored and when	
12.1 General requirements	B	Contamination	Pests	Hygiene, cleaning, and incoming material inspection in place as well as pathogen and environmental monitoring procedures	Hygiene, cleaning, and raw material monitored through good hygiene practices inspections and audits (monthly) Pathogen monitoring program in place (weekly)	

Instructions for Completing PRP Work Sheet 2: PRP Management (*continued*)	
Column G	Describe the job role or title of the department/function within the FBO responsible for monitoring the relevant hazard measurement parameters.
Column H	Describe the correction and corrective action aimed at preventing a reoccurrence of a rise above the allowable or permitted hazard measurement parameters.
Column I	Indicate the monitoring and hazard measurement parameter records to be maintained.
Column J	Describe the verification activities necessary to confirm the accuracy of the monitoring and hazard measurement parameters.
Column K	Describe the FBO documents and relevant external documents, for example, statutory and regulatory requirements.

	G. Who is responsible	**H.** Correction/ corrective action	**I.** Records	**J.** Verification activities	**K.** Reference documents
	Dairy plant quality assurance laboratory Dairy plant sanitization Dairy plant food safety	Training Cleaning of area where deviation was found Raw material is sent back to supplier or discarded if not compliant	Good hygiene practices inspection Audit reports Pathogen monitoring Raw material monitoring	Pest Control records no pest activity	Dairy Plant Design and Construction of Buildings PRP Dairy Plant Site Location and Standards PRP Dairy Plant Layout of Premises and Workspace PRP Dairy Plant Internal Structure PRP Dairy Plant Environment Suitability, Cleaning, and Maintenance PRP Raw Material Handling Procedure Dairy Plant Product Inspection Procedure Dairy Plant Cleaning and Sanitizing Procedures Dairy Plant Awareness and Training Procedure Dairy Plant Environmental and Pathogen Monitoring Procedure Pest Control Folder/Manual (external pest control company)

continued

CHAPTER 3

Table 3.20 PRP 12, Work Sheet 2: PRP Management *(Continued)*

A. PRP (go step-by-step through ISO/TS 22002-1) 12 Pest control	Hazards			**E.** Control measures	**F.** What is monitored and when
	B. Agent(s)	**C.** Presence, growth, survival, increase, contamination	**D.** Origin, cause, source, vector, condition		
12.2 Pest control program	B, C	Contamination	Pests, chemicals used	Pest control program in place, outsourced to an external company Dairy plant designed site contact is the sanitizing supervisor Contact person is the dairy plant sanitizing supervisor Documents and records are with the dairy plant sanitizing supervisor List of approved pesticide chemicals used is available on a USB stick that is with the Pest Management Program folder/manual The food safety manager approves all dairy plant chemical pesticides	Pest activity, infestation Pest activity is frequently monitored according to the Pest Management Program
12.3 Preventing access	B, C	Contamination	Holes, cracks, open doors, ventilation openings	Building maintenance in place Pest access points are sealed All doors to the outside have closures, windows cannot be opened, ventilation openings are designed to minimize the potential entry of pests Dairy plant–approved pesticides maintained Material safety data sheets for dairy plant–approved pesticides maintained	Pest activity, infestation Pest activity is frequently monitored according to the Pest Management Program
12.4 Harborage and infestations	B	Contamination	Raw material Bad housekeeping Pallets, and so on	GMP and good housekeeping in place throughout the dairy plant Material found to be infested is separated or discarded Outside space is not used for storage	Pest activity, infestation Pest activity is frequently monitored according to the Pest Management Program, monthly pest prevention audit

	G. Who is responsible	**H.** Correction/ corrective action	**I.** Records	**J.** Verification activities	**K.** Reference documents
	Dairy plant quality assurance laboratory Dairy plant sanitization Dairy plant food safety	Contain-ment during construction Eliminate source of pest entry	Pest manage-ment service report (external provider)	Pest Control records no pest activity	Dairy Plant Pest Control Program Dairy Plant Pest Control Map Pest Control Folder/Manual (external pest control company)
	Dairy plant quality assurance Dairy Plant maintenance Dairy plant sanitization	Closing entry point of pests	Pest manage-ment service report (external provider)	Pest Control records no pest activity	Dairy Plant Design and Construction of Buildings PRP Dairy Plant Site Location and Standards PRP Dairy Plant Layout of Premises and Workspace PRP Dairy Plant Internal Structure PRP Dairy Plant Environment Suitability, Cleaning, and Maintenance PRP
	Dairy plant sanitation	Cleaning of infested area Root cause analysis Training	Inspection/audit report Training Destruction of nonconforming product	Pest Control records no pest activity Audit	Dairy Plant Raw Material Handling Procedure Dairy Plant Product Inspection Procedure

continued

Table 3.20 PRP 12, Work Sheet 2: PRP Management *(Continued)*

	Hazards					
A. PRP (go step-by-step through ISO/TS 22002-1) 12 Pest control	**B.** Agent(s)	**C.** Presence, growth, survival, increase, contamination	**D.** Origin, cause, source, vector, condition	**E.** Control measures	**F.** What is monitored and when	
12.5 Monitoring and detection	B	Contamination	Pests	Pest control program in place, outsourced to an external company Pest Control map of detectors and traps included in the Pest Control Folder/Manual Detectors and traps conform to ISO/TS 22002-1 Detectors and traps are frequently inspected according to Pest Management Program	Pest activity, infestation Pest activity is frequently monitored according to the Pest Management Program	
12.6 Eradication	B, C	Contamination	Pests	Eradication measures shown in pest management service report Only authorized and trained dairy plant personnel handle pesticides Records of dairy plant-approved pesticides are maintained in the pest control service report	Pest activity, infestation Pest activity is frequently monitored according to the Pest Management Program	

CHAPTER 3

	G. Who is responsible	**H.** Correction/ corrective action	**I.** Records	**J.** Verification activities	**K.** Reference documents
	Dairy plant sanitization	Review Pest Management Program	Pest management service report	Pest Control records no pest activity Audit	Dairy Plant Pest Control Folder/ Manual (external pest control company) Dairy Plant Environmental and Pathogen Monitoring Procedure Dairy Plant Good Hygiene Practices Inspection Procedure Dairy Plant Awareness/Training Procedure
	Dairy plant sanitization Dairy plant food safety	Review Pest Management Program	Pest management service report	Pest Control records no pest activity	Dairy Plant Pest Control Folder/ Manual (external pest control company) Dairy Plant Awareness and Training Procedure

Instructions for Completing PRP Work Sheet 3: PRP Verification Action Plan	
Column A	The establishment should provide details on the number and title of the PRP. It is recommended that the number should match the number of the relevant part in the appropriate FSMS scheme standard, for example, in ISO/TS 22002-1, 12—Pest control.
Column B	The establishment should provide details on the verification actions associated with the PRP and on the individual or entity responsible for reviewing these verification actions.

Table 3.21 PRP 12, Work Sheet 3: PRP Verification Action Plan

A. PRP	**B.** Verification action
12 Pest control	Reviewed by the laboratory manager and PRP pest control team
	Review of pest sighting log Review of pest management service reports Food safety management system audits Internal GMP/hygiene audits Review of approved chemical pesticide Review of material safety data sheets Frequency and criticality review

Instructions for Completing PRP Work Sheet 4: PRP Meeting Summary	
Column A	List meeting dates.
Column B	List attendees among the team and other invitees.
Column C	Provide the reason for the meeting.
Column D	Record decisions and next steps.
Column E	Identify the individuals or entities responsible for executing any decisions.
Column F	Record deadlines.
Column G	Indicate the dates of relevant actions.

Table 3.22 PRP 12, Work Sheet 4: PRP Meeting Summary

A. Date	**B.** Participants	**C.** Purpose	**D.** Outcome (decisions/actions)	**E.** Responsibility	**F.** Deadline	**G.** Deadline reached
April 20, 2018	G Moran, O Brown, M Rodrigues, B White, D Collins, O Murphy, C Flack	Initial review of PRP	Update PRP management work sheet Review related PRPs	G Moran to complete verification sheet	May 15, 2018	May 15, 2018
April 28, 2018	G Moran, O Brown, M Rodrigues, B White, D Collins, O Murphy, C Flack	Complete GAP sheet Review PRP management work sheet Appoint designated person	Completed and approved Reviewed and approved Appointed designated person	G Moran to update PRP work sheets	May 15, 2018	May 15, 2018
February 17, 2019	G Moran, O Brown, M Rodrigues, B White, D Collins, O Murphy, C Flack	Review pesticide chemicals and material safety data sheets	Reviewed/approved pesticide chemical specification Updated material safety data sheets folder	PRP team to complete	February 17, 2019	February 17, 2019
February 17, 2019	G Moran, O Brown, M Rodrigues, B White, D Collins, O Murphy, C Flack	Review and update based on changes in ISO 22000:2018	Current PRPs underwent a comprehensive review for compliance with ISO/TS 22002-1 and ISO 22000:2018 starting on February 15, 2019, and completed on February 20, 2019	PRP team to complete	February 20, 2019	February 20, 2019

Instructions for Completing PRP Work Sheet 5: PRP Gap Registration and Resolution	
Column A	Provide a description of the FSMS scheme requirement.
Column B	Provide a description of the requirement arising from the FSMS scheme where the gap exists.
Column C	Provide a short description of the specific requirement where the gap exists within the FBO.
Column D	Detail the relevant FSMS policy.
Column E	Describe the gap.
Column F	Provide the action to be taken to address the requirement identified as not having been fulfilled.
Column G	Provide details of the actions taken to address the gap and the date of the completion of the actions.
Column H	Add any additional relevant comments as required.

Table 3.23 PRP 12, Work Sheet 5: PRP Gap Registration and Resolution

Fill out this work sheet only if gaps have been identified.

A. ISO/TS 22002-1, 12 Pest control	**B.** Description (of the requirement of the standard)	**C.** Specific requirement	**D.** Associated dairy policy	**E.** Gap	**F.** Action plan (including time frame for completion)	**G.** Gap resolution (actions completed, with date)	**H.** Comments
12.2 Pest control programs	The establishment shall have a nominated person to manage pest control activities and deal with appointed expert contractors	Nominated person to manage pest control activities	Food safety policy	No clear designated person	Agree to nominated person by next PRP team meeting	Nominated person appointed; see PRP team meeting May 15, 2018	Dairy plant sanitizing supervisor appointed designated person

Instructions for Completing PRP Work Sheet 6: Hazard Agent	
Column A	Classify food safety hazard agents, for example, biological, chemical, or physical hazard agents.
Column B	Indicate the food safety hazard agent code, for example, allergen = A, biological = B, chemical = C, physical = P.

Table 3.24 PRP 12, Work Sheet 6: Hazard Agent

A. Hazardous agents	**B.** Hazard class
Biological (for example, vegetative or spores, depending on circumstances)	B
Chemical (for example, cleaning chemicals, nonfood-grade lubricants, oils and greases, and chemical residues)	C
Physical (for example, various types of foreign material, including metal, wood, plastic, or other foreign bodies)	P
Allergens (for example, milk, soy, wheat, eggs, fish, shellfish, tree nuts, peanuts)	A

PRP 13: Personnel hygiene and employee facilities

Sample completed work sheets for PRP 13 follow (tables 3.25–3.30). For instructions on filling out each PRP work sheet, see the boxes that precede each sample completed work sheet.

Instructions for Completing PRP Work Sheet 1: PRP Scope	
A. PRP study scope	Provide the PRP title from the standard or scheme (for example, Personnel hygiene and employee facilities). Provide the standard PRP number (for example, in ISO/TS 22002-1, 13—Personnel hygiene and employee facilities). Provide the facility name, product category, processes, product, PRP start date, status of the PRP (for example, draft, approved), and end date.
B. PRP review history	In this section, record information about the history of the PRP revision, with an explanation of the reason why this update has been done: "according to plan" or "unscheduled." For an unscheduled revision, why has this revision been undertaken? (What reason?)
C. PRP team members	For every PRP study, the organization needs to establish an HACCP team with specific responsibilities and roles. Names within the company, department name, and responsibilities should be detailed. The competence of each team member should also be documented.
D. Specialist input	To establish PRP studies, companies may need advice from an outsourced expert (consultant/subject matter expert). The expert's role should be explained: input/specialist advice.
E. Authorization	Team members must indicate their approval of the document by providing their names, positions, responsibilities held, and signature. The authorized team member should provide his/her signature and the date signed.

Table 3.25 PRP 13, Work Sheet 1: PRP Scope

PRP	13 Personnel hygiene and employee facilities		
A. PRP study scope			
Facility	Joe Bloggs Dairy Plant	Start date	February 17, 2019
Product category	Grade A Interstate Milk Shippers registered whole milk	Status	Draft
Processes	High-temperature/short-time pasteurizer, aseptic filling, retort	End date	Ongoing
Products	Grade A aseptically processed and packaged milk		
B. PRP review history	Check as appropriate	Notes/reason for unscheduled review	Dates of last three reviews
New PRP study	✓	Current PRPs underwent a comprehensive review for compliance with ISO/TS 22002-1 and ISO 22000:2018 starting on February 15, 2019, and completed on February 17, 2019. These management sheets describe each PRP in place at the dairy plant facility.	
Scheduled review	December 20, 2019		
Unscheduled review			
C. PRP team members			
Name	Position	Department	Responsibility/role
G Moran	Food safety manager	Food safety	Food safety/quality assurance
O Brown	Hygienist/microbiologist	Food safety	Hygienist/ microbiologist
M Rodrigues	Milk processing manager	Milk processing	Milk processing
B Murphy	Laboratory manager	Quality assurance	Laboratory
D Small	Warehouse manager	Warehousing	Warehousing
O Murphy	Engineering manager	Engineering	Engineering
C Flack	Factory manager	Management	Management

D. Specialist input		
Name	Location/job title	Input/specialist advice
Angela Yard	Consultant	PRP team facilitator
E. Authorization		
Food safety team leader/quality assurance manager	Signature: *G Moran*	Date: *February 17, 2019*
Management team member	Signature: *C Flack*	Date: *February 17, 2019*

CHAPTER 3

Instructions for Completing PRP Work Sheet 2: PRP Management	
Column A	Describe the ISO/TS 22002-1 requirements.
Column B	Describe the hazard agent, for example, biological (B), chemical (C), physical (P), or a combination.
Column C	Describe how the hazard is manifest as a threat, including presence, increase, or survival.
Column D	Describe the cause, origin, condition, source, or vector of a hazard.
Column E	Describe the control measures the FBO has in place to control relevant hazards.
Column F	Describe the hazard measurement parameters and the monitoring frequency of the measurement parameters.

Table 3.26 PRP 13, Work Sheet 2: PRP Management

A. PRP (go step-by-step through ISO/TS 22002-1) 13 Personnel hygiene and employee facilities	Hazards					
	B. Agent(s)	**C.** Presence, growth, survival, increase, contamination	**D.** Origin, cause, source, vector, condition	**E.** Control measures	**F.** What is monitored and when	
13.1 General requirements	B, C, P (see below)	Presence, contamination	Contamination by pathogens Contamination by cleaning and sanitizing residues Contamination by extraneous material	Dairy plant hygiene policy Dairy plant hygiene awareness and training	Pathogen monitoring, daily Good hygiene practices, weekly	
13.2 Personnel hygiene facilities and toilets	B	Presence, contamination	Contamination by vegetative pathogens	Provision of personnel hygiene facilities Hygienic design of personnel hygiene facilities Location and cleaning/maintenance of personnel hygiene facilities	Pathogen monitoring, daily Cleaning/sanitizing, daily Temperature of water Maintenance, weekly Supply of soap and/or sanitizer	

Instructions for Completing PRP Work Sheet 2: PRP Management (*continued*)	
Column G	Describe the job role or title of the department/function within the FBO responsible for monitoring the relevant hazard measurement parameters.
Column H	Describe the correction and corrective action aimed at preventing a reoccurrence of a rise above the allowable or permitted hazard measurement parameters.
Column I	Indicate the monitoring and hazard measurement parameter records to be maintained.
Column J	Describe the verification activities necessary to confirm the accuracy of the monitoring and hazard measurement parameters.
Column K	Describe the FBO documents and relevant external documents, for example, statutory and regulatory requirements.

G. Who is responsible	**H.** Correction/ corrective action	**I.** Records	**J.** Verification activities	**K.** Reference documents
All personnel Hygienist Quality assurance/ laboratory	Pathogen monitoring Retraining, if required Disciplinary action, if required	Personnel hygiene Good hygiene practices inspections Audits Pathogen monitoring	Good hygiene practices inspections Audit Document/ record review	Dairy Plant Measures for Prevention of Cross Contamination PRP Dairy Plant Hygiene Policy Dairy Plant Cleaning/Sanitizing Procedures Dairy Plant Awareness and Training Procedure Dairy Plant Environmental and Pathogen Monitoring Procedure
Facilities management Hygienist Quality assurance/ laboratory Cleaning operators/ service providers Maintenance	Preventive maintenance Retraining, if required Reclean/ resanitize	Good hygiene practices inspections Audits Personnel hygiene facil- ities cleaning logbook	Good hygiene practices inspections Audit Document/ record review	Dairy Plant Measures for Prevention of Cross Contamination PRP Dairy Plant Construction and Layout of Building PRP Dairy Plant Environment Suitability, Cleaning, and Maintenance PRP Dairy Plant Management of Purchased Materials PRP Dairy Plant Utilities PRP Dairy Plant Cleaning/Sanitizing Procedures Dairy Plant Awareness and Training Procedure Dairy Plant Product Traceability Procedure Dairy Plant Environmental and Pathogen Monitoring Procedure

continued

Table 3.26 PRP 13, Work Sheet 2: PRP Management *(Continued)*

A. PRP (go step-by-step through ISO/TS 22002-1) 13 Personnel hygiene and employee facilities	Hazards			**E.** Control measures	**F.** What is monitored and when	
	B. Agent(s)	**C.** Presence, growth, survival, increase, contamination	**D.** Origin, cause, source, vector, condition			
13.2 Personnel hygiene facilities and toilets *(continued)*	C	Presence, contamination	Cleaning and sanitizing solution residues	Material safety data sheets for cleaning and/or sanitizing chemicals Approved cleaning and sanitizing chemicals Chemical storage	Toxic residues, daily/weekly	
	P	Presence, contamination	Extraneous material arising from poor personnel facility maintenance and/or cleaning, for example, paint	Preventive maintenance Cleaning log	Maintenance Cleaning daily/weekly	

	G. Who is responsible	**H.** Correction/ corrective action	**I.** Records	**J.** Verification activities	**K.** Reference documents
	Quality assurance/ laboratory Sanitizing operators Cleaning service providers	Environmental monitoring frequency review Quality assurance training, if required Reclean/ resanitize	Good hygiene practices inspections Audits Cleaning/ sanitizing	Good hygiene practices inspections Audit Document/ record review	Dairy Plant Measures for Prevention of Cross Contamination PRP Dairy Plant Construction and Layout of Building PRP Dairy Plant Environment Suitability, Cleaning, and Maintenance PRP Dairy Plant Management of Purchased Materials PRP Dairy Plant Utilities PRP Dairy Plant Cleaning/Sanitizing Procedures Dairy Plant Awareness and Training Procedure Dairy Plant Product Traceability Procedure Dairy Plant Environmental and Pathogen Monitoring Procedure
	Maintenance Cleaning service provider	Quality assurance training, if required Reclean/ resanitize	Good hygiene practices inspections Audits Cleaning/ sanitizing Maintenance	Good hygiene practices inspections Audit Document/ record review	Dairy Plant Construction and Layout of Building PRP Dairy Plant Environment Suitability, Cleaning, and Maintenance PRP Dairy Plant Management of Purchased Materials PRP Dairy Plant Utilities PRP Dairy Plant Cleaning/Sanitizing Procedures Dairy Plant Awareness and Training Procedure Dairy Plant Product Traceability Procedure Dairy Plant Environmental and Pathogen Monitoring Procedure

continued

CHAPTER 3

Table 3.26 PRP 13, Work Sheet 2: PRP Management *(Continued)*

A. PRP (go step-by-step through ISO/TS 22002-1) 13 Personnel hygiene and employee facilities	Hazards			**E.** Control measures	**F.** What is monitored and when	
	B. Agent(s)	**C.** Presence, growth, survival, increase, contamination	**D.** Origin, cause, source, vector, condition			
13.3 Staff canteens and designated eating areas	B	Presence, contamination	Contamination by vegetative pathogens	Hygienic storage of prepared food Cooking and holding temperatures	Cleaning/sanitizing, daily Pathogen monitoring, daily Temperature and time limitations, daily	
	C	Presence, contamination	Cleaning and sanitizing solution residues	Material safety data sheets for cleaning and/or sanitizing chemicals Approved cleaning and sanitizing chemicals Chemical storage	Toxic residues, daily/weekly	
	P	Presence, contamination	Extraneous material arising from poor personnel facility maintenance and/or cleaning, for example, paint	Preventive maintenance Cleaning log	Maintenance Cleaning, daily/weekly	
13.4 Workwear and protective clothing	B	Presence, contamination	Contamination by pathogens Glove use, where specified Improper footwear	Personal hygiene policy (hair, dirt, personnel perspiration, and so on) Hair restraints/beard snoods Dedicated dairy plant footwear Properly maintained food foamers Specification for laundering of uniforms/lab coats Adequate supply of laundered uniforms/lab coats Lockers provided for uniform storage Clean uniforms to be worn	Temperature Pathogen monitoring	

	G. Who is responsible	**H.** Correction/ corrective action	**I.** Records	**J.** Verification activities	**K.** Reference documents
	Hygienist Canteen staff	Cleaning/ sanitizing schedule/ program Ingredient/ product disposal	Good hygiene practices inspections Audits Environmental and pathogen monitoring Cleaning/ sanitizing Cooking and holding temperature Waste disposal	Good hygiene practices inspections Audit Document/ record review	Dairy Plant Environment Suitability, Cleaning, and Maintenance PRP Dairy Plant Cleaning/Sanitizing Procedures Dairy Plant Awareness and Training Procedure Dairy Plant Environmental and Pathogen Monitoring Procedure
	Quality assurance/ laboratory Sanitizing operators Cleaning service providers	Environmental monitoring frequency review Quality assurance training, if required Reclean/ resanitize	Good hygiene practices inspections Audits Cleaning/ sanitizing	Good hygiene practices inspections Audit Document/ record review	Dairy Plant Environment Suitability, Cleaning, and Maintenance PRP Dairy Plant Cleaning/Sanitizing Procedures Dairy Plant Awareness and Training Procedure Dairy Plant Environmental Monitoring Procedure
	Maintenance Cleaning service provider	Quality assurance training, if required Reclean/ resanitize	Good hygiene practices inspections Audits Cleaning/ sanitizing Maintenance	Good hygiene practices inspections Audit Document/ record review	Dairy Plant Environment Suitability, Cleaning, and Maintenance PRP Dairy Plant Cleaning/Sanitizing Procedures Dairy Plant Awareness and Training Procedure
	Quality assurance laboratory Cleaning operator	Reclean	Clean-in-place charts for all dairy plant processing equipment	Good hygiene practices inspections Audit Document/ record review	Dairy Plant Measures for Prevention of Cross Contamination PRP Dairy Plant Construction and Layout of Building PRP Dairy Plant Environment Suitability, Cleaning, and Maintenance PRP Dairy Plant Management of Purchased Materials PRP Dairy Plant Personal Hygiene Policy Dairy Plant Awareness and Training Procedure

continued

Table 3.26 PRP 13, Work Sheet 2: PRP Management *(Continued)*

A. PRP (go step-by-step through ISO/TS 22002-1) 13 Personnel hygiene and employee facilities	Hazards			**E.** Control measures	**F.** What is monitored and when
	B. Agent(s)	**C.** Presence, growth, survival, increase, contamination	**D.** Origin, cause, source, vector, condition		
13.4 Workwear and protective clothing *(continued)*	C	None			
	P	Presence, contamination	Extraneous material arising from personnel jewelry, false fingernails, fingernail polish, buttons, pens, and so on	Personal hygiene policy (jewelry, fingernails, pens, and so on)	Good hygiene practices, daily
13.5 Health status	B	Presence, contamination	Contamination by pathogens because of personnel ill health, minor cuts, or infectious disease	Dairy plant personal hygiene policy Dairy plant hygiene awareness and training Supervisor notification Glove use after minor cuts and handwashing Personnel prohibition to work handling food products	Personnel health status Pathogen monitoring frequency, daily/weekly
	C	None			
	P	Presence, contamination	Contamination from Band-Aid/plaster	Use of Band-Aids reported to management	Use of Band-Aids, if allowed
13.6 Illness and injuries	B	Presence, contamination	Contamination by pathogens because of personnel injury on hands and lower portions of the arms	Dairy plant personal hygiene policy Dairy plant hygiene awareness and training Supervisor notification Glove use after minor cuts and handwashing Personnel prohibition to work handling food products	Personnel health status Pathogen monitoring frequency, daily/weekly
	C	None			

	G. Who is responsible	**H.** Correction/ corrective action	**I.** Records	**J.** Verification activities	**K.** Reference documents
	All personnel, including visitors and contractors	Retraining, if required Disciplinary action, if required	Good hygiene practices inspections Audits Cleaning/ sanitizing Maintenance	Good hygiene practices inspections Audit Document/ record review	Dairy Plant Personal Hygiene Policy Dairy Plant Awareness and Training Procedure
	All personnel Hygienist Medical health nurse, if available	Personnel prohibition to work handling food products	Personnel hygiene/health Good hygiene practices inspections Audits Pathogen monitoring	Good hygiene practices inspections Audit Document/ record review	Dairy Plant Measures for Prevention of Cross Contamination PRP Dairy Plant Personal Hygiene Policy Dairy Plant Cleaning/Sanitizing Procedures Dairy Plant Awareness and Training Procedure Dairy Plant Product Traceability Procedure Dairy Plant Environmental and Pathogen Monitoring Procedure
	Food safety manager	Use of gloves	Band-Aids use	Good hygiene practices inspections Audit Document/ record review	Dairy Plant Personal Hygiene Policy Dairy Plant Awareness and Training Procedure
	All personnel Hygienist Medical health nurse, if available	Personnel prohibition to work handling food products	Personnel hygiene/health Good hygiene practices inspections Audits Pathogen monitoring	Good hygiene practices inspections Audit Document/ record review	Dairy Plant Measures for Prevention of Cross Contamination PRP Dairy Plant Personal Hygiene Policy Dairy Plant Awareness and Training Procedure Dairy Plant Environmental and Pathogen Monitoring Procedure

continued

Table 3.26 PRP 13, Work Sheet 2: PRP Management *(Continued)*

A. PRP (go step-by-step through ISO/TS 22002-1) 13 Personnel hygiene and employee facilities	Hazards			**E.** Control measures	**F.** What is monitored and when	
	B. Agent(s)	**C.** Presence, growth, survival, increase, contamination	**D.** Origin, cause, source, vector, condition			
13.6 Illness and injuries *(continued)*	P	Presence, contamination	Contamination from Band-Aid/ plaster	Use of Band-Aids reported to management	Use of Band-Aids, if allowed	
13.7 Personal cleanliness	B	Presence, contamination	Contamination by pathogens because of lack of personal hygiene by personnel	Dairy plant personal hygiene policy Dairy plant hygiene awareness and training Gloves, where required	Pathogen monitoring frequency Good hygiene practices inspections/ observations, daily/ weekly	
	C	None				
	P	None				
13.8 Personal behavior	B	Presence, contamination	Contamination by pathogens	Dairy plant personal hygiene policy Dairy plant hygiene awareness and training Gloves, where required	Pathogen monitoring frequency, daily/weekly	
	P	None				
	P	Presence, contamination	Extraneous material arising from personnel behavior, for example, smoking, chewing gum, jewelry, pens exposed, false nails, eyelashes, medicines, and so on	Dairy plant personal hygiene policy Dairy plant smoking policy Dairy plant hygiene awareness and training Designed areas for storing smoking materials, medicines Maintenance of personal lockers (cleaned and kept free of soiled clothing, storage of religious/ cultural objects, and so on); personal effects Handwashing signs	Good hygiene practices, weekly	

CHAPTER 3

	G. Who is responsible	**H.** Correction/ corrective action	**I.** Records	**J.** Verification activities	**K.** Reference documents
	Food safety manager	Use of gloves	Band-Aid use	Good hygiene practices inspections Audit Document/ record review	Dairy Plant Personal Hygiene Policy Dairy Plant Awareness and Training Procedure
	All personnel Hygienist Quality assurance/ laboratory	Pathogen monitoring Quality assurance training, if required Disciplinary action, if required	Personnel hygiene Good hygiene practices inspections Audits Pathogen monitoring	Good hygiene practices inspections Audit Document/ record review	Dairy Plant Measures for Prevention of Cross Contamination PRP Dairy Plant Personal Hygiene Policy Dairy Plant Awareness and Training Procedure Dairy Plant Environmental and Pathogen Monitoring Procedure
	Cleaning/ sanitizing supervisor Quality assurance/ laboratory	Pathogen monitoring Quality assurance training, if required Disciplinary action, if required	Personnel hygiene Good hygiene practices inspections Audits Pathogen monitoring	Good hygiene practices inspections Audit Document/ record review	Dairy Plant Measures for Prevention of Cross Contamination PRP Dairy Plant Personal Hygiene Policy Dairy Plant Awareness and Training Procedure Dairy Plant Environmental and Pathogen Monitoring Procedure
	All personnel Hygienist Quality assurance/ laboratory	Pathogen monitoring Quality assurance training, if required Disciplinary action, if required	Personnel hygiene Good hygiene practices inspections Audits Pathogen monitoring	Good hygiene practices inspections Audit Document/ record review	Dairy Plant Measures for Prevention of Cross Contamination PRP Dairy Plant Environment Suitability, Cleaning, and Maintenance PRP Dairy Plant Personal Hygiene Policy Dairy Plant Smoking Policy Dairy Plant Awareness and Training Procedure Dairy Plant Environmental and Pathogen Monitoring Procedure

Instructions for Completing PRP Work Sheet 3: PRP Verification Action Plan	
Column A	The establishment should provide details on the number and title of the PRP. It is recommended that the number should match the number of the relevant part in the appropriate FSMS scheme standard, for example, in ISO/TS 22002-1, 13—Personnel hygiene and employee facilities.
Column B	The establishment should provide details on the verification actions associated with the PRP and on the individual or entity responsible for reviewing these verification actions.

Table 3.27 PRP 13, Work Sheet 3: PRP Verification Action Plan

A. PRP	**B.** Verification action
13 Personnel hygiene and employee facilities	Reviewed by hygienist and personnel hygiene PRP team
	Review of environment, pathogen, and foreign objects monitoring Review of good hygiene practices inspections Food safety management system audits Internal GMP/hygiene audits Review of visitors/contractors hygiene Review of personal hygiene Review of personal protective equipment/workwear Review of employee prohibition to work (under special conditions) Review of training Frequency and criticality review

Instructions for Completing PRP Work Sheet 4: PRP Meeting Summary	
Column A	List meeting dates.
Column B	List attendees among the team and other invitees.
Column C	Provide the reason for the meeting.
Column D	Record decisions and next steps.
Column E	Identify the individuals or entities responsible for executing any decisions.
Column F	Record deadlines.
Column G	Indicate the dates of relevant actions.

Table 3.28 PRP 13, Work Sheet 4: PRP Meeting Summary

A. Date	**B.** Participants	**C.** Purpose	**D.** Outcome (decisions/actions)	**E.** Responsibility	**F.** Deadline	**G.** Deadline reached
April 20, 2018	G Moran, O Brown, M Rodrigues, B White, D Collins, O Murphy, C Flack	Initial PRP review	Update PRP management work sheet Review related PRPs	G Moran to complete verification sheet	May 15, 2018	May 15, 2018
April 28, 2018	G Moran, O Brown, M Rodrigues, B White, D Collins, O Murphy, C Flack	Complete GAP sheet Review PRP management work sheet	Completed and approved Reviewed and approved	G Moran to update PRP work sheets	May 15, 2018	May 15, 2018
February 17, 2019	G Moran, O Brown, M Rodrigues, B White, D Collins, O Murphy, C Flack	Review and update of utility specifications	Complete the update of the water supply specification	PRP team to complete	February 17, 2019	February 17, 2019
February 17, 2019	G Moran, O Brown, M Rodrigues, B White, D Collins, O Murphy, C Flack	Review and update based on changes to ISO 22000:2018	The current PRPs underwent comprehensive reviews of compliance with ISO/TS 22002-1 and ISO 22000:2018 starting on February 17, 2019, and completed on February 20, 2019	PRP team to complete	February 20, 2019	February 20, 2019

Instructions for Completing PRP Work Sheet 5: PRP Gap Registration and Resolution	
Column A	Provide a description of the FSMS scheme requirement.
Column B	Provide a description of the requirement arising from the FSMS scheme where the gap exists.
Column C	Provide a short description of the specific requirement where the gap exists within the FBO.
Column D	Detail the relevant FSMS policy.
Column E	Describe the gap.
Column F	Provide the action to be taken to address the requirement identified as not having been fulfilled.
Column G	Provide details of the actions taken to address the gap and the date of the completion of the actions.
Column H	Add any additional relevant comments as required.

Table 3.29 PRP 13, Work Sheet 5: PRP Gap Registration and Resolution

Fill out this work sheet only if gaps have been identified.

A. ISO/TS 22002-1, 13 Personnel hygiene and employee facilities	**B.** Description (of the requirement of the standard)	**C.** Specific requirement	**D.** Associated dairy policy	**E.** Gap	**F.** Action plan (including time frame for completion)	**G.** Gap resolution (actions completed, with date)	**H.** Comments
13.5 Health status	Medical examinations, where permitted, shall be carried out at intervals defined by the organization	Health screening of personnel	Food safety policy	Health screening policy not in compliance with country regulations and not effectively communicated to personnel	Review/update health screening policy and communicate effectively within the FBO as soon as practical	Review/approved policy and reenforced policy/practice with relevant personnel; see PRP team meeting May 15, 2018	Need to continue to monitor for next 12 months
13.8 Personnel behavior	Prohibition of storage of product contact tools and equipment in personal lockers	Product contact tools and equipment to be stored in FBO-supplied toolbox	Food safety policy	Practice does not match the requirements of the standard	Reenforce policy/practice and include in good hygiene practices inspections	Review/approved new health screening policy and communicate it to all personnel; see PRP team meeting May 17, 2019	Need to continue to monitor for next six months to sustain improvements to date

CHAPTER 3

Instructions for Completing PRP Work Sheet 6: Hazard Agent	
Column A	Classify food safety hazard agents, for example, biological, chemical, or physical hazard agents.
Column B	Indicate the food safety hazard agent code, for example, allergen = A, biological = B, chemical = C, physical = P.

Table 3.30 PRP 13, Work Sheet 6: Hazard Agent

A. Hazardous agents	**B.** Hazard class
Biological (for example, vegetative or spores, depending on circumstances)	B
Chemical (for example, cleaning chemicals, nonfood-grade lubricants, oils and greases, and chemical residues)	C
Physical (for example, various types of foreign material, including metal, wood, plastic, or other foreign bodies)	P
Allergens (for example, milk, soy, wheat, eggs, fish, shellfish, tree nuts, peanuts)	A

PRP 14: Rework

Sample completed work sheets for PRP 14 follow (tables 3.31–3.36). For instructions on filling out each PRP work sheet, see the boxes that precede each sample completed work sheet.

Instructions for Completing PRP Work Sheet 1: PRP Scope	
A. PRP study scope	Provide the PRP title from the standard or scheme (for example, Rework). Provide the standard PRP number (for example, in ISO/TS 22002-1, 14—Rework). Provide the facility name, product category, processes, product, PRP start date, status of the PRP (for example, draft, approved), and end date.
B. PRP review history	In this section, record information about the history of the PRP revision, with an explanation of the reason why this update has been done: "according to plan" or "unscheduled." For an unscheduled revision, why has this revision been undertaken? (What reason?)
C. PRP team members	For every PRP study, the organization needs to establish an HACCP team with specific responsibilities and roles. Names within the company, department name, and responsibilities should be detailed. The competence of each team member should also be documented.
D. Specialist input	To establish PRP studies, companies may need advice from an outsourced expert (consultant/subject matter expert). The expert's role should be explained: input/specialist advice.
E. Authorization	Team members must indicate their approval of the document by providing their names, positions, responsibilities held, and signature. The authorized team member should provide his/her signature and the date signed.

CHAPTER 3

Table 3.31 PRP 14, Work Sheet 1: PRP Scope

PRP	14 Rework		
A. PRP study scope			
Facility	Joe Bloggs Dairy Plant	Start date	February 17, 2019
Product category	Grade A Interstate Milk Shippers registered whole milk	Status	Draft
Processes	High-temperature/short-time pasteurizer, aseptic filling, retort	End date	Ongoing
Products	Grade A aseptically processed and packaged milk		
B. PRP review history	Check as appropriate	Notes/reason for unscheduled review	Dates of last three reviews
New PRP study	✓	Current PRPs underwent a comprehensive review for compliance with ISO/TS 22002-1 and ISO 22000:2018 starting on February 15, 2019, and completed on February 17, 2019. These management sheets describe each PRP in place at the dairy plant facility.	
Scheduled review	December 20, 2019		
Unscheduled review			
C. PRP team members			
Name	Position	Department	Responsibility/role
G Moran	Food safety manager	Food safety	Food safety/quality assurance
O Brown	Hygienist/microbiologist	Food safety	Hygienist/ microbiologist
M Rodrigues	Milk processing manager	Milk processing	Milk processing
B Murphy	Laboratory manager	Quality assurance	Laboratory
D Small	Warehouse manager	Warehousing	Warehousing
O Murphy	Engineering manager	Engineering	Engineering
C Flack	Factory manager	Management	Management

D. Specialist input		
Name	Location/job title	Input/specialist advice
Angela Yard	Consultant	PRP team facilitator
E. Authorization		
Food safety team leader/quality assurance manager	Signature: *G Moran*	Date: *February 17, 2019*
Management team member	Signature: *C Flack*	Date: *February 17, 2019*

Instructions for Completing PRP Work Sheet 2: PRP Management	
Column A	Describe the ISO/TS 22002-1 requirements.
Column B	Describe the hazard agent, for example, biological (B), chemical (C), physical (P), or a combination.
Column C	Describe how the hazard is manifest as a threat, including presence, increase, or survival.
Column D	Describe the cause, origin, condition, source, or vector of a hazard.
Column E	Describe the control measures the FBO has in place to control relevant hazards.
Column F	Describe the hazard measurement parameters and the monitoring frequency of the measurement parameters.

Table 3.32 PRP 14, Work Sheet 2: PRP Management

	Hazards					
A. PRP (go step-by-step through ISO/TS 22002-1) 14 Rework	**B.** Agent(s)	**C.** Presence, growth, survival, increase, contamination	**D.** Origin, cause, source, vector, condition	**E.** Control measures	**F.** What is monitored and when	
14.1 General requirements	B, C, P	Contamination	Microbiological, chemical, or extraneous matter contamination	Hygiene, cleaning, product inspection Pathogen, environmental monitoring Extraneous material procedures Traceability	Hygiene, cleaning, storage monitored through good hygiene practices inspections and audits, monthly Pathogen monitoring program in place, weekly	

Instructions for Completing PRP Work Sheet 2: PRP Management (*continued*)	
Column G	Describe the job role or title of the department/function within the FBO responsible for monitoring the relevant hazard measurement parameters.
Column H	Describe the correction and corrective action aimed at preventing a reoccurrence of a rise above the allowable or permitted hazard measurement parameters.
Column I	Indicate the monitoring and hazard measurement parameter records to be maintained.
Column J	Describe the verification activities necessary to confirm the accuracy of the monitoring and hazard measurement parameters.
Column K	Describe the FBO documents and relevant external documents, for example, statutory and regulatory requirements.

G. Who is responsible	**H.** Correction/ corrective action	**I.** Records	**J.** Verification activities	**K.** Reference documents
Dairy plant warehousing Dairy plant quality assurance laboratory Dairy plant food safety	Training Product rework Product disposal	Good hygiene practices inspection Audit reports Pathogen monitoring Product inspection Traceability	Good hygiene practices inspections Audits Product inspection Environmental/ pathogen monitoring	Dairy Plant Design and Construction of Buildings PRP Dairy Plant Site Location and Standards PRP Dairy Plant Layout of Premises and Workspace PRP Dairy Plant Internal Structure PRP Dairy Plant Environment Suitability, Cleaning, and Maintenance PRP Warehousing PRP Rework Procedure Dairy Plant Product Inspection Procedure Dairy Plant Awareness and Training Procedure Dairy Plant Product Traceability Procedure Dairy Plant Environmental and Pathogen Monitoring Procedure Pest Control Folder/Manual (external pest control company)

continued

CHAPTER 3

Table 3.32 PRP 14, Work Sheet 2: PRP Management *(Continued)*

	Hazards				
A. PRP (go step-by-step through ISO/TS 22002-1) 14 Rework	**B.** Agent(s)	**C.** Presence, growth, survival, increase, contamination	**D.** Origin, cause, source, vector, condition	**E.** Control measures	**F.** What is monitored and when
14.2 Storage, identification, and traceability	B	Contamination	Reclaimed or reworked product may have been handled, stored, or used in a way to subject it to contamination with pathogens	Product that has not been continuously in control of the dairy plant is assumed to contain pathogens and to be reclaimed or reworked If product is no longer under the control of the dairy plant, it cannot be assumed to have been held to preclude temperature abuse or adulteration Only product that has not left the control of the dairy plant should be used, kept segregated, handled, protected, and cooled as appropriate for the product, with the exception of product approved by the regulatory agency Reworking is done in a clean area and in a manner that will not contaminate the product being salvaged	Environmental and pathogen monitoring Good warehousing practices Product segregation Product protection (temperature), daily/weekly
	C	Contamination	Allergens mixed with products on which the labels do not indicate the presence of allergens	Foods containing undeclared allergens may cause life-threatening reactions in sensitive individuals	Reworked product segregation Product labeling

	G. Who is responsible	**H.** Correction/ corrective action	**I.** Records	**J.** Verification activities	**K.** Reference documents
	Dairy plant warehousing Dairy plant milk processing Dairy plant quality assurance laboratory Dairy plant food safety	Training Product rework Product disposal	Good hygiene practices/good warehousing practice Inspection Audit reports Rework (classification) Pathogen monitoring Product inspection Traceability	Good hygiene practices/good warehousing practice inspections Audits Environmental/ pathogen monitoring	Dairy Plant Design and Construction of Buildings PRP Dairy Plant Site Location and Standards PRP Dairy Plant Layout of Premises and Workspace PRP Dairy Plant Internal Structure PRP Warehousing PRP Rework Procedure Product Traceability Procedure Dairy Plant Product Inspection Procedure Dairy Plant Awareness and Training Procedure Dairy Plant Environmental and Pathogen Monitoring Procedure
	Dairy plant warehousing Dairy plant milk processing Dairy plant food safety	Training Product rework Product disposal	Good hygiene practices/good warehousing practice inspection Audit reports Rework (classification) Traceability	Good hygiene practices/good warehousing practice inspections Audits Environmental monitoring	Dairy Plant Design and Construction of Buildings PRP Dairy Plant Site Location and Standards PRP Dairy Plant Layout of Premises and Workspace PRP Dairy Plant Internal Structure PRP Warehousing PRP Measures of Prevention of Cross Contamination PRP Rework Procedure Allergen Management Procedure Dairy Plant Product Inspection Procedure Dairy Plant Awareness and Training Procedure Dairy Plant Environmental and Pathogen Monitoring Procedure

continued

Table 3.32 PRP 14, Work Sheet 2: PRP Management *(Continued)*

	Hazards					
A. PRP (go step-by-step through ISO/TS 22002-1) 14 Rework	**B.** Agent(s)	**C.** Presence, growth, survival, increase, contamination	**D.** Origin, cause, source, vector, condition	**E.** Control measures	**F.** What is monitored and when	
14.2 Storage, identification, and traceability *(continued)*	P	Contamination	Extraneous material may result in choking or other physical harm to consumers	Opening of products is conducted in a manner that will minimize the opportunity for bits, packaging, cutting tools, and so on, from entering the product Verification that, at some point in the process, the ingredient or the milk product to which the ingredient is added will pass through a filter, screen, or small orifice	Foreign objects contamination, each batch	
14.3. Rework usage	B,C,P	Contamination	Microbiological, chemical, or extraneous matter contamination	Rework procedure and additional documentation specifying the conditions of rework, the process step, the acceptable quantity, type, conditions of rework, any preprocessing steps, and so on Opening of products is conducted in a manner that will minimize the opportunity for bits, packaging, cutting tools, and so on from entering the product Verification that, at some point in the process, ingredient or the milk product to which the ingredient is added will pass through a filter, screen, or small orifice	Hygiene Cleaning foreign object contamination, each batch Good hygiene practices inspections Audits Environment and pathogen monitoring program in place, weekly	

	G. Who is responsible	**H.** Correction/ corrective action	**I.** Records	**J.** Verification activities	**K.** Reference documents
	Dairy plant warehousing Dairy plant milk processing Dairy plant quality assurance Dairy plant food safety	Training Product rework Product disposal	Good hygiene practices/good warehousing practice inspection Audit report Rework (classification) Foreign objects monitoring Product inspection Traceability	Good hygiene practices/good warehousing practice inspections Audits Foreign objects monitoring	Dairy Plant Design and Construction of Buildings PRP Dairy Plant Site Location and Standards PRP Dairy Plant Layout of Premises and Workspace PRP Dairy Plant Internal Structure PRP Warehousing PRP Rework Procedure Dairy Plant Product Inspection Procedure Dairy Plant Awareness and Training Procedure Dairy Plant Environmental and Pathogen Monitoring Procedure
	Dairy plant quality assurance Dairy plant maintenance Dairy plant sanitization	Training Product rework Product disposal	Good hygiene practices inspection Audit report Rework (classification) Environment, pathogen, and foreign objects monitoring Product inspection Traceability Waste disposal	Good hygiene practices/good warehousing practice inspections Audits Environment, pathogen, and foreign object monitoring Product inspection	Dairy Plant Design and Construction of Buildings PRP Dairy Plant Site Location and Standards PRP Dairy Plant Layout of Premises and Workspace PRP Dairy Plant Internal Structure PRP Dairy Plant Environment Suitability, Cleaning, and Maintenance PRP Warehousing PRP Rework Procedure Dairy Plant Product Inspection Procedure Dairy Plant Awareness and Training Procedure Dairy Plant Environmental and Pathogen Monitoring Procedure

Instructions for Completing PRP Work Sheet 3: PRP Verification Action Plan	
Column A	The establishment should provide details on the number and title of the PRP. It is recommended that the number should match the number of the relevant part in the appropriate FSMS scheme standard, for example, in ISO/TS 22002-1, 14—Rework.
Column B	The establishment should provide details on the verification actions associated with the PRP and on the individual or entity responsible for reviewing these verification actions.

Table 3.33 PRP 14, Work Sheet 3: PRP Verification Action Plan

A. PRP	**B.** Verification action
14 Rework	Reviewed by laboratory manager and pest control PRP team
	Review of environment, pathogen, and foreign objects monitoring Review of good hygiene practices/good warehousing practice inspections Food safety management system audits Internal GMP/hygiene audits Review of cleaning/sanitizing program/schedule/records Review of product disposal Review of traceability Review of training Frequency and criticality review

Instructions for Completing PRP Work Sheet 4: PRP Meeting Summary	
Column A	List meeting dates.
Column B	List attendees among the team and other invitees.
Column C	Provide the reason for the meeting.
Column D	Record decisions and next steps.
Column E	Identify the individuals or entities responsible for executing any decisions.
Column F	Record deadlines.
Column G	Indicate the dates of relevant actions.

Table 3.34 PRP 14, Work Sheet 4: PRP Meeting Summary

A. Date	**B.** Participants	**C.** Purpose	**D.** Outcome (decisions/actions)	**E.** Responsibility	**F.** Deadline	**G.** Deadline reached
April 20, 2018	G Moran, O Brown, M Rodrigues, B White, D Collins, O Murphy, C Flack	Initial PRP review	Update PRP management work sheet Review related PRPs	G Moran to complete verification sheet	May 15, 2018	May 15, 2018
April 28, 2018	G Moran, O Brown, M Rodrigues, B White, D Collins, O Murphy, C Flack	Complete GAP sheet Review PRP management work sheet	Completed and approved Reviewed and approved	G Moran to update PRP work sheets	May 15, 2018	May 15, 2018
February 17, 2019	G Moran, O Brown, M Rodrigues, B White, D Collins, O Murphy, C Flack	Review and update of utility specifications	Complete the update of the water supply specification	PRP team to complete	February 17, 2019	February 17, 2019
February 17, 2019	G Moran, O Brown, M Rodrigues, B White, D Collins, O Murphy, C Flack	Review and update based on changes to ISO 22000:2018	The current PRPs underwent comprehensive reviews of compliance with ISO/TS 22002-1 and ISO 22000:2018 starting on February 17, 2019, and completed on February 20, 2019	PRP team to complete	February 20, 2019	February 20, 2019

CHAPTER 3

Instructions for Completing PRP Work Sheet 5: PRP Gap Registration and Resolution	
Column A	Provide a description of the FSMS scheme requirement.
Column B	Provide a description of the requirement arising from the FSMS scheme where the gap exists.
Column C	Provide a short description of the specific requirement where the gap exists within the FBO.
Column D	Detail the relevant FSMS policy.
Column E	Describe the gap.
Column F	Provide the action to be taken to address the requirement identified as not having been fulfilled.
Column G	Provide details of the actions taken to address the gap and the date of the completion of the actions.
Column H	Add any additional relevant comments as required.

Table 3.35 PRP 14, Work Sheet 5: PRP Gap Registration and Resolution

Fill out this work sheet only if gaps have been identified.

A. ISO/TS 22002-1, 14 Rework	**B.** Description (of the requirement of the standard)	**C.** Specific requirement	**D.** Associated dairy policy	**E.** Gap	**F.** Action plan (including time frame for completion)	**G.** Gap resolution (actions completed, with date)	**H.** Comments
14.2 Storage, identification, and traceability	The rework classification or the reason for the rework designation shall be recorded (for example, product name, date of production, shift, line of origin, shelf life)	Recording of rework classification	Food safety policy	Rework procedure does not fully meet the requirements of ISO/TS 22002-1	Review/update rework procedure	Rework procedure updated; see PRP team meeting May 15, 2018	Need to complete training and verify effectiveness of implementation

CHAPTER 3

Instructions for Completing PRP Work Sheet 6: Hazard Agent	
Column A	Classify food safety hazard agents, for example, biological, chemical, or physical hazard agents.
Column B	Indicate the food safety hazard agent code, for example, allergen = A, biological = B, chemical = C, physical = P.

Table 3.36 PRP 14, Work Sheet 6: Hazard Agent

A. Hazardous agents	**B.** Hazard class
Biological (for example, vegetative or spores, depending on circumstances)	B
Chemical (for example, cleaning chemicals, nonfood-grade lubricants, oils and greases, and chemical residues)	C
Physical (for example, various types of foreign material, including metal, wood, plastic, or other foreign bodies)	P
Allergens (for example, milk, soy, wheat, eggs, fish, shellfish, tree nuts, peanuts)	A

General information on HACCP

History of HACCP

In the 1960s, the Pillsbury Corporation developed the HACCP system with the U.S. National Aeronautics and Space Administration to ensure food safety aboard the first manned space missions. The HACCP system and application guidelines were defined by the CAC, which implements the Food Standards Program of the Food and Agriculture Organization of the United Nations and the World Health Organization.[4]

Following an outbreak of *Escherichia coli* 0157 in Scotland in 1996, the Pennington Report recommended that HACCP be adopted by all food businesses to ensure food safety (Pennington 1997). All global food safety initiative scheme standards—BRC Global Standards, FSSC 22000, Safe Quality Food Programs, GLOBALG.A.P., and so on—have established specific requirements for the incorporation of HACCP into FSMSs. An effective HACCP has become invaluable in supporting any food safety due diligence defense.

HACCP principles

An FSMS represents a systematic approach to identifying and controlling hazards, whether microbiological, chemical, or physical, that could pose a threat to the production of safe food. This involves identifying what could go wrong in a food system and planning how to prevent this occurrence.

An FSMS must be based on HACCP principles, which enable FBOs to identify and control hazards before they threaten the safety of food or of consumers. There are seven principles of HACCP, as follows:

First, *identify the hazards*. This requires that FBOs examine each stage—purchasing, delivery, storage, preparation, cooking, refrigeration, and so on—in their food operations and identify what might go wrong. This might involve *Salmonella* in a cooked chicken product because of cross contamination with raw meat (biological hazard), the contamination of uncovered food by detergent (chemical hazard), or a piece of broken glass that has fallen into an uncovered food (physical hazard).

Second, *determine the critical control points*. FBOs need to identify the points in their operations that ensure their control over the hazards, For instance, cooking raw meat thoroughly will kill pathogens, such as *Escherichia coli* O157.

Third, *establish critical limits*. FBOs must set limits to enable them to identify when a CCP is out of control. Thus, during cooking, the center of a beef burger patty must reach a minimum temperature of 75°C or an equivalent time temperature combination, such as 70°C for two minutes, to ensure that pathogens are destroyed.

Fourth, *establish a system to monitor the control over the CCP*. In identifying CCPs and critical limits, FBOs should possess a method to monitor and record what is happening at each CCP. Typically, monitoring involves measuring parameters, such as temperature and time. However, the method and frequency of this monitoring often depends on the size and nature of the FBO operations. However, in any case, the monitoring process should be simple, clear, and easy. For example, refrigerated food might be probed to ensure that the temperature is maintained at 5°C or less.

Fifth, *establish the corrective action to be taken if monitoring indicates that a particular CCP is not under control*. Thus, if the temperature of a food in a refrigerator rises to 10°C because of a technical failure in the appliance, the corrective action might be to discard the food and repair the unit according to the manufacturer's instruction manual to ensure the correct temperature of 5°C is achieved.

Sixth, *establish verification procedures to confirm that the HACCP system is effective*. FBOs should review and correct their FSMSs periodically and whenever they alter their operations. For instance, after replacing an oven, an FBO should, by probing food, determine that the time and temperature settings of the new appliance are accurate and achieve the minimum safe cooking temperature for the particular dish.

Seventh, *establish documentation on all procedures and records appropriate to these principles and to the application of these principles*. For the successful implementation of an FSMS based on HACCP, appropriate documentation and records must be kept and be readily available. It is unrealistic to operate HACCP or to demonstrate compliance with current legislation without providing evidence, such as written records. As with the FSMS itself, the complexity of the recordkeeping will depend on the nature and complexity of the FBO's business. The aim should be to ensure that control is maintained without generating excessive paperwork.

The benefits of hazard control and HACCP

Hazard control or HACCP provides businesses with a cost-effective system for controlling food safety, from ingredients through production, storage, and distribution to sales and service among final consumers. The preventive approach of hazard control or HACCP not only improves food safety management, but also complements other quality management systems. It promotes the following main benefits:

- Saves the business money in the long run
- Avoids poisoning among customers of the business
- Food safety standards are enhanced
- Ensures that the business complies with the law
- Food quality standards are improved
- Organizes the processes of the business to produce safe food
- Organizes business staff by fostering teamwork and efficiency
- Due diligence becomes a defense in court

The International Finance Corporation (IFC) has developed a comprehensive cost-benefit analysis tool to enable FBOs to establish the benefits of adopting hazard controls, HACCP, or an FSMS (see chapter 6).

Preliminary steps in developing a hazard control or HACCP plan

When the FBO sets about establishing and developing a hazard control or HACCP plan, the FBO needs to develop the processes required for the production of safe products. Drawing up a hazard control or HACCP plan starts with the collection of information. The fact-finding process involves several preliminary steps. ISO 22000:2018 requires that all relevant information needed to conduct the hazard analysis be collected, maintained, updated, and documented. Records should also be kept.

THE PURPOSE OF THE PRELIMINARY STEPS

HACCP systems and FSMSs are systematic preventive approaches to ensuring the safe production of food products.

Prior to the application of a hazard control or HACCP plan, an FBO should be operating according to the CAC (2003) General Principles of Food Hygiene, the appropriate Codex Alimentarius codes of practice, and relevant food safety legislation. FBOs must understand the food sector requirements that apply to their food products and processes. They are obligated to implement, operate, and ensure the effectiveness of the planned activities and any changes to these activities.

THE FIVE PRELIMINARY STEPS

The CAC (2003) outlines five preliminary steps that need to be completed before a hazard control or HACCP plan is developed. These preliminary steps must be addressed in sequence. They are (1) assemble the hazard control or HACCP team, (2) describe the food and the distribution of the food, (3) describe the intended use and the consumers of the food, (4) develop a flow diagram that describes the process, and (5) verify the flow diagram.

Preliminary step 1: Assemble the hazard control or HACCP team

To ensure that all likely hazards and CCPs are identified, a multidisciplinary team of people must be assembled to develop, implement, and maintain the HACCP system. The hazard control or HACCP team should consist of people who have operational experience, product-specific knowledge, and a good understanding of the production process. The hazard control or HACCP team should include the following types of employees: quality assurance managers, technical staff, production supervisors and managers, laboratory personnel, engineering staff, and sanitation staff.

If the FBO is small, the hazard control or HACCP team may be supported by an external FSMS consultant. In this case, there should be a written agreement or contract in place between the FBO and the FSMS consultant clearly defining roles and responsibilities. Given the risk associated with the product or commodity being produced or processed, the FBO has a duty to ensure that the consultant is suitably competent and can perform the assigned tasks.

A hazard control or HACCP team leader needs to be designated to oversee the development, implementation, and maintenance of the hazard control or HACCP system. The team leader must have a good understanding of hazard control or HACCP and a working knowledge of the product and the relevant production process. It is desirable that the team leader should also have proven competence in the design and delivery of training. Attendance of the team leader at a recognized training-the-trainer course is recommended.

Preliminary step 2: Describe the food and the distribution of the food

A full description of the product needs to be prepared to provide a profile of the product and to help determine the food safety hazards associated with the production of the product. A key element is the collection of relevant information on associated food safety hazards and acceptable limits. The hazard control or HACCP team needs to collect food safety hazard identification data and information on acceptance levels that are defined and documented by (1) statutory and regulatory agencies, (2) the CAC, (3) customers, and (4) scientific studies.

Product descriptions must cover relevant food safety information, such as (1) available water process parameters, for example, pH, heavy metals, and so on; (2) end product characteristics, including, for example, shape, size, color, texture, and smell; (3) details on the method of preservation; (4) packaging; (5) storage conditions; (6) shelf life; (7) special labeling; (8) customer preparation; and (9) details on the method of distribution.

Preliminary step 3: Describe the intended use and the consumers of the food

Information on the expected use of a product by end users and consumers should be identified because the intended use of a product will affect hazard analysis decisions. This might include, for example, information on whether the food must be cooked before consumption, or is ready to eat without cooking. Intended use information should also identify whether the end user is the general public or a specific consumer group, particularly vulnerable groups, such as infants, the elderly, pregnant women, ill people, immuno-compromised persons, or cancer patients.

Preliminary step 4: Develop a flow diagram that describes the process

The hazard control or HACCP team should draw up a flow diagram that provides a clear, simple outline of all inputs, steps, and outputs in the food production process. All steps in the process must be set out, including any reworking or recycling of materials. The flow diagram provides the foundation for a systematic hazard analysis.

Preliminary step 5: Verify the flow diagram

An on-site review of the flow diagram must be carried out to check that the diagram accurately reflects the production process associated with the product. The hazard control or HACCP team should follow the production process on-site and check that the flow diagram includes all steps that are carried out. When verifying the accuracy of the flow diagram, consideration needs to be given to the number of work shifts and hours of operation, batch sizes, optional ingredients, and nonroutine steps, such as equipment maintenance.

The completion of the five preliminary steps in the development of a hazard control or HACCP plan represents a solid foundation for the successful application of the seven HACCP principles.

Hazard control or HACCP plan workbook

The following workbook details a sample FBO hazard control plan that is based on Codex Alimentarius and ISO 22000 requirements. It provides information on the implementation of a dairy sector HACCP system and the development of the associated FSMS documentation. These methodologies can be applied in the case of any food product.

ISO 22000:2018 AND HACCP

The hazard control plan workbook is recommended for use in conjunction with ISO 22000:2018. ISO 22000:2018 introduced two new terms, namely, "hazard control plan" and "action criterion." A hazard control plan is equivalent to an HACCP plan, with one major difference: it identifies both CCP and operational PRP (OPRP) control measures. ISO 22000:2018 covers both the HACCP plan, which identifies CCP control measures, and the OPRP plan, which defines the OPRP control measures. In effect, the hazard control plan combines these two categories of control measures into one plan.

The second new term is "action criterion," which is associated only with an OPRP control measure. An action criterion is defined as a measurable or observable specification for the monitoring of an OPRP. It has been established to determine whether an OPRP remains in control and distinguishes between what is acceptable and unacceptable. Refer to table 3.41, later in this chapter, for details.

The workbook overview (table 3.37) chronicles the 13 sample HACCP work sheets (tables 3.38–3.49, figure 3.2), which are similar to ones to be filled out by designated HACCP teams. The 13 sample work sheets include 10 main work sheets (tables 3.38–3.46, figure 3.2) and 3 supplementary work sheets (tables 3.47–3.49). Each work sheet contains brief descriptions of the information to be inserted in each field as well as sample completed fields. Editable work sheets and templates can be found at the following location: http://www.ifc.org/foodsafety/handbook/templates.

Table 3.37 Overview and Guide: HACCP Work Sheets

Main work sheets	Supplementary work sheets	Comments
Work sheet 1: HACCP scope		Registration and approval of the HACCP study
Work sheet 2: product/ingredient descriptions		Product and process description, including characteristics of raw materials and end products
Work sheet 3: flow diagram		Simplified process flow diagram with OPRP and CCP location
	Work sheet A: hazardous agent codes and classification	Guidance for food safety/HACCP team: assessing hazards controlled by the HACCP system
Work sheet 4: hazard identification and description	← (from Work sheets A and B)	Each potential hazard is listed and the significance is determined according to the severity of the health effects and the likelihood of the hazard's occurrence
	Work sheet B: hazard assessment table	Coding and classifying potentially hazardous agents that need to be considered during the study
Work sheet 5: control measure selection and categorization		With help of the decision tree, the control measures are categorized as CCP, OPRP, or modification
Work sheet 6: validation of control measures		Evidence that the control measure can achieve the targeted limits
Work sheet 7: HACCP plan, including OPRPs		List and overview of all identified CCPs and OPRPs with control measures, limits, corrective actions, and responsibilities
Work sheet 8: verification plan		Overview of verification activities that shows that the CCPs and OPRPs have been implemented properly
Work sheet 9: modification(s) and follow-up		List of modifications, with details
Work sheet 10: meeting summary		Recording meetings, attendance, and decisions made by the team
	Work sheet C (optional): list of supporting documents	Supporting information, recording, and filing

MAIN WORK SHEETS

Work sheet 1: HACCP scope

Work sheet 1 defines and documents the scope of the hazard control study, along with the revision history. It also lists the HACCP team members conducting the study. It has two sections. The first is to be completed before the start of the study, while the second is to be completed following the end of the study. The work sheet consists of eight sections. Completion instructions are included within the sample work sheet (table 3.38) to the right of the relevant rows.

Table 3.38 HACCP Work Sheet 1: HACCP Scope		**Instructions**
Complete the section below at the start of the hazard control plan/HACCP study.		
HACCP study no.	*Version no.*	**HACCP study** Supply the HACCP study number, version number, study details (may involve checking one of several descriptions), and the study start date.
#122015	V1.0	
HACCP study details	*Check as appropriate*	
New HACCP study	✓	
Scheduled review		
Unscheduled review		
Date study started	February 1, 2018	
HACCP study scope		**HACCP study scope** Complete the HACCP study scope information, including factory name, plant/line description or number, brand name, product name, product code, and FSMS reference.
Factory	Joe Bloggs LLC	
Plant/line	2211	
Brand	Bloggs	
Product name	Whole milk	
Product code	Interstate Milk Shippers #1	
FSMS reference	ISO 22000:2018	
Description of scope of study (for example, module, start and end point, or products included)		**Description of scope of study** Complete the description of scope of study information by offering a short description of the process undergone and the product.
Grade A aseptically processed and packaged milk		
Scheduled or unscheduled review: main changes/reasons/causes		**Scheduled or unscheduled review** Provide information on hazard control or HACCP review history, including the type, "scheduled" or "unscheduled." For unscheduled reviews, also indicate the cause or reason for the review.
ISO 22000:2018/FSSC 22000 review		

continued

<table>
<tr><th colspan="4">Table 3.38 HACCP Work Sheet 1: HACCP Scope (Continued)</th><th>Instructions</th></tr>
<tr><td colspan="4">HACCP team members</td><td rowspan="10">HACCP team members
Supply details on hazard control or HACCP team members.</td></tr>
<tr><td>Name</td><td>Responsibility/role/ expertise</td><td colspan="2">Department/company</td></tr>
<tr><td>G Moran</td><td>Food safety manager</td><td colspan="2">Food safety/quality assurance</td></tr>
<tr><td>O Brown</td><td>Hygienist/ microbiologist</td><td colspan="2">Hygienist</td></tr>
<tr><td>M Rodrigues</td><td>Milk processing manager</td><td colspan="2">Milk processing</td></tr>
<tr><td>B Murphy</td><td>Laboratory manager</td><td colspan="2">Laboratory</td></tr>
<tr><td>D Small</td><td>Warehouse manager</td><td colspan="2">Warehousing</td></tr>
<tr><td>O Murphy</td><td>Engineering manager</td><td colspan="2">Engineering</td></tr>
<tr><td>C Flack</td><td>Factory manager</td><td colspan="2">Management</td></tr>
<tr><td>N Williams</td><td>Veterinary</td><td colspan="2">Food safety/quality assurance</td></tr>
<tr><td colspan="4">Authorization for new HACCP study or update to new version</td><td rowspan="2">Authorization of HACCP study
The authorized person should sign and date the authorization.</td></tr>
<tr><td>Factory manager</td><td>C Flack</td><td colspan="2">Date: February 15, 2018</td></tr>
<tr><td colspan="5">Complete the section below upon completion of the HACCP study.</td></tr>
<tr><td colspan="4">Planned modification(s) according to HACCP study</td><td rowspan="8">Planned modification(s)
Identify modification number, provisional control measures and deadlines, next review date, and the date the current study was issued.</td></tr>
<tr><td>Modification no.</td><td>Provisional control measure(s) for immediate application</td><td colspan="2">Deadline</td></tr>
<tr><td></td><td></td><td colspan="2">Date:</td></tr>
<tr><td></td><td></td><td colspan="2">Date:</td></tr>
<tr><td></td><td></td><td colspan="2">Date:</td></tr>
<tr><td colspan="2">HACCP study review</td><td colspan="2">HACCP study issue date</td></tr>
<tr><td>Next scheduled review (date):</td><td>December 20, 2018</td><td>Study issued</td><td>Date: February 15, 2018</td></tr>
<tr><td colspan="4">Authorization of completed study</td></tr>
<tr><td>Food safety team leader</td><td>G Moran</td><td colspan="2">Date: February 15, 2018</td><td rowspan="3">Authorization of completed study
Authorized persons should sign and date the study.</td></tr>
<tr><td>Hygienist/microbiologist</td><td>O Brown</td><td colspan="2">Date: February 12, 2018</td></tr>
<tr><td>Factory manager</td><td>C Flack</td><td colspan="2">Date: February 12, 2018</td></tr>
</table>

Work sheet 2: Product and ingredient description

Work sheet 2 defines and documents the characteristics of the product, which may include details on the production process and product category (table 3.39). The description of the safety of the product should encompass the sensitivity to and potential for safety risks. Traceability should be facilitated by clarifying the supply chain, ranging from the raw materials used to the distribution of the finished product. An extensive specification of the end product is required to ensure a comprehensive assessment of the relevant food safety procedures.

The end product information specified on the work sheet must clearly reflect the following product details: product name; type; general product specifications, such as appearance and weight; specific requirements, such as relevant legislation or customer requirements; raw materials and ingredients used (composition); safety indicators (chemical, microbiological and physical, allergens); product packaging; main steps and processing conditions (production method); shelf life and storage conditions; safety-related product labeling; intended use by consumers and proper use; transportation conditions and distribution methods; potential for mishandling or misuse of the product; target consumer groups; and other characteristics having an impact on food safety.

The description of raw and auxiliary materials that have come into contact with the food product should concisely indicate the following: the names of these raw materials, ingredients, and auxiliary materials; composition; high-risk ingredients; safety indicators (chemical, microbiological and physical, allergens); origin or supplier; main stages and processing conditions (production method); methods of packaging and transportation; storage conditions and shelf life; preparation or processing before use or reprocessing; and acceptance criteria related to food safety.

All indicators on the sample form are provided solely for illustrative purposes. In designing its own specifications, the FBO should consider all indicators in light of relevant legislation, regulations, required technological specifications, customer requirements, and other requirements.

Table 3.39 HACCP Work Sheet 2: Product and Ingredient Description		Instructions
End product characteristics		**End product characteristics** Provide details on the product or product family name, type, physical and chemical characteristics, key processing steps, and other characteristics. Indicate details on raw materials, high-risk ingredients, packaging materials, rework, and other characteristics.
Name (product[s], product group[s], line)	Grade A aseptically processed and packaged milk	
Composition	Cow's milk	
Type (e.g., raw, cooked, ready to eat)	Ready to eat	
Key chemical, biological, and physical characteristics	Chemical parameters *Heavy metals* • Lead, mg/kg, not more than 0.1 • Arsenic, mg/kg, not more than 0.05 • Cadmium, mg/kg, not more than 0.03 • Mercury, mg/kg, not more than 0.005 *Antibiotics* • Chloramphenicol is not allowed • Tetracycline group is not allowed • Streptomycin is not allowed • Penicillin is not allowed • Inhibitory substances are not allowed • Melamines are not allowed *Radionuclides* • Cs-137, Bq/kg, not more than 100 • Sr-90, Bq/kg, not more than 37	
	Biological parameters • Mesophilic aerobic and facultative anaerobic microorganisms—no more than -100,000 CFU/g • Coliforms in 0.1 CFU/mL—not allowed • Pathogens including *Salmonella* spp 25.0 g—not allowed • *Staphylococcus aureus* in 1.0 g—not allowed • *Listeria* in 25.0 g—not allowed	
	Physical parameters • Group purity—not less than 1 • Particles of mechanical impurities are not allowed	
Key processing steps (e.g., drying, heat treatments, freezing)	Storage, clarifier/separator, normalization, pasteurization, filler, storage, distribution/logistics	
Other		

continued

Table 3.39 *(Continued)*		Instructions
Specifications and regulatory requirements (food safety related)		**Specifications and regulatory requirements** Indicate details on product specifications and regulatory requirements.
Product specifications	JB-0346-7654-A	
Product-specific regulatory requirements	PMO 2005	
Filling and packing		**Filling and packing** Supply details on packaging and packaging system requirements.
Packaging description (e.g., size)	High-density polyethylene gallon container with a polypropylene snap-on screw tamper-evident cap	
Packaging system (e.g., modified atmosphere)	Aseptic packaging	
Claims and label information		**Claims and label information** Complete details on claims about product and label information.
Instructions for use by consumers (including use or storage after opening)	Keep refrigerated, Grade A pasteurized, homogenized, vitamin A and D added, 30% less fat than regular milk	
Statements for safe use (e.g., allergen information, special instructions for safe handling)	Shelf life seven days; storage temperature not to exceed +6°C—24 hours	
Other	Date of manufacture	
Distribution/storage/description		**Distribution/storage/description** Fill in details on distribution, storage, shelf life, and other conditions.
Distribution instructions (e.g., ambient, chilled, frozen)	Product is cased in standard milk cases—four units per case, using refrigerated trucks from 0°C to +20°C	
Storage instructions (e.g., ambient, chilled, frozen)	Distributed using refrigerated trucks from 0°C to +20°C in a vehicle fitted out for the shipment of food for the wholesale and retail trade	
Shelf life conditions	Storage conditions at temperature from 0°C to +20°C. Shelf life seven days	
Other	Not applicable	
Use by consumers		**Use by consumers** Supply details on intended use, special consumer groups, and reasonably expected mishandling and misuse.
Intended use	Ready-to-serve product. May also be used as an ingredient in preparing meals	
Target group of users and special consumer considerations (e.g., infants, the elderly)	Consumers of all ages consume this product	
Reasonably expected mishandling and misuse	Not stored under proper refrigeration	
Incoming material characteristics		**Incoming material characteristics**
Name of raw materials, ingredients	Cow's milk	Define all raw materials, ingredients, and materials coming into contact with the food.
Composition	Cow's milk	Specify ingredients, including food additives and processing aids.
High-risk ingredients	Cow's milk: a hospitable environment for the development of microorganisms (lactic acid bacteria, streptococci, coliforms, putrefaction bacteria, *Salmonella* spp, among others)	Provide a list of high-risk ingredients, including allergens (celery, corn, eggs [usually a protein], citrus, pumpkin, legumes, peanuts, soybeans, milk, seafood, sesame, tree nuts, wheat), microbiological hazards (*Salmonella* spp., *Clostridium botulinum*, *Staphylococcus aureus*, *Yersinia enterocolitica*, *Listeria monocytogenes*, *Vibrio* spp., *Escherichia coli* O157:H7, *Clostridium perfringens*, *Bacillus cereus*, *Campylobacter* spp., *Shigella* spp.), and sources of foreign bodies (packing material, transport, material type).

continued

Table 3.39 HACCP Work Sheet 2: Product and Ingredient Description *(Continued)*		**Instructions**
Incoming material characteristics *(continued)*		**Incoming material characteristics** *(continued)*
Key chemical, biological, and physical characteristics	Chemical parameters *Toxic elements* • Lead, mg/kg, not more than 0.1 • Arsenic, mg/kg, not more than 0.05 • Cadmium, mg/kg, not more than 0.03 • Mercury, mg/kg, not more than 0,005 *Pesticides* • Hexachloran α, β, γ isomers, mg/kg, not more than 1.25 (in terms of fat) • DDT and its metabolites, mg/kg, not more than 1.0 (in terms of fat) *Radionuclides* • Cs-137, Bq/kg, not more than 100 • Sr-90, Bq/kg, not more than 3.7 • Inhibiting substances are not allowed *Antibiotics* • Chloramphenicol is not allowed • Tetracycline group is not allowed • Streptomycin is not allowed • Penicillin is not allowed	Properties or characteristics important in determining safety may include physical characteristics (particle size, porosity, weight), chemical characteristics (pH, water activity, acidity), or microbiological characteristics (content colony-forming unit per g).
	Biological parameters • Number of somatic cells, 1,000/cm^3 • Mesophilic aerobic and facultative anaerobic microorganisms—no more than 100,000 CFU/g • Coliforms in 0.1 CFU/mL are not allowed • Pathogens, including *Salmonella* spp. 25.0 g—not allowed • *Staphylococcus aureus* in 1.0 g—not allowed • *Listeria* in 25.0 g—not allowed	
	Physical parameters • Density, kg/m^3, at least 1,028 • Group of purity—not less than 1 • Particle mechanical impurities not allowed	
Supplier	World of Milk dairy farm	**Supplier** Specify the raw material supplier.

continued

Table 3.39 (Continued)		Instructions
Processing main steps and conditions (production method)	Obtained during the mechanical milking of cattle, followed by cooling to +6°C	**Processing main steps and conditions** Specify processes to block the occurrence, reproduction, or survival of microorganisms.
Packing and transportation containers	Closed tightly sealed transportation containers (stainless steel tanks); food rubber gaskets used in sealing the lids	**Packing and transportation containers** Specify the type of material that is in contact with the food product.
Storage conditions and shelf life	Storage temperature not to exceed +6°C. 24 hours	**Storage conditions and shelf life** Specify the shelf life and appropriate storage conditions for the raw materials.
Preparation or processing before use	Filtering, cooling	**Preparation or processing before use** Specify the stages of preparation or processing of raw materials prior to use to minimize food hazards.
Acceptance criteria related to safety	Temperature when accepted of not more than +10°C Availability of veterinary certificate Test for the absence of antibiotics (chloramphenicol, tetracycline group, streptomycin, penicillin) Group of purity—not less than 1 Particles of mechanical impurities are not allowed	**Acceptance criteria related to safety** Specify safety criteria for the raw materials checked by the company at acceptance.
Other (e.g., preservatives, processing aids, services)	Not applicable	**Other** Specify any other information.

Work sheet 3: Flow diagram

Work sheet 3 illustrates all production steps for the product or similar products within a hazard control or HACCP system (CAC 2003). It takes the form of a flow diagram (figure 3.2). The flow diagram should be constructed by the hazard control or HACCP team and should cover all operational steps pertaining to a specific product. The same flow diagram may be used for any number of products manufactured through a similar process.

Prepare flow diagrams for the products or process categories covered by the hazard control or HACCP system. Flow diagrams should provide a basis for evaluating the possibility of an occurrence, increase, or introduction of food safety hazards. The flow diagrams need to take into account the relevant process steps, their sequence, and how they relate to each other. If work is subcontracted or outsourced, this should be indicated in the flow diagram. The flow diagram should detail the introduction of raw materials, ingredients, and so on. If rework is an option in the process or in recycling, these steps need to be included. The realization of waste, by-products, intermediate products, and end products should also be included.

Figure 3.2 HACCP Work Sheet 3: Flow Diagram

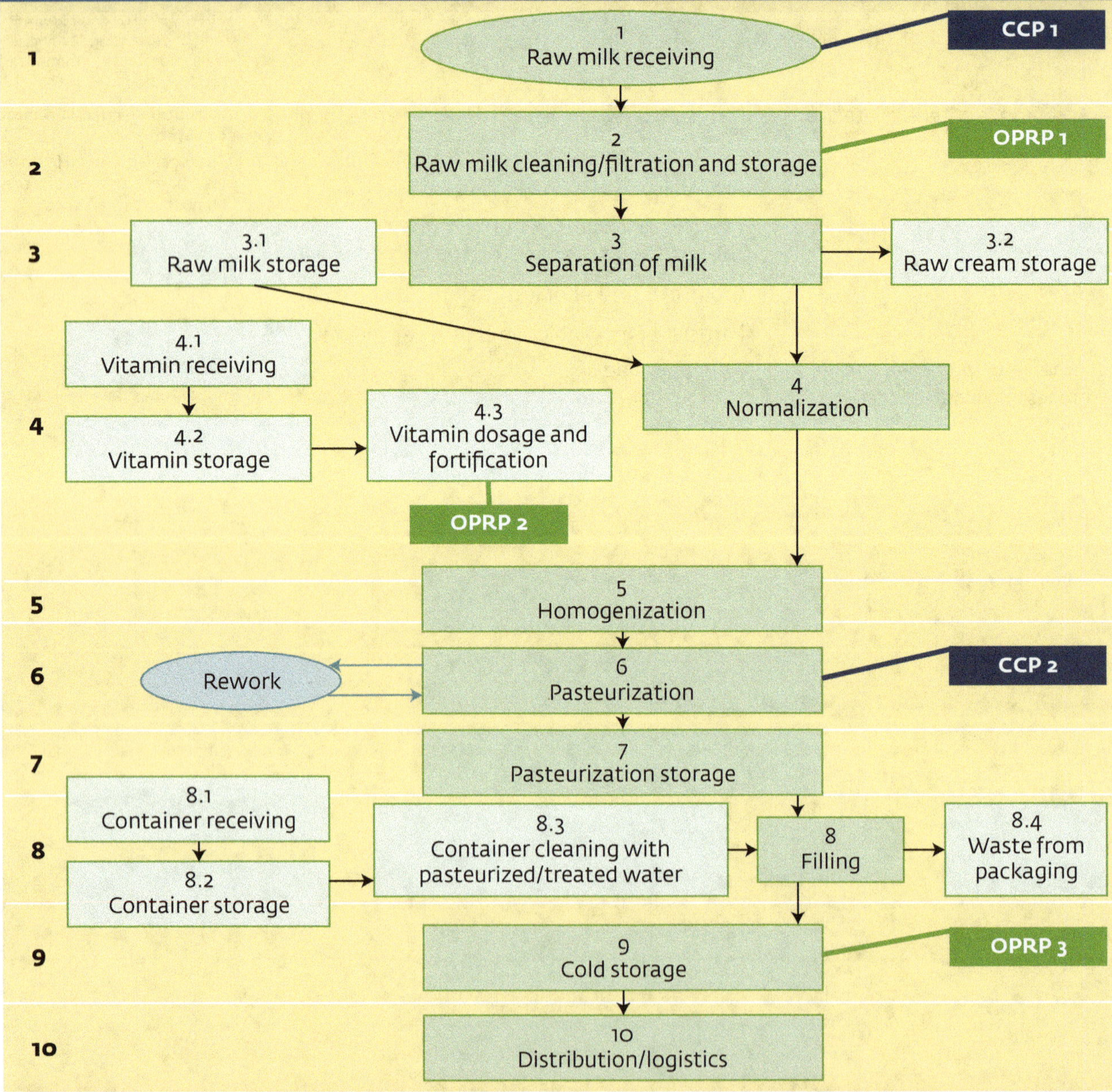

Authorization			
Name	Position/responsibilities on the team	Signature	Date
G Moran	Food safety manager, HACCP team leader	*G Moran*	*February 1, 2018*

The accuracy and actuality of the flow diagrams and layout should be verified by the hazard control, HACCP, or food safety team for compliance with the documented situation. This verification should be repeated periodically (at least annually) to identify and document modifications in the process installation and layout. The FBO needs to make a diagram for all process steps, including all control steps (CCP) with specific parameters. An individual should be designated to be responsible for most steps in the creation of the flow diagram. The main steps are as follows:

- Construct a flow diagram of the process
- Number each step in the process
- Indicate the CCP when the hazard control plan or HACCP system study is completed
- Indicate the OPRP when the hazard control plan or HACCP system study is completed
- Record the on-site verification of the flow diagram

1	Input control of raw and auxiliary materials is carried out in the enterprise laboratory in accordance with the guidelines on technochemical and microbiological control at dairy industry companies; these are duly approved and consistent with the standards of research methods specified in the technical specifications for this product.
2	Milk selected based on safety indicators is purified by mechanical filters, then immediately cooled to 4°C (±2°C) and fed to the intermediate storage tanks. The duration of raw milk storage at a temperature up to 4°C is 12 hours, and at up to 6°C, 6 hours.
3	Part of the milk is separated in cream separators to select the cream.
4	The milk is normalized in mass fractions of fat and protein in such a way that these shares of the normalized mass fractions correspond to milk fat and protein shares in the end product. ■ As for fat, milk is normalized by adding cream. ■ As for protein, milk can be normalized by mixing milk batches with content of varying protein weight percentages. ■ For dosage of vitamins, the responsible person must weigh the required number of vitamins and prepare a solution as recommended by the technological instruction in compliance with the safety requirements. The required amount of the complex needs to be taken, trying not to raise dust, and wearing protective gloves and goggles to avoid contact with skin and eyes. After use of the vitamins, the package must be tightly closed. Vitamins should be stored in a dry, dark place, with limited access, at a temperature not higher than +25°C. They can be stored in a sealed package without air and light for one year.
5	Milk is homogenized at a pressure of 12–18 bars.
6	The milk is pasteurized at a temperature not less than 85°C, and the time of pasteurization should not be less than 20 seconds (this time is conditioned by constructive features of the equipment) and cooled to 4°C (+2°C). If the pasteurization temperature does not reach the required level, the milk should be pasteurized again (rework).
7	Pasteurized refrigerated milk enters the tank for intermediate storage before further processing. The maximum shelf life of pasteurized milk to sterilization is 24 hours.
8	Pasteurized milk bottling is carried out under hygienic conditions. The packaging material is cleaned with pasteurized or treated water before bottling. Packaging material is supplied only by approved suppliers. The certificate of analysis or compliance is provided. The laboratory provides incoming inspection according to company requirements.
9	Packets put in shrink film or cardboard trays are stacked on pallets for foodstuff transportation and fed to the dry, clean chamber at a temperature of 0°C to +20°C. Here the pasteurized milk is cooled to a temperature of +20°C or less in under 24 hours, after which the process is considered complete. Products in storage must be protected from direct sunlight.
10	The shelf life of pasteurized milk with a fat content of 5.0% in a package of composite material with a nominal volume of 1 liter at a temperature ranging from 0°C to +20°C is four months from the date of manufacture.

Instructions

The document should be approved on-site after the flow diagram check is completed.

CHAPTER 3

Work sheet 4: Hazard identification and description

Work sheet 4 defines and documents each potential hazard identified during the food production process by the HACCP team and determine its significance according to the severity of the health effect and the likelihood of occurrence (table 3.40). The FBO hazard control, HACCP, or food safety team should identify, analyze, and evaluate all potential biological, chemical, and physical hazards that can have an adverse effect on the safety of the products.

The identification should include all aspects of FBO operations within the scope of the hazard control, HACCP, or FSMS system, such as raw materials and ingredients (specifications, process control at suppliers, and so on); characteristics of interim and end products (intrinsic product specifications, for example); characteristics of the processes used, including by subcontracted services; PRPs (layout of the facility, production

Table 3.40 HACCP Work Sheet 4: Hazard Identification and Description

Location of potential hazard	Hazard description						
Indicate the step (e.g., raw material, processing, or distribution) at which the hazard may be introduced.	Describe clearly and specifically the hazards that are reasonably expected to occur at each step: class (B, P, C, or A), agent, size, origin, nature, etc.						
Step no.	Step (description)	Hazard class	Hazardous agent description	Hazard no.	Origin or source of the hazard	Nature of the hazard	Acceptable level in end product
1	Raw milk receiving	C	Therapeutic drugs (antibiotics)	C1	Primary milk production (farm)	Presence	Absence
1	Raw milk receiving	C	Toxic elements (heavy metals)	C3	Primary milk production (farm)	Presence, introduction	Lead, mg/kg, not more than 0.1 Arsenic, mg/kg, not more than 0.05 Cadmium, mg/kg, not more than 0.03
Instructions							
Step no. Defines a sequential number for each process step.	**Step description** Defines the title or description of the process step.	**Hazard class** Defines the hazard agent class: B (biological), C (chemical), P (physical), A (allergen).	**Hazardous agent description** Defines the hazard controlled by the measure.	**Hazard no.** Defines the hazard agent code: B1, C1, P, A.	**Origin or source of the hazard** Defines where and how the product or environment can become contaminated.	**Nature of the hazard** Defines particular hazard threats, such as availability, capacity for growth, survival, allocation of toxins or toxic chemicals, or migration of chemicals.	**Acceptable level in end product** Defines the acceptable level of the hazard as required by law or customer specifications.

lines, installations, and equipment; location of rooms, routing, storage, and separation of raw materials, interim products, end products, ventilation, and so on; production processes, including purchasing, cleaning and disinfection, packaging, maintenance, pest control, waste management, and so on; personnel, including arrangements for visitors and external service providers, for example, mechanics (hygiene, knowledge on food hygiene and food safety, requirement to notify diseases and infections, and so on).

The FBO hazard control, HACCP, or food safety team shall conduct a hazard analysis to identify which hazards are of such a nature that their elimination, reduction, or control at acceptable levels is essential to the production of safe food. The hazard analysis should cover the likely occurrence of hazards and the severity of their adverse health effects. Whenever the FBO changes procedures in a manner that could adversely affect food safety, all relevant steps of the hazard analysis must be updated.

Hazard assessment			Justification of hazard selection and assessment
Q1: Based on the hazard description, likelihood of occurrence (before applying the control measure), and severity of health effects, does this hazard need to be controlled, i.e., is it a significant hazard?			Provide supporting data/references on likelihood of occurrence, information on severity of health effects, and acceptable level in end product.
Likelihood of occurrence	Severity of adverse health effect	Significant hazard? (Yes/No)	For each hazard, document why it is or why it is not likely to occur or to cause adverse health effects. For a nonsignificant hazard, document if it is managed, e.g., by a PRP, through a specification or major allergen declaration. Make sure that all hazards likely to occur are considered. Justify why a certain hazard has been disregarded.
Frequent (4)	Can lead to serious illness (4)	Significant (16)	Hazard likelihood is frequent, antibiotics used to treat animals
Could occur (2)	Can cause illness (3)	Insignificant (6)	Last two years, there were no heavy metals identified in incoming milk. This hazard is controlled by prerequisite programs for the analysis of incoming raw materials and finished products
Instructions			
Likelihood of occurrence Defines the likelihood of hazard occurrence.	**Severity of adverse health effect** Defines the severity of any adverse health effect arising from the hazard.	**Significant hazard?** Defines whether the hazard is significant. For significant hazards, select and categorize control measures(s) on work sheet 5 (table 3.41).	**Justification of hazard selection and assessment** Defines why it is or is not likely to occur and to cause, or not cause, adverse health effects

continued

Table 3.40 HACCP Work Sheet 4: Hazard Identification and Description *(Continued)*

Location of potential hazard	Hazard description
Indicate the step (e.g., raw material, processing, or distribution) at which the hazard may be introduced.	Describe clearly and specifically the hazards that are reasonably expected to occur at each step: class (B, P, C, or A), agent, size, origin, nature, etc.

Step no.	Step (description)	Hazard class	Hazardous agent description	Hazard no.	Origin or source of the hazard	Nature of the hazard	Acceptable level in end product
1	Raw milk receiving	B	*Salmonella, Staphylococcus aureus, L monocytogenes, Listeria, Shigella*	B1	Primary milk production (farm), transportation	Presence, introduction	Absence
1	Raw milk receiving	P	Extraneous material (e.g., stone, glass)	P1	Primary milk production (farm), transportation	Presence	Absence
1	Raw milk receiving	A	Allergen	A1	Primary milk production (farm), transportation	Presence	Always present
6	Pasteurization	B	Pathogenic micro organisms *Salmonella, S. aureus, L monocytogenes*	B1	Primary milk production (farm), personnel, work environment	Survival	Absence
		C	Absence	—	—	—	—
		P	Absence	—	—	—	—

Instructions

Step no.	**Step description**	**Hazard class**	**Hazardous agent description**	**Hazard no.**	**Origin or source of the hazard**	**Nature of the hazard**	**Acceptable level in end product**
Defines a sequential number for each process step.	Defines the title or description of the process step.	Defines the hazard agent class: B (biological), C (chemical), P (physical), A (allergen).	Defines the hazard controlled by the measure.	Defines the hazard agent code: B1, C1, P, A.	Defines where and how the product or environment can become contaminated.	Defines particular hazard threats, such as availability, capacity for growth, survival, allocation of toxins or toxic chemicals, or migration of chemicals.	Defines the acceptable level of the hazard as required by law or customer specifications.

	Hazard assessment			Justification of hazard selection and assessment
	Q1: Based on the hazard description, likelihood of occurrence (before applying the control measure), and severity of health effects, does this hazard need to be controlled, i.e., is it a significant hazard?			Provide supporting data/references on likelihood of occurrence, information on severity of health effects, and acceptable level in end product.
	Likelihood of occurrence	Severity of adverse health effect	Significant hazard? (Yes/No)	For each hazard, document why it is or why it is not likely to occur or to cause adverse health effects. For a nonsignificant hazard, document if it is managed, e.g., by a PRP, through a specification or major allergen declaration. Make sure that all hazards likely to occur are considered. Justify why a certain hazard has been disregarded.
	Rare (1)	Can lead to serious illness (4)	Insignificant (4)	Hazard is controlled by PRP (incoming raw material and finished product)
	Could occur (2)	Can cause illness (3)	Insignificant (6)	Taking into account a moderate level of hygiene in primary milk production on the farm, there is a remote probability of encountering foreign objects in milk
	Rare (1)	Can lead to serious illness (4)	Insignificant (4)	This hazard is controlled by the prerequisites programs–allergen control procedure and mentioned on the label as cow's milk. This hazard is not insignificant for a consumer who may suffer from the allergy
	Could occur (2)	Can lead to serious illness (4)	Significant (8)	Pasteurization can be violated by the survival probability of microorganisms in the milk, causing a severe health hazard
	—	—	—	—
	—	—	—	—
	Instructions			
	Likelihood of occurrence Defines the likelihood of hazard occurrence.	**Severity of adverse health effect** Defines the severity of any adverse health effect arising from the hazard.	**Significant hazard?** Defines whether the hazard is significant. For significant hazards, select and categorize control measures(s) on work sheet 5 (table 3.41).	**Justification of hazard selection and assessment** Defines why it is or is not likely to occur and to cause, or not cause, adverse health effects

CHAPTER 3

Work sheet 5: Control measure selection and categorization

Work sheet 5 defines and documents the selection and categorization of control measures related to the hazards that have been identified (see work sheet B). The work sheet helps in determining whether the control measures need to be managed by the hazard control plan through OPRPs or CCPs. The hazard control, HACCP, or food safety team should identify and document the control measures that are to be applied or implemented if the hazard identification and hazard analysis conclude that the risk of an identified hazard is significant and needs to be eliminated, reduced, or controlled at an acceptable level. The team should conduct an assessment of every step in the process using a decision tree. The assessment should be based on several

Table 3.41 HACCP Work Sheet 5: Control Measure Selection and Categorization

Step and hazard				Control measures
Transfer the hazards considered significant in work sheet 4 to this work sheet (5)				Select and describe a control measure or combination of control measures capable of preventing, eliminating, or reducing the hazard to an acceptable level Document the rationale for the selection, e.g., effectiveness of applied control measures alone or in combination against identified hazard (refer to documents if possible)
Step no.	Step description	Hazard no.	Hazardous agent description	Description of control measures
1	Raw milk receiving	C1	Therapeutic drugs (antibiotics: chloramphenicol, tetracycline family, streptomycin, penicillin)	Control of raw milk to assure the absence of antibiotics using express method (Delvotest)
2	Raw milk filtration	P1	Extraneous foreign material—glass	PRP (incoming raw material) -filtering and purity control of raw milk
—	—	—	—	—
6	Pasteurization	B1	Pathogenic microorganisms	Pasteurization
—	—	—	—	—
8.3	Container cleaning with pasteurized/ treated water		*E.coli*	There are no control measures
Instructions				
Step no. Defines a sequential number for each process step.	**Step description** Defines the title or description of the process step.	**Hazard no.** Defines the hazard agent code: B1, C1, P, A.	**Hazardous agent description** Defines the hazard that is controlled by the measure.	**Description of control measures** Describes the control measure or combination of control measures taken to prevent, eliminate, or reduce hazards to an acceptable level.

factors, including the differing expertise within the team and external and internal information. For each step, including all products and processes and all parts of the PRPs, the assessed aspects should be identified. The reasons for deciding whether CCPs are required or not should be documented and traceable. More than one control measure may be required to control a hazard, and more than one hazard may be controlled by a control measure. Control measures may be classified as PRPs, OPRPs, or part of hazard control or HACCP plan.

Each field on the work sheet contains instructions and guidance on the information or rating to be entered into the relevant fields (table 3.41). The work sheet also contains questions with answer options. In this case, the significance of the selection of each answer is explained.

Categorization of control measures in OPRPs and CCPs (answer questions Q1 to Q5 as necessary)

Q1: Based on the likelihood of occurrence (before applying the control measure) and the severity of adverse health effects (work sheet 4), is this hazard significant (needs to be controlled)? YES: This is a significant hazard. Go to Q2. NO: This is not a significant hazard.

Q2: Will a subsequent processing step, including the expected use by the consumer, guarantee the removal of this significant hazard, or its reduction to an acceptable level? YES: Identify and name the subsequent step. NO: Go to Q3.

Q3: Are control measures or practices in place at this step, and do they exclude, reduce, or maintain this significant hazard to/at an acceptable level? YES: Go to Q4. NO: Modify the process or product and go to Q1.

Q4: Is it possible to establish critical limits for the control measure at this step? YES: Go to Q5. NO: This hazard is managed by an OPRP and action criteria.

Q5: Is it possible to monitor or observe the control measure in such a way that corrections can be made immediately if there is a loss of control? YES: This hazard is managed by the hazard control/HACCP plan (CCP). NO: This hazard is managed by an OPRP and action criteria.

Q1	Q2	Q3	Q4	Q5	CCP, OPRP, or process modification	Decision justification: provide supporting evidence that selected control measure(s) and target/critical limits or action criteria will adequately control the hazard
Yes	No	Yes	Yes	Yes	CCP 1	Express method allows testing for each batch of raw materials and detection of antibiotics in dairy raw materials
Yes	No	Yes	Yes	No	OPRP 1	Filtration of milk by filter with a cell diameter of 0.01 mm enables prevention of impurities in milk
—	—	—	—	—	—	—
Yes	No	Yes	Yes	Yes	CCP 2	Pasteurization destroys some pathogenic microorganisms in milk or, at a minimum, reduces their number to an acceptable level—absence in 25 mg
—	—	—	—	—	—	—
Yes	No	No	—	—	Process modification	Process change needed; use pasteurized or additionally purified water

Instructions

Categorization of control measures in OPRPs and CCPs	CCP, OPRP, or process modification	Decision justification
Provides questions that the hazard control or HACCP team should answer, giving a range of possible responses.	Identifies the category of the control measure selected.	Notes the rationale behind the choice of a control measure or a combination of control measures.

CHAPTER 3

Work sheet 6: Validation of control measures

Work sheet 6 defines and documents the FBO validation of the control measures identified in work sheet 5. Its purpose is to provide evidence that the control measure can achieve the targeted limits. The work sheet contains several questions that prompt for the type of information required. This is designed to elicit from the organization information about the effectiveness of the controls that the FBO has established to address each hazardous agent. Each field in the work sheet contains instructions or guidance on the information or rating to be entered (table 3.42).

Table 3.42 HACCP Work Sheet 6: Validation of Control Measures

The hazard control/HACCP team must provide or ask for evidence that selected control measures are capable of achieving the intended control over the hazards identified.

The hazard control/HACCP team leader shall provide answers to the following questions:

- Have potential hazards been correctly identified as significant or not?
- Are applied control measures capable of reducing the significant hazards to acceptable levels?
- Are the critical limits correct and appropriate?
- Will the corrections restore product safety control?

CCP no. or OPRP no.	Step	Hazardous agent description	Control measure	Justification for the selection of control measures
CCP 1	1	Therapeutic drugs: antibiotics: tetracycline group, penicillin, streptomycin, chloramphenicol	Control of raw milk for the absence of antibiotics using the Delvotest	Rapid test allows quick determination of the presence of antibiotics in raw materials. This methodology is approved and ensures test accuracy and reliability
OPRP 1	2	Extraneous foreign material	Filtration and purity control of raw milk	Filtration of milk on a filter cell with a diameter of 0.01 mm enables the prevention of impurities in the finished product
CCP 2	6	Pathogenic microorganisms, including *Salmonella*, *S. aureus*, *L monocytogenes*	Pasteurization	Pasteurization destroys some pathogenic microorganisms in the milk or reduces their numbers to an acceptable level
Instructions				
CCP no. or OPRP no. Defines numbers for the CCP and OPRP.	**Step** Defines a sequential number for each process step.	**Hazardous agent description** Defines the hazard, that is controlled by the measure.	**Control measure** Defines the control measures selected for this hazard.	**Justification for the selection of control measures** Defines whether the control measure functions in practice.

Checking control measure effectiveness	Critical limits (for CCP only)	Justification for the selection of critical limits	Corrections
Monthly check using the enzyme-linked immuno-sorbent assay or high-performance liquid chromatography method	Absence	Legislation for raw milk	Return to supplier or disposal of milk
Determination of purity according to the standard	Not applicable	Not applicable	Not applicable
Monthly microbiological analysis of the product	Pasteurization temperature not less than 85°C; time, not less than 20 seconds	Technological instruction of pasteurized milk	Flow diversion and repasteurization
Instructions			
Checking control measure effectiveness Defines the extent to which the control measure is effective.	**Critical limits** Defines the critical limits determined for this CCP.	**Justification for the selection of critical limits** Defines the basis for determining the relevant critical limits.	**Corrections** Defines the actions necessary to prevent a negative effect on food safety if a critical limit is exceeded; it also indicates the person responsible.

Work sheet 7: Hazard control plan

Work sheet 7 (table 3.43) defines and documents the details of all CCPs and OPRPs and indicates the control measures, critical limits, action criterion, and corrections taken, plus the verification events detailed in work sheet 8.

Table 3.43 HACCP Work Sheet 7: Hazard Control Plan

CCP no. or OPRP no.	Hazard class	Step no.	Step description	Hazardous agent description	Control measure(s)	Critical limits and targets (or limits, if applicable) that measure effectiveness
CCP 1	C	1	Raw milk receiving	Therapeutic drugs: antibiotics: chloramphenicol, tetracycline family, streptomycin, penicillin	Control of raw milk for the absence of antibiotics using the express method (Delvotest)	100% absence
OPRP 1	P	2	Raw milk filtration	Extraneous foreign material—glass	Raw milk filtering and purity control	Not applicable
CCP 2	B	6	Pasteurization	Pathogenic microorganisms, including *S. aureus*, *L. monocytogenes*	Control of temperature and pasteurization timing	Pasteurization temperature not less than 85°C, time not less than 20 seconds
Instructions						
CCP no. or OPRP no. Defines the CCP and OPRP numbers.	**Hazard class** Defines the hazard agent class: B (biological), C (chemical), P (physical), A (allergens).	**Step no.** Defines a sequential number for each process step.	**Step description** Defines the title or description of the process step.	**Hazardous agent description** Defines the hazard controlled by the measure.	**Control measure(s)** Defines the control measures selected for this hazard.	**Critical limits and targets (or limits, if applicable) that measure effectiveness** Defines the critical limits determined for this CCP.

	Monitoring: how, frequency, who?	Corrections, responsibilities	Corrective actions, responsibilities	Records	Verification (details in work sheet 8)
	Delvotest, each batch, by quality specialist	Return of milk to supplier or environmental disposal of product/ procurement manager	Inform dairy farm and veterinary service provider, identifying reasons for therapeutic drugs use/quality manager	Raw milk receiving log	Control by immunofluorescence methods monthly from each supplier, laboratory technician
	Determination of purity according to standard, each batch, quality specialist	Repeated filtering by quality specialist	Unannounced audit of supplier coordinated by the quality manager	Filtering and cooling log	Checking of cooling log by laboratory manager
	Automatic registration of pasteurization temperature and time, visual inspection of temperature indicator, continuously, by the pasteurization operator	Stopping milk supply for filling, backflow, and repasteurization of milk, pasteurization operator	Checking technical condition of the device; checking monitoring, and metering the instrument; pasteurization training for operator, mechanical engineer, human resources manager	Pasteurization log, thermogram	Parameter control of reference thermometer hourly by shift foreman and control of thermometer every shift by microbiologist
	Instructions				
	Monitoring: how, frequency, who? Defines the monitoring method, its frequency, and the person responsible.	**Corrections, responsibilities** Defines the actions necessary to prevent a negative effect on food safety if a critical limit is exceeded; it also cites the person responsible.	**Corrective actions, responsibilities** Defines the actions necessary to eliminate the reasons for exceeding critical limits, thereby preventing a repeated occurrence.	**Records** Defines the records to be maintained.	**Verification** Defines the verification of the actions conducted.

Work sheet 8: Verification plan

Work sheet 8 defines and documents verification activities intended to substantiate the effectiveness of a hazard control plan in a particular case (table 3.44). The purpose is to provide evidence to show that the CCPs and OPRPs have been implemented properly.

FBOs must establish, document, and implement procedures for the verification of the hazard control or HACCP system. The main purpose of the verification is to determine compliance with the specifications of the systems and to confirm that the systems are effective. Auditing methods, procedures, tests (including random sampling and analysis), and other evaluations, in addition to monitoring, are applied to accomplish this purpose.

The verification procedures should be established and documented and should include, at a minimum, a purpose; methods, including standard operating procedures or the tests applied; tasks and responsibilities; frequency, and records.

The procedures should address, at a minimum, the following topics: a review of the HACCP system and its corresponding records; an analysis of any product recalls and product dispositions; an assessment of all general control measures, nonconformities, and corrective actions taken to confirm effective control of the CCPs; an assessment of all general control measures to seek confirmation of implementation and to demonstrate effective control of associated hazards; conformity of the actual flow diagrams and layout with the documented situation; conformity of OPRP and CCP documents with the operational situation; analysis of customer and consumer complaints related to hygiene and food safety; a review of analytical outcomes of random sampling and analysis of products; evaluation of compliance in the context of applicable legislation and regulations (as well as with foreseeable changes in legislation and regulations), identification of changes in the legislation and regulations concerning food safety; a review of the gaps between current and targeted levels of knowledge, awareness, and staff training with respect to hygiene and food safety, and the results in terms of effective on-the-job training sessions; and consistency of the current documentation.

Table 3.44 HACCP Work Sheet 8: Verification Plan					
CCP no. or OPRP no.	Verification activity	Verification procedure	Frequency	Person responsible	Records
CCP 1	Verify the input and efficiency control of raw milk in the absence of therapeutic drugs	Selective periodic monitoring Control of records	Monthly for each supplier Weekly	Food safety manager Laboratory manager	Register of input control Laboratory technician workbook
OPRP 1	Monitor implementation of the raw material filtration procedure and its effectiveness	Periodic control of cleaning process and records for cleaning and cooling	Weekly	Laboratory manager	Cleaning and cooling register
CCP 2	Verify milk pasteurization and its effectiveness and efficiency	Periodic control of pasteurization temperature and time	Control of reference thermometer parameters—hourly	Shift supervisor	Milk pasteurization register
		Periodic control of thermograms	Control of thermograms—every shift	Microbiologist	Thermogram
		Peroxidase test	Peroxidase test—every shift	Quality specialist	Peroxidase test register
Instructions					
CCP no. or OPRP no. Defines CCP and OPRP numbers.	**Verification activity** Defines the purpose of the verification.	**Verification procedure** Defines the methods or procedures to be used, the observations to be made, or the measurements and actions to be taken if there is a deviation, or follow-up.	**Frequency** Defines the frequency with which verification should be conducted.	**Person responsible** Defines the individual, department, or function responsible for conducting verification.	**Records** Defines the records to be maintained.

Work sheet 9: Modification(s) and follow-up

Work sheet 9 defines and documents all modifications to the plan and tracks any follow-up steps resulting from these modifications. It references details on the process step and the hazard (table 3.45).

Table 3.45 HACCP Work Sheet 9: Modification(s) and Follow-Up

Production process step		Hazard description		Modification			
Step no.	Step description	Hazard no.	Hazardous agent description	Modification no.	Recommended modification and confirmation of transfer for action	Limit date	Provisional control measure(s)
8	Filling	P1	Foreign body	2	Implement control of the packaged milk with x-ray detector to reveal foreign bodies	February 20, 2018	None
8.3	Handling containers with water	B1	*E. Coli*	1	Used for rinsing containers, pasteurized or additionally purified water	February 20, 2018	Increased to weekly the frequency of microbiological control of water used
Instructions							
Step no. Defines a sequential number for each process step.	**Step description** Defines the title or description of the process step.	**Hazard no.** Defines the hazard agent code: B1, C1, P, A.	**Hazardous agent description** Defines the hazard that is controlled by the measure.	**Modification no.** Defines the modification number.	**Recommended modification and confirmation of transfer for action** Defines the recommended modification and confirmation of information to be transferred to the relevant department or group for action.	**Limit date** Defines the planned date for corrective action.	**Provisional control measure(s)** Defines the immediate provisional (containment) control measure to be applied if modifications have not yet been implemented.

Work sheet 10: Meeting summary

Work sheet 10 defines and documents meetings held by the hazard control, HACCP, or food safety team. It records attendance and the decisions of the meetings (table 3.46). Meetings of the team are an important form of information transfer, updating the entire team about the implementation and effectiveness of their food safety system. The team should have a defined plan for meetings, but, in the case of some unpredicted event, the team may hold unscheduled meetings.

Table 3.46 HACCP Work Sheet 10: Meeting Summary

Date	Participants	Purpose	Outcome (decisions/ actions)	Persons responsible	Planned deadline, completion date	Performed
February 1, 2018	G Moran O Brown M Rodrigues B Jackson D Smith O Murphy C Flack N Williams	Review/ update the product description	Updated the product description	G Moran	February 10, 2018	February 5, 2018
December 12, 2018	G Moran O Brown M Rodrigues B Jackson D Smith O Murphy C Flack N Williams	Verify the flow diagram, compare document versus practice	No action required	G Moran	December 22, 2018	December 20, 2018
Instructions						
Date Defines the meeting date.	**Participants** Defines the team members and invitees attending.	**Purpose** Defines details on the reason for the meeting.	**Outcome** Defines the decisions made at the meeting, for example, next steps.	**Persons responsible** Defines the individuals responsible for executing decisions.	**Planned deadline, completion date** List the planned completion date.	**Performed** Defines the actual completion date.

CHAPTER 3

SUPPLEMENTARY WORK SHEETS

Work sheet A: Hazardous agent codes and classification

Work sheet A defines the guidelines for the food safety, hazard control, or HACCP team in assessing the hazards controlled through the hazard control plan (table 3.47). This is an optional activity in the implementation of the hazard control or HACCP plan.

Table 3.47 HACCP Work Sheet A: Hazardous Agent Codes and Classification

Ingredient or process	Hazard no.	Hazard class	Hazardous agent description
Raw milk	B1	Biological	Presence of vegetative pathogens (*Salmonella, Staphylococcus aureus, L monocytogenes, Listeria, Shigella*)
	C1	Chemical	Presence of therapeutic drugs—antibiotics: chloramphenicol, tetracycline family, streptomycin, penicillin
	C2	Chemical	Presence of mycotoxins
	C3	Chemical	Presence of toxic elements (heavy metals)
	P1	Physical	Extraneous material (not less then 2 mm—glass, stone, and so on)
	A1	Allergen	Allergy to cow milk protein
Pasteurized milk	B1	Biological	Presence of vegetative pathogens
	B2	Biological	Contamination of vegetative pathogens
Other ingredients/ packaging materials	B1	Biological	Presence of vegetative pathogens
	C1	Chemical	Presence of toxic or carcinogenic substances
	P1	Physical	Extraneous material
Water	B1	Biological	*E. coli*
Instructions			
Ingredient or process Details the ingredient or process.	**Hazard no.** Defines the hazard agent code: B1, C1, P, A.	**Hazard class** Defines the hazard agent class: B (biological), C (chemical); P (physical, and A (allergen).	**Hazardous agent description** Defines the hazard that is controlled by the measure.

Work sheet B: Hazard assessment table

Work sheet B defines and documents the hazard assessment or risk assessment (table 3.48). Its purpose is to offer guidance to the FBO hazard control, HACCP, or food safety team in assigning risks associated with each hazard type. This guidance table is for reference or guidance purposes only; hence, there is no blank template.

Table 3.48 HACCP Work Sheet B: Hazard Assessment Table

Severity of health effect						
Can cause fatality	5					
Can lead to serious illness	4			Significant (controlled by OPRPs or CCPs)		
Can cause illness	3					
Can cause inconvenience	2		Insignificant (controlled by PRPs)			
Almost no significance	1					
	Score	1	2	3	4	5
		Rare (1/year)	Could occur (1/6 months)	Likely (1/month)	Frequent (1/week)	Very frequent (1/day)
		Likelihood of occurrence				

Note: The hazard assessment table helps separate significant from nonsignificant hazards and to document the decision.

Work sheet C: HACCP list of supporting documents

Work sheet C cites details of reference documents (procedures and work instructions) associated with an FBO hazard control plan (table 3.49).

Table 3.49 HACCP Work Sheet C: HACCP List of Supporting Documents

No.	Document title or designation	Status and issue of the document	Document developer	Filing location
1	ISO 22000:2018	Valid from September 1, 2005, first edition	ISO	Standardization and Certification Office
2	ISO/TS 22002-1:2009	Valid from 2009	ISO	Standardization and Certification Office
3	Enterprise standard Interstate Milk Shippers 008, Purchases of raw and auxiliary materials	Valid from January 1, 2011, first edition	Head of procurement and logistics	Standardization and Certification Office
4	Ministry of Health (2009)	Valid from September 6, 2009	Ministry of Health	Standardization and Certification Office
Instructions				
No. Provides the sequential number assigned to each document in the register.	**Document title or designation** Defines the document number and title.	**Status and issue of the document** Indicates the date published and, if needed, the document issue.	**Document developer** Identifies the document author or publisher.	**Filing location** Records the storage location and where the document can be found.

Two other analysis and critical control point systems

Threat assessment and critical control point (TACCP) and vulnerability assessment and critical control point (VACCP) are relatively new programs based on the HACCP program. They address threats and vulnerabilities, respectively.

The FSMSs developed by industry and regulators and based on HACCP principles have been proven effective against unintended food safety hazards. The principles have not been routinely used to detect or mitigate deliberate attacks, however, and are therefore not as relevant to food defense. The aim of food defense is to control intentional food safety hazards that may cause harm to consumers or companies.

Thus, threats, in the case of TACCP, signify, for example, food tampering, intentional adulteration of food, and food defense. Although, at several points, TACCP and HACCP overlap, such as in recommendations on the use of tamper-proof seals and various quality control checks, TACCP systems generally require more substantial employee involvement relative to HACCP because the former cover issues that arise in food manufacturing or that entail transportation security, information technology security, and employee background checks.

VACCP systems, meanwhile, also focus on food fraud, but widen the scope of analysis and assessment to include the systematic prevention of any potential adulteration of food, whether intentional or not, by identifying the vulnerable points in a supply chain. It is especially concerned with economically motivated adulteration. Examples of topics of interest in a VACCP system include product substitution, unapproved product enhancements, counterfeiting, and trade in stolen goods.

TACCP and VACCP, similar to HACCP, each require a control plan that covers mitigation strategies and correction procedures. The programs may also require audits of an entire supply chain, assessments of suppliers, and extensive quality control checks on ingredients.

Annex 3A furnishes more information on TACCP, along with sample TACCP work sheets. Annex 3B provides a table highlighting brief explanations and definitions of various issues, concerns, and initiatives typically involved in a VACCP system. Editable work sheets and templates can be found at the following location: http://www.ifc.org/foodsafety/handbook/templates.

Chapter 4 supplies a sample food defense procedure that additionally clarifies HACCP, TACCP, and VACCP. The procedure is labeled SOP-044, that is, standard operating procedure 044.

Annex 3A. Instructions and sample work sheets: Threat assessment and critical control point

WS 1 Overview and Guide: TACCP Work Sheets

Main work sheets	Supplementary work sheets	Comments
WS 1 Overview and guide: TACCP work sheets		Registration and approval of the TACCP study
WS 2 TACCP instructions and content		Details of TACCP instructions
WS 3 Threat and vulnerability scope		Scope of the TACCP study and SOP
WS 4 Terms and definitions		Most common definitions related to a TACCP study
WS 5 Threat assessment critical control point (TACCP)		Aim and process for the TACCP study
WS 6 Types of threats and case study examples		Categories of TACCP threats with examples
WS 7 Understanding the attacker		General outline of the attacker relevant to the TACCP study
WS 8 Assessing threats, vulnerabilities, and risk		Outline of process for assessing threats, vulnerabilities, and risk, plus risk assessment rating system
WS 9 Critical controls for consideration		Overview of current controls related to the site
WS 10 Response to an incident		High-level overview of roles and responsibilities in responding to an incident
	WS A Site team	Details of TACCP team
	WS B Site plan	Computer-aided design (CAD) drawing showing FBO site map layout including boundaries
	WS C Flow diagram	Product process flow diagram
	WS D Types of threats	Categories of threats relevant to the site, with examples
	WS E TACCP site self-assessment	Site self-assessment. This is for illustrative purposes only. It is not complete.
	WS F Site threat identification	Catalog and list of risks related to the site
	WS G Risk register	High-level site risk register
	WS H Threat decision tree	Threat decision tree

WS 2 TACCP Instructions and Content			
No.	**Item**	**Description**	**Link to sheet in WS 1**
2.1	TACCP team	Identify the TACCP team and responsibilities. The TACCP team should be separate from the HACCP team.	A. SITE TEAM
2.1.1	Training material	Ensure all TACCP team members are trained in the following principles:	
		• Scope of the assessment	Scope
		• Terms and definitions	Definitions
		• Aim of TACCP	TACCP aim
		• Types of threats to consider	Threats
		• Understanding the attacker	The attacker
		• Assessing threats and risk assessment procedures	Assessing threats
		• Critical controls in relation to TACCP	Critical controls
		• Response to an incident	Response to an incident
2.2	Site plan	Insert a site plan, including access/entry points and external site perimeter.	B. SITE PLAN
2.3	Process flow	Update the process flow from raw material purchase to delivery to customer. Site must include ALL steps where there is a threat or risk to the product. This is not an HACCP flowchart—consider only the steps involving interventions that may pose a threat to the product. Consider the supply chain prior to entering the site and post-site, including third-party storage and haulage.	C. FLOW DIAGRAM
2.4	Site types of threats	The site must identify all types of threats relevant to the site and any raw materials. The update, as necessary, is to be completed by the TACCP team.	D. TYPES OF THREAT
2.5	Site self-assessment	The site TACCP is to complete a self-assessment based on the systems currently in place at the site. Each point to be rated as compliant, improved, or a weakness. All areas of improvement and weakness must be detailed on the site threat page, and controls must be detailed on the risk register. The site is to update this page as gaps are closed.	E. TACCP SITE SELF-ASSESSMENT
2.6	Site threat identification and risk assessment	Update with risks that are applicable to the site. This list is not exhaustive. Consider and insert on this sheet any additional threats relevant to the site.	F. SITE THREAT IDENTIFICATION
2.7	Risk register	Detail all risks relevant to site following assessment and short-, medium-, and long-term controls to mitigate risks.	G. RISK REGISTER
2.8	Threat decision tree	The threat decision tree is used to determine if a threat is controlled via the prerequisite program or a critical control.	H. THREAT DECISION TREE

WS 3 Threat and Vulnerability Scope

TACCP background

A core element to the defense of food is a systematic evaluation of vulnerable links in the supply chain carried out by an experienced and trusted team. This is the threat assessment critical control point (TACCP). The evaluation reflects established procedures in risk management, and it is likely that organizations will increasingly incorporate TACCP into crisis and business continuity management frameworks.

One of the major guiding documents for TACCP is the PAS 96:2014, which was originally developed in 2008 by the Centre for the Protection of National Infrastructure (CPNI) in consultation with food manufacturers (such as Heinz, Kellogg, and Kraft), organizations (including the Food Standards Agency, National Farmers Union, and Food and Drink Federation), and retailers (such as Sainsbury's, Tesco, and Marks & Spencer). At the 2013 GFSI conference in Barcelona, Spain, Terry Donohoe from the Food Standards Agency gave a talk on the identification of future food safety risks. He specifically mentioned TACCP, the PAS 96, and the need to look for threat points, hazard points, and value points in the process of ensuring safe food.

Scope

TACCP, avoidance and mitigation of threats to food and food supply: the aim of this study is to identify and manage the potential threats and vulnerabilities associated with sites and products that could have an impact on the consumer and thereby a consequential impact on business. The assessment includes the following:

1.0 Food security programs

2.0 Outside grounds and roof

3.0 Employee and visitor programs

4.0 Material receiving

5.0 Facility operations

6.0 Finished goods storage/shipping

Procedure

Assessment of the risks through the process to identify potential threats and vulnerabilities and the implementation of controls on raw materials, packaging, finished products, processes, premises, distribution networks, and business systems

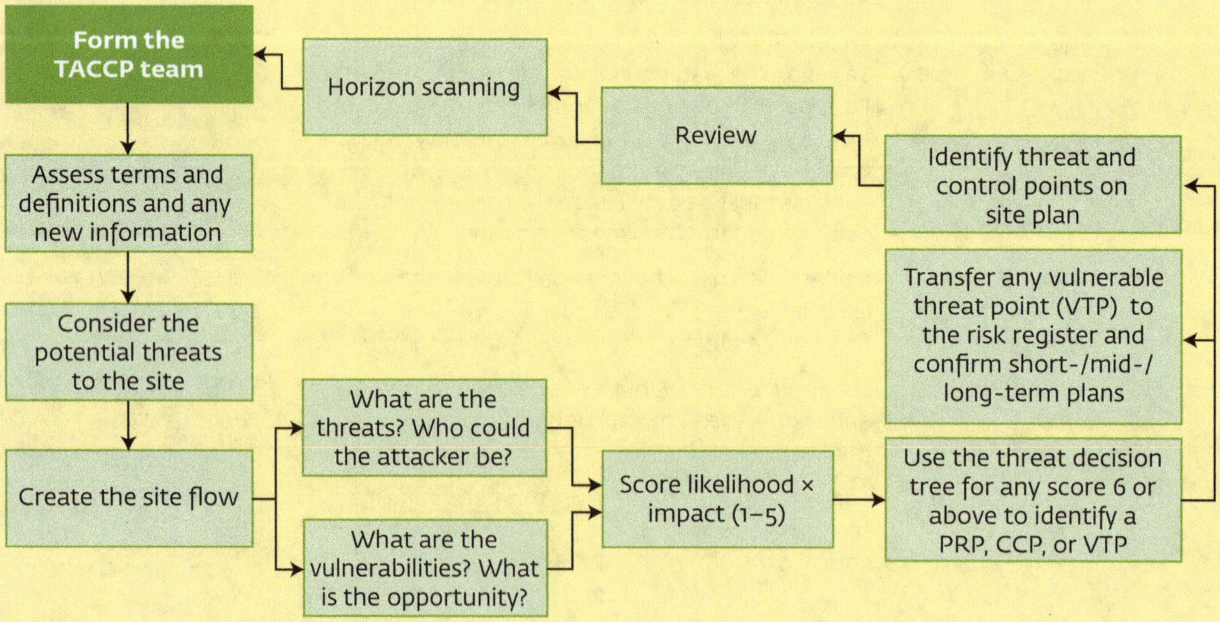

Sources: BSI (British Standards Institution), 2014, "PAS 96:2014: Guide to Protecting and Defending Food and Drink from Deliberate Attack," BSI, London; AIB International, 2010, "Food Defense Guidelines 2010," AIB International, Manhattan, KS.

CHAPTER 3

WS 4 Terms and Definitions	
Term	Definition
4.1 Cybersecurity	Procedures used to protect electronic systems from sources of threat; examples of threats include malware and hackers intent on misusing, corrupting, or shutting down information technology systems
4.2 Food defense	Procedures adopted to assure the security of food and drink and associated supply chains from malicious and ideologically motivated attack leading to contamination or supply disruption
4.3 Food fraud	Involves deliberately placing food on the market, for financial gain, with the intention of deceiving consumers. There are many sorts of food fraud, but the two main types are • Selling food that is unfit and potentially harmful, such as — Recycling animal by-products back into the food chain — Packing and selling beef and poultry of unknown origin — Knowingly selling goods that are past their "use by" date • Deliberately mislabeling food, such as by — Substituting products with cheaper alternatives, for example, selling farmed salmon as wild salmon or selling basmati rice that has been adulterated with cheaper varieties — Making false statements about the source of ingredients, that is, their geographic, plant, or animal origin Food fraud may also involve the sale of meat from animals that have been stolen and/or illegally slaughtered, as well as wild game animals, such as deer, that may have been poached.
4.4 Food protection	Procedures adopted to deter and detect fraudulent attacks on food
4.5 Food supply	Elements of what is commonly called a food supply chain
4.6 Hazard	A naturally occurring or accidental event that may cause loss or harm or that results from the incompetence or ignorance of individuals
4.7 Hazard analysis critical control point	A system that identifies, evaluates, and controls hazards
4.8 Insider	An individual within or associated with an organization who has access to its assets but who may misuse that access and become a threat to its operations
4.9 Personnel security	Procedures used to (a) confirm an individual's identity, qualifications, experience, and right to work and (b) monitor an individual's conduct as an employee or contractor. Personnel security should not be confused with personal security. Personnel security principles are applied to assure the trustworthiness of the staff of an organization, but may be applied as vendor accreditation to certify the staff of suppliers involved in processes.
4.10 Threat	A naturally occurring or accidental event that may cause loss or harm or that results from the incompetence or ignorance of individuals. The term threat is not used in the sense of threatening behavior or a promise of unpleasant consequences for a failure to comply with a malicious demand.
4.11 Threat assessment critical control point (TACCP)	The systematic management of risk through the evaluation of threats, the identification of vulnerabilities, and the implementation of controls with respect to materials and products, purchasing, processes, premises, distribution networks, and business systems by a knowledgeable and trusted team with the authority to implement changes to procedures

WS 5 Threat Assessment Critical Control Point (TACCP)	
5.1 TACCP aim	• Reduce the likelihood (chance) of a deliberate attack • Reduce the consequences (impact) of an attack • Protect organizational reputation • Reassure customers, the press, and the public that proportionate steps are being taken to protect food • Satisfy international expectations and support the work of trading partners • Demonstrate that reasonable precautions are being taken and that due diligence is being exercised in protecting food by, in broad terms: — Identifying specific threats to the company's business — Assessing the likelihood of an attack by considering the motivation of the prospective attacker, the vulnerability of processes, and the opportunity and the capability potential attackers have of carrying out the attack — Assessing the potential impact by considering the consequences of a successful attack — Judging the priority to be given to different threats by comparing their likelihood and impact — Identifying the proportionate controls needed to discourage the attacker and give early notification of an attack — Maintaining information and intelligence systems to enable revision of priorities
5.2 Process	TACCP must be a team activity and consider the following: • Who might want to attack? • How might they do it? • Where are the vulnerabilities? • How can attackers be stopped? The TACCP team should: • Evaluate all new information that has become available. • Identify individuals or groups that may represent a threat to the organization and assess their motivation, capability, and determination. • Identify individuals or groups that may represent a threat to specific operations (e.g., premises, factories, other sites). • Select products that are representative of specific processes and undertake the following: — Identify individuals or groups that may want to target the specific products; — Draw up a process flowchart for the products; — Examine each step of the processes to identify vulnerable points where an attacker might hope for success and identify the people who would have access to these points; — Identify possible threats associated with the products at each step and assess the impact that the process may have in mitigating the threats.
	NOTE 1: *Model adulterants include low-cost alternative ingredients to premium components; model contaminants could include highly toxic agents, toxic industrial chemicals, readily available noxious materials, and inappropriate substances like allergens or ethnically unwholesome foodstuffs.* NOTE 2: *For example, cleaning may remove the contaminant, heat treatment may destroy it, and other food components may neutralize it.*

continued

WS 5 *(Continued)*	

5.2 Process *(continued)*	• Select the points in the process where an attack would have the most effect, and where the threat might be most readily detected.
	• Assess the likelihood that routine control procedures may detect a threat.
	
	NOTE: For example, routine laboratory analysis could detect added water or unusual fats and oils; effective management of purchasing operations would challenge unusual purchase orders.
	
	• Score the likelihood of the realization of the threat and the impact this would have and chart the results to show the priority the threat should be assigned.
	
	NOTE: The TACCP team might ask, "If we were trying to undermine our business, what would be the best way?" It might consider how an attacker might select the materials needed to carry out an attack, including
	  • *Availability*
	  • *Cost*
	  • *Toxicity*
	  • *Physical form*
	  • *Safety in use, for example, pesticides on farms and aggressive flavor materials in factories may be convenient contaminants*
	• Where the priority is high, identify who has unsupervised access to the product or process and whether they are trustworthy, and if that trust can be justified.
	• Identify, record confidentially, agree on, and implement proportionate preventive action (critical controls). The TACCP team should have a confidential reporting and recording procedure that allows management action on decisions but does not expose weaknesses to those without a need to know.
	• Determine the review and revision arrangements for TACCP evaluations.
	
	NOTE: Review of the TACCP evaluation should take place after any alert or annually, and at points where new threats emerge or when there are changes in good practice.
	
	• Maintain a routine watch of official and industry publications that may offer an early warning of changes that may become new threats or that alter the priority among existing threats, including emerging local issues.

WS 6 Types of Threats and Case Study Examples		
6.1 General		Deliberate attacks on food and food supply take several forms. Clause 3 describes the characteristics of the main threats to food authenticity and safety—economically motivated adulteration (EMA) and malicious contamination—and outlines the nature of other threats.
6.2 Economically motivated adulteration (EMA)		A core element in food defense is the systematic evaluation of vulnerable links in the supply chain carried out by an experienced and trusted team. This threat assessment critical control point (TACCP) evaluation reflects established procedures in risk management. Organizations will likely increasingly incorporate such evaluations into crisis management and business continuity management.
	Case 1	In 2013, allegations were reported that a food factory in Asia was labeling cooking oil as peanut, chili, and olive oil when it contained none of these oils.
	Case 2	A 2013 report suggested that one-third of retail fish in the United States was mislabeled. Examples included tilapia sold as red snapper and tilefish sold as halibut.
	Case 3	In 2010, some producers of buffalo mozzarella in Italy were accused of adulteration of their product with cow's milk.
	Case 4	Staff for a European meat packer felt, mistakenly, that, by covering it with disinfectant, they could avoid removal of a product because it harbored the virus associated with foot and mouth disease.
6.3 Malicious contamination		The motivation for malicious contamination may be to cause localized (see case 5) or widespread (see case 6) illness or death. In case 6, the attacker did not want the contamination to be detected before the food was consumed, therefore the contaminant had to be an effective toxin with little effect on the palatability of the food. The motivation in case 7 was publicity. Public opinion would have been against the attackers if harm had been caused to members of the public, but the supplier could not take that risk. Materials that may be used by an attacker to gain publicity, or to extort money, are more readily available than those needed to cause widespread harm. The case of allergens (see case 8) shows the harm, impact, and cost that can be caused to a business with little risk for the attacker. Contamination close to the point of consumption or sale, as in case 6, is more likely to cause harm to health than is an attack on crops or primary ingredients.
	Case 5	In 2005, a major British bakery reported that several customers had found glass fragments and sewing needles inside the wrappers of loaves of bread.
	Case 6	In 1984, the Rajneeshee sect in Oregon attempted to affect the result of a local election by contaminating food in 10 different salad bars, causing 751 people to become infected with *Salmonella*.
	Case 7	In 2013, a major soft drink supplier was forced to withdraw products from a key market when it received a bottle in which the contents had been replaced with mineral acid. The attackers included a note indicating that more would be distributed to the public if the company did not comply with their demands.
	Case 8	In 2007, a bakery found piles of peanuts in the factory. It withdrew its products and closed for a week-long deep clean to reestablish its nut-free status.

continued

WS 6 *(Continued)*		
6.4 Extortion		The motivation for extortion by an individual or group is financial, to obtain money from the victim organization. Such activity is attractive to the criminal mind if the product, such as baby food (see case 9), is sensitive or if a company is seen as rich (see case 10). A small number of samples can be used to show the company that the attacker has the capability to attack, and is enough to cause public concern and media interest.
	Case 9	In 1990, a former police officer was convicted of extortion after contaminating baby food with glass and demanding money from the multinational manufacturer.
	Case 10	In 2008, a man was jailed in the United Kingdom after being convicted for threatening to bomb a major supermarket and contaminate its products.
6.5 Espionage		The primary motivation of espionage is to seek commercial advantage by gaining access to intellectual property. Attackers may infiltrate using insiders to report, or they may attack remotely through IT systems. Alternatively, organizations may try to entice executives to reveal confidential information or use covert recording to capture such material, or they may simply steal the material, as case 12 suggests.
	Case 11	A business consultancy uses the theft of the intellectual property of a fictitious innovative snack product as an example of commercial espionage.
	Case 12	In July 2014, Reuters reported that a woman was charged in the United States with attempting to steal patented U.S. seed technology as part of a plot to smuggle out types of specialized corn for use in China.
6.6 Counterfeiting		The motivation for counterfeiting is financial gain, by fraudulently passing off inferior goods as established and reputable brands. Both organized and petty crime can cause companies financial loss and harm to their reputation. The former, for example, can use sophisticated printing technologies to produce product labels that are indistinguishable from the genuine ones. Petty criminals may steal genuine packs or even refill single-use containers for resale. Organized criminals may try to mimic the food contents closely to delay detection and investigation. Petty criminals may be tempted by the opportunity to make a "quick killing" and be less concerned about the safety of the food.
	Case 13	In 2013, enforcement officers seized 9,000 bottles of fake Glen's Vodka from an illegal factory.
	Case 14	In 2011, 340 bottles of a famous Australian brand of wine were seized, following complaints of poor quality to the owner, which had no link with Australia.
6.7 Cybercrime		Modern information and communications technologies provide new opportunities for malpractice. In the United Kingdom for the year to February 2013, Action Fraud received 58,662 cyber-enabled fraud reports and 9,898 computer misuse crime reports representing 41 percent of all of its reports, with an average loss of £3,689,16. In case 15 the fraudster aims to defraud both the business and consumers. It is common for the attacker to try and exploit the individual's ignorance of the technologies involved. Identity theft is perhaps more familiar to the public, but organizations may be aware of their identity being stolen to enable procurement fraud, in which goods are ordered in their name but diverted to the fraudsters' premises, leaving the organizations to carry the cost and litigation.
	Case 15	In 2014, Financial Fraud Action UK advised restaurant managers to stay vigilant because fraudsters were attempting to target their customers in a new phone scam. They phoned restaurants claiming there was a problem with their card payments system; the restaurant was then told to redirect any card payments to a phone number provided by the fraudster.

WS 7 Understanding the Attacker	
7.1 General	The success of a deliberate attack on food or food supply depends on several things: • Does the attacker have the motivation and drive to overcome the obvious, and less obvious, obstacles to their actions? If the obstacles seem massive and success seems unlikely, many would-be attackers would seek an easier target. • Does the attacker have the capability to carry out the attack? A group is more likely to find the resources and learn the skills needed. • Does the attacker have the opportunity to carry out the attack? A physical attack requires physical access to the target, but a cyberattack may only require access to a computer. • Would the attacker be deterred by the chance of detection or any potential penalties?
7.2 The extortionist	The extortionist wants to gain financially from an attack but does not want to be caught, and concentrates on avoiding detection. Their target is more likely to be a high-profile business with lots to lose from negative publicity. They may work alone and be resourceful, secretive, and self-interested. Some individuals may claim to be able to take action against a business while lacking the capability to carry it out; the business may judge the claim as not credible but still report and respond seriously.
7.3 The opportunist	The opportunist may hold an influential position within an operation and be able to evade internal controls. They may have some technical knowledge but their main asset is access. They are likely to be discouraged by the chance of detection, so unannounced visits by customers or auditors, or ad hoc sampling for analysis, may deter their actions. A supplier who cannot risk failing to deliver to a customer may take the chance that occasional adulteration would not be detected. Success on one occasion may make it easier to attempt a repeat. Opportunists may persuade themselves that the adulteration is legitimate, for example, chicken in a pork sausage would still be meat.
7.4 The extremist	The extremist takes their cause or campaign so seriously that they distort its context and overlook wider issues. The dedication to their cause may have no limits, and their determination to progress it can be great. Extremists may want to cause harm and are likely to enjoy publicity after the event. It may not matter, and may be a benefit, if they themselves are harmed. The risk of failure is a deterrent, but the risk of capture after the event is not. They are typically resourceful and innovative in devising ways to attack. Some single-issue groups may want to disrupt business operations and reputation but fear that mass harm to the public would damage their cause and lead them to lose support.
7.5 The irrational individual	Some individuals have no rational motive for their actions. Their priorities and preoccupations have become distorted so they are unable to take a balanced view of the world. Some may have clinically diagnosed mental health issues. They may be readily deterred by simple steps that prevent them from gaining access to their target or make detection easy.
7.6 The disgruntled individual	The disgruntled individual may believe that an organization has been unfair to them and seek revenge. For example, they may be an aggrieved employee or former employee, supplier, or customer. They may have expert knowledge of the operation and access to it. This attacker is likely to act alone rather than as part of a group. If an insider, they could be dangerous, but are more likely to want to cause embarrassment and financial loss than harm to the public. If not an insider, this individual is more likely to claim or boast of having done something than actually being able to do it.

continued

WS 7 *(Continued)*	
7.7 The hacktivist and other cybercriminals	A hacktivist or other cybercriminal aims to subvert controls on computerized information and communications systems in order to stop them from working effectively, to steal or to corrupt data which they hold, or to disrupt Internet business. Their motivation may be criminal, but may also be to demonstrate their expertise and ability to beat any protective system devised to stop them. This type of attacker has information and communications technology expertise that can cause commercial harm and may pose an increasing threat to food safety as Internet activity increases.
7.8 The professional criminal	Organized crime may see food fraud as a relatively simple crime, with big gains in prospect, little chance of apprehension, and modest penalties if convicted. The global trade in food in which food materials move, often with little notice, across enforcement area borders appears to encourage the professional criminal. They may be deterred by close collaboration between food operations and national and international police authorities.

WS 8 Assessing Threats, Vulnerabilities, and Risk	
8.1 Assessing threats	The product, the premises, and the organization can be the target of an attack from a range of groups and individuals. The TACCP team should consider suppliers under financial stress, alienated employees and former employees, single-issue groups, local pressure groups, commercial competitors, media organizations, terrorists, criminals, and individuals (see Clause 4), and each element should be assessed separately. Commonly, a short supply chain involving fewer people may be less risky than a longer supply chain. The TACCP team could ask the following questions to analyze the threats. A consideration of the responses to these questions can provide an understanding of the impact of a successful attack and the likelihood an attack will take place. This informs a judgment on the proportionate level of protection required. Relative to the product: • Have significant cost increases affected the product? • Does the product have particular religious, ethical, or moral significance for some people? • Does the product contain ingredients or other material sourced from overseas? Relative to the premises: • Are the premises located in a politically or socially sensitive area? • Do the premises share access or key services with controversial neighbors? • Are new recruits, especially agency and seasonal staff, appropriately screened? • Are services to the premises adequately protected? • Are external utilities adequately protected? • Are hazardous materials, which could be valuable to hostile groups, stored on-site? • Are large numbers of people (including the general public) using the location? • Do any employees have reason to feel disgruntled or show signs of dissatisfaction? • Are internal audit arrangements independent? • Have key roles been occupied by staff for many years with little supervision?

continued

CHAPTER 3

WS 8 Assessing Threats, Vulnerabilities, and Risk *(Continued)*	
8.1 Assessing threats *(continued)*	Relative to the organization: • Is the organization under foreign ownership by nations involved in international conflict? • Does the organization have a celebrity or high-profile chief executive or proprietor? • Does the organization have a reputation for having significant links with customers, suppliers, and so on, in unstable regions of the world? • Are the brands of the organization regarded as controversial by some? • Does the organization or its customers supply high-profile customers or events?
8.2 Assessing vulnerabilities	Economically motivated adulteration (EMA) A typical feature of EMA is the substitution of a low-cost item in place of a relatively high-cost component/ingredient. The TACCP team needs to be alert to the availability of such alternatives. An example where this may happen is when added value is claimed (e.g., such as for organic, non-GMO, locally grown, free range, or with protected designations of origin). The attacker is likely to have ready access to lower-value equivalents that are almost indistinguishable. The TACCP team needs to be confident that its own operations and those of its suppliers are in trustworthy hands. This can be achieved using advice on personnel security.
	Questions that the TACCP team could ask include • Are low-cost substitute materials available? • Have there been significant material cost increases? • Has pressure increased on suppliers' trading margins? • Does the organization trust their suppliers and their suppliers' managers? • Do key suppliers use personnel security practices? • Do suppliers think that the organization monitors their operation and analyze their products? • Is the organization supplied through remote, obscure chains? • Are major materials becoming less available (e.g., from crop failure) or alternatives becoming more plentiful (e.g., from overproduction)? • Have there been unexpected increases or decreases in demand? • How do suppliers dispose of excessive amounts of waste materials? • Is the organization aware of shortcuts in the process that may affect the organization? • Are staff and the staff of suppliers encouraged to report concerns (whistleblowing)? • Are accreditation records, certificates of conformance, and analysis reports drafted independently?
	Malicious contamination Questions that the TACCP team could ask about its own operations and the operations of its suppliers include • Are food safety audits rigorous and up-to-date? • Are personnel security procedures in use? • Is access to products restricted to those with a business need?

continued

<table>
<tr><th colspan="2">WS 8 *(Continued)*</th></tr>
<tr><td>8.2 Assessing vulnerabilities (continued)</td><td>Malicious contamination (continued)
• Do storage containers have tamper-evident seals?
• Is the organization involved with controversial trade?
• Is the organization owned by nationals from conflict areas?
• Is there opportunity for access by sympathizers of single-issue groups?
• Do any employees bear a grudge against the organization?
• Is staff boredom, discipline, or recruitment a problem?
• Have business competitors been accused of espionage or sabotage?</td></tr>
<tr><td>8.3 Assessing risks</td><td>Organizations need to understand the threats that they face, but should focus attention on the priority threats. For each identified threat, the TACCP team might give a score for the likelihood of each threat occurring and for its impact.</td></tr>
</table>

Likelihood of threat happening/detection	Score	Impact of threat
Rare	1	Trivial
Unlikely	2	Minor
Possible	3	Moderate
Likely	4	Major
Almost certain	5	Severe

The likelihood of a threat occurring can be judged by considering

- Whether attackers would achieve their aims if successful
- Whether attackers would be able to gain access to the product or process
- Whether attackers would be deterred by protective measures
- Whether attackers would prefer other targets
- Whether an attack would be detected before it had any impact

The impact might be assessed in financial terms or in terms of the seniority of the staff needed to deal with it.

The level of risk is determined using the table below, and appropriate controls should be put in place on the risk register to mitigate the risk.

Impact of threat	5	C	B	A	A	A
	4	D	C	B	B	A
	3	E	D	C	C	B
	2	E	D	D	C	B
	1	E	E	D	C	C
		1	2	3	4	5
		Likelihood of threat happening/detection				

Very high risk	Threat A
High risk	Threat B
Moderate risk	Threat C
Low risk	Threat D
Trivial risk	Threat E

CHAPTER 3

WS 9	Critical Controls for Consideration
9.1 Controlling access	If prospective attackers have no access to their target, then their attack cannot take place. It is not possible or desirable to prevent all access, but physical measures may limit access to certain individuals and those with a legitimate need.
	For consideration:
	Access to premises: • Access to people on business only • Vehicle parking outside perimeter • Premises zoned to restrict access to those with a business need • Visible and comprehensive perimeter fencing • Closed-circuit television (CCTV) monitoring/recording of perimeter vulnerabilities
	Access to vehicles: • Monitored access points • Traffic calming on approach roads • Scheduled deliveries • Documentation checked before admittance • Missed deliveries investigated
	Access to people: • Controlled access • Facilities for changing clothes should maintain control over workwear • Screening of visitors • Entry by appointment only • Proof of identity required for entry • Visitors accompanied throughout • Positive identification of staff and visitors • CCTV monitoring/recording of sensitive areas
	Other aspects: • Secure handling of mail • Restrictions on portable electronic and camera equipment • Limitations on access to main services • Compliance with the International Organization for Standardization (ISO)/International Electrotechnical Commission 27000 standards on information security management systems on cybersecurity
9.2 Tamper detection	Much raw material storage, some product storage, most distribution vehicles, and all packaged foods should be tamper evident. Should an attacker gain access, tamper evidence gives some chance that the attack may be detected in time to avoid the impact.
	Detecting tampering: • Numbered seals on bulk storage silos • Numbered seals on stores of labels and labeled packs • Effective seals on retail packs • Numbered seals on hazardous materials • Tight stock control on key materials • Recording of seal numbers on delivery vehicles • Secure usernames and passwords for electronic access • Incursion reporting by cybersystems
9.3 Assuring personnel security	Personnel security guidance can mitigate the insider threat to organizations. Its principles can also be used by food businesses to judge whether key staff within the organizations that supply goods and services can be trusted to comply with specifications and procedures, and to work in the best interest of both the supplier and customer.
	Preemployment checks: • Proof of identity • Proof of qualifications • Verification of contractors • More sensitive roles identified and accompanied by appropriate recruitment
	Ongoing personnel security: • Staff in critical roles motivated and monitored • Whistleblowing arrangements • Temporary staff supervised • Individuals able to work alone • Favorable security culture
	End of contract arrangements: • Access and ID cards and keys recovered • Computer accounts closed or suspended • Termination interview to assess security implications

WS 10 Response to an Incident	
10.1 Management of a food protection crisis	Food protection and defense procedures aim to reduce the risk of an attack but cannot eliminate it, so emergency response and business continuity protocols are essential.
	Food protection sits within the business crisis management system and shares its general objectives: • To minimize physical and financial harm to consumers, customers, employees, and others • To collaborate with investigative and enforcement authorities • To gain public support for the organization • To minimize the cost—financial, reputational, and personal—of the incident • To prevent recurrence • To identify offenders Where contamination is implicit, quarantine and the withdrawal or recall of products may be expected. Each incident will be assessed with the on-site TACCP team, the quality assurance cluster lead, and other teams such as the supplier quality and food safety team, as required. All raw material nonconformances should be logged into the raw materials nonconformance database.

WS A Site Team

Name	Position	Knowledge, experience, and training
G Moran	Food safety manager	BA in English 15 years' experience in the pharmaceutical and food industry 4+ years in dairy nutrition HACCP level 6 ISO 22000/FSSC 22000 auditor/lead auditor PCQI auditor training TACCP, food defense, biovigilance, and bioterrorism training
O Brown	Hygienist/ microbiologist	MS in analytical chemistry 20 years in the food industry 4+ years in dairy nutrition HACCP level 5 ISO 22000/FSSC 22000 internal auditing
M Rodrigues	Milk processing manager	BA in English 8 years' experience in the food industry 4+ years in dairy nutrition HACCP level 6 ISO 22000/FSSC 22000 internal auditing
B Murphy	Laboratory manager	PhD in microbiology 8 years' experience in the food industry 4+ years in dairy nutrition HACCP level 6 ISO/IEC 17025:2017 ISO 22000/FSSC 22000 internal auditing
D Small	Warehouse manager	BA in English 20 years' experience in the food industry HACCP level 5 ISO 22000/FSSC 22000 internal auditing
O Murphy	Engineering manager	BA in English 12 years' experience in the food industry HACCP level 5 ISO 22000/FSSC 22000 internal auditing
C Flack	Factory manager	BA in English 15 years' experience in the food industry HACCP level 5 ISO 22000/FSSC 22000/BRC internal auditing

WS B Site Plan

Insert a site plan, including access/entry points and the external site perimeter.

WS C Flow Diagram

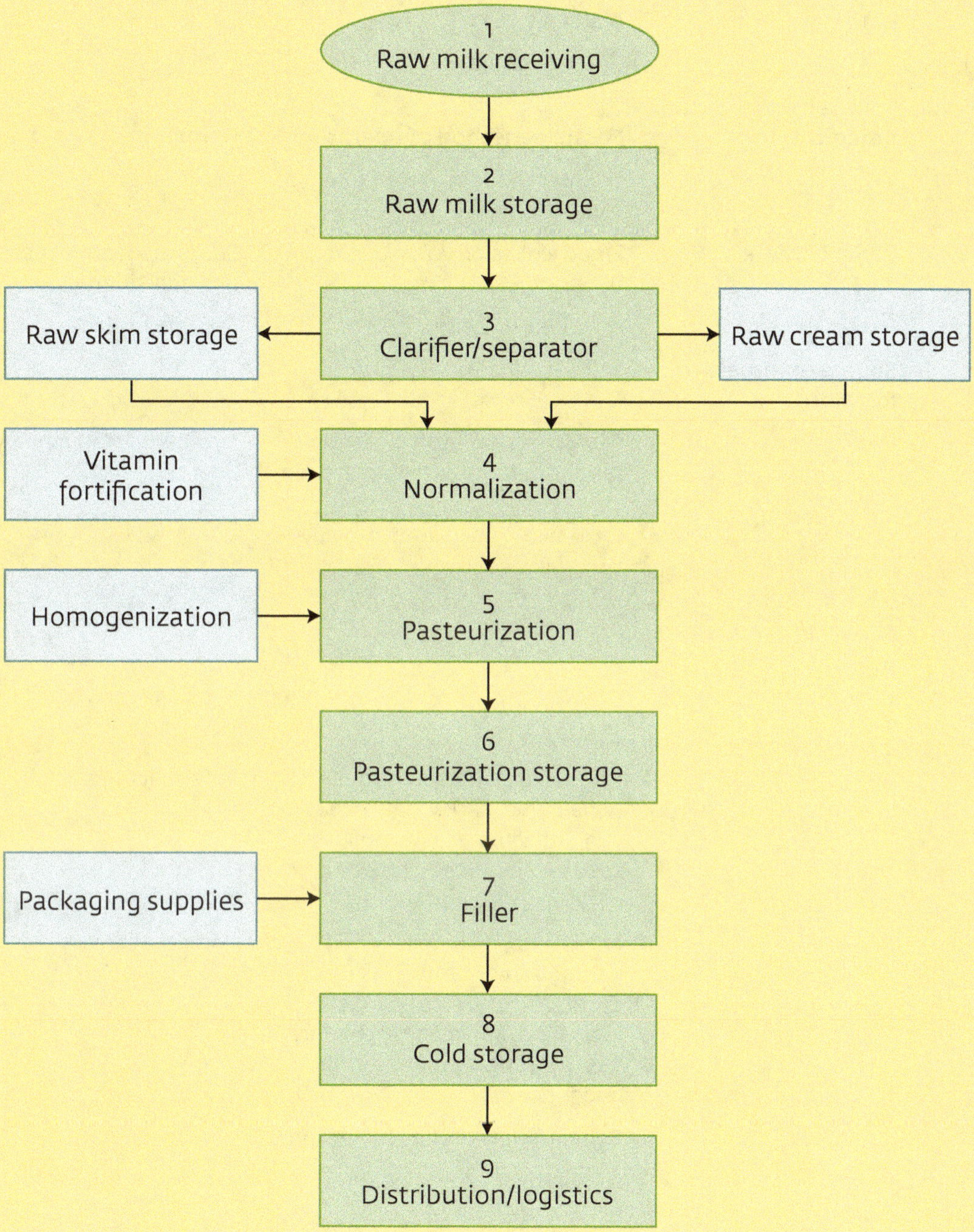

WS D Types of Threat			
Site information	Site address	Joe Bloggs LLC	
	Date	March 17, 2019	
Product details		Grade A whole milk	
Number	**Source of threats to organization**	**Possible method of operation**	**Comments**
	Criminals	Counterfeiting, misappropriation of packaging	Risk to brand
	Company buyers	Fraud, collusion with suppliers	Supplier assessments, audits, and questionnaires
	Hactivists	Attack on website	Security in place
	Source of threats to location		
	Disgruntled staff	Petty contaminaton, potential serious malicious contamination	No agency staff employed on-site
	Frontline staff	Theft, collusion with customers	Security checks on entering and exiting the site
	Neighbors		Secured access
	Source of threats to product		
	Microbiological	Pathogens, viruses, toxins, parasites	Potential threats
	Allergens	Per EU legislation, milk, gluten (wheat, rye, barley, oats), celery, egg, fish, crustacean, mollusks, sesame, soya, lupin, mustard, nuts, peanuts, and sulfites	Potential threats
	Chemical	Antibiotics, mycotoxins, packaging contaminants, pesticides, cleaning chemicals, lubricants, ink, medical chemicals, fertilizers, heavy metals, coloring, and flavors	Potential threats
	Radiological	Radioactive material, for example	Potential threats
	Legal	Not of the proper substance, nature, or quality	Potential threats
	Physical	Glass and shatterable materials, metal, wood, hard plastics, packaging materials, stones, personnel-related, bone and gristle, flexible plastic, and intrinsic food foreign bodies such as stones in fruit or shell in nut products	Potential threats

WS E	TACCP Site Self-Assessment				
	Site name: 2211	Date: 3/17/2019			Completed by: G Moran
	Section	***Compliant***	***Improvement***	***Weakness***	
1.0	Food security programs	3	2	3	
2.0	Outside grounds and roof	6	0	0	
3.0	Employee and visitor programs	5	3	2	
4.0	Material receiving	8	7	1	
5.0	Facility operations	7	6	3	
6.0	Finished goods storage/shipping	2	7	0	
	TOTAL	31	25	9	65
	% Compliance	48%	39%	14%	

	Criteria	**Rating (Insert "1" in relevant field)**				
1.0	**Food security programs**	*Compliant*	*Improvement*	*Weakness*	**Comments (input document reference data where documentation is in place)**	**Actions required**
1.1	Operational risk management program completed for facility. (Documented)	1		1	Risk management program last updated in 2011 (Revision 3)	Updated risk assessment of site needed
1.2	Site TACCP team identified and all stakeholders trained in TACCP principles. Crisis management team established. (Documented)			1	TACCP team identified but training has not taken place	Train TACCP team
1.3	Program to ensure security of incoming mail and packages. (Documented)		1		Documentation of procedure is poor	Update procedure to include controls

continued

WS E TACCP Site Self-Assessment *(Continued)*

1.0	Food security programs *(Continued)*	*Compliant*	*Improvement*	*Weakness*	**Comments**	**Actions required**
1.4	Program to protect and backup computer systems and documentation critical to food safety (Documented)			1	Documentation not backed up	Devise a backup system for food-safety-critical documentation
1.5	Company controlled off-site warehousing, manufacturing, and distribution included in food security programs	1			Not applicable	Not applicable
1.6	Customer/consumer complaint program established and procedures to investigate alleged tampering issues (Documented)		1		Procedure for complaint handling is outdated	Update procedure
1.7	Written procedures and policies in place for a contracted security service (Documented)	1			SOP-017 site defense policy	None
2.0	**Outside grounds and roof**	*Compliant*	*Improvement*	*Weakness*	**Comments**	**Actions required**
2.1	Secured perimeters to restrict access to the facility and related outbuildings	1			Perimeter secured by physical barriers	
2.2	Security cameras utilized at key locations around facility and outbuildings	1			Netwatch-operated cameras in place	
2.3	Regular patrols conducted of outside grounds and roof area (Documented)	1			Not currently documented	Contact contract cleaners re: documentation
2.4	Program in place to address any unusual security issues noted on outside grounds (Documented)	1			24-hour security on-site	

continued

WS E *(Continued)*						
2.5	Entrances to facility are minimized and monitored	1			24-hour security on-site	
2.6	Metal or metal-clad doors utilized on entrances to facility	1			In place	Confirmed
3.0	Employee and visitor programs	*Compliant*	*Improvement*	*Weakness*	**Comments**	**Actions required**
3.1	Formal pre-hire screening program in place for all employees and contracted persons (Documented)		1		Documented process not available	Confirm with HR
3.2	No employees or contracted individuals working without pre-hire screening program completed and approved (Documented)		1		Documented process not available	Confirm with HR
3.3	No evidence of personal belongings outside of designated areas	1			Production areas inspected daily	None
3.4	Formal uniform or outer garment program (Documented)	1			SOP-018 gowning procedure	None
3.5	Employees not allowed outside of facility or designated outside break areas during work hours			1	Currently no restriction on employee movement during break times	Discuss with TACCP team if required
3.6	Employee lockers in locker rooms and other personal storage areas inspected on a regular basis			1	Inspections occurring only prior to audits, not being documented	SOP and schedule for locker inspections required

continued

CHAPTER 3

WS E TACCP Site Self-Assessment *(Continued)*

3.0	Employee and visitor programs *(Continued)*	*Compliant*	*Improvement*	*Weakness*	**Comments**	**Actions required**
3.7	Visitors, contractors, guests, etc. report to a designated entrance and sign in	1			All visitors report to security, all contractors report to maintenance manager	None
3.8	Facility policies provided to visitors, contractors, guests, etc. and plant-issued identification provided with date of issue and expiration		1		System needs to be updated	Update system to ensure information provided is concise and accurate
3.9	Visitors, contractors, guests, etc. comply with the company dress policy/protective clothing policy	1			Everyone entering medium- and high-care areas must comply with gowning requirements	None
3.10	Formal program to accompany visitors in facility and verify access to food sensitive areas	1			SOP-004	None
4.0	**Material receiving**	*Compliant*	*Improvement*	*Weakness*	**Comments**	**Actions required**
4.1	Suppliers provide evidence of food security programs (Documented)	1			Evidence of supplier food security programs not documented	Confirm on vendor approval system if information is there and decide what suppliers we need this for
4.2	Supplier guarantees on file for all ingredients and packaging	1			Available on supply quality portal	
4.3	Formalized ingredient and packaging testing programs are in place (in-house testing, outside testing, or certificates of analysis) (Documented)		1		Raw material/ packaging testing programs are poor	Define requirements for each raw material/ packaging item

continued

WS E	(Continued)					
4.4	Unloading equipment (hoses, pipes, caps, augers, etc.) is secured and inspected prior to use			1	Documented procedure for securing and inspecting unloading equipment not in place	Discuss with TACCP team if required
4.5	Unloading process is conducted in a secured area or monitored during entire process	1			All unloading occurs in specified areas with supervision by relevant personnel	None
4.6	Trailer is inspected after unloading and all unloading equipment re-secured		1		Procedure for securing of unloading equipment not documented	Discuss with TACCP team if required
4.7	Amount of product received is verified against the receiving document	1			Done in stores/ at weighbridge	None
4.8	Written procedures in place to cover receipt of all received materials (Documented)	1			SOP-002	None
4.9	Arrival of truck at facility verified and driver identification verified (Documented)	1			SOP-001	None
4.10	Bill of lading and receiving documents verified. Should include material name, amount of material, amount of seals, seal numbers, lot numbers, etc.		1		Done upon receipt, check SOP	Review SOP
4.11	Truck and/or trailer inspection conducted by trained facility personnel before and after unloading (Documented)		1		Warehouse manual	Review SOP

continued

WS E TACCP Site Self-Assessment *(Continued)*

4.0	Material receiving *(Continued)*	*Compliant*	*Improvement*	*Weakness*	**Comments**	**Actions required**
4.12	Product(s), amount, labels, lot numbers, etc. verified at time of receipt (Documented)		1		Warehouse manual	Review SOP
4.13	Procedures in place for handling damaged or rejected materials (Documented)	1			SOP-003	Review SOP
4.14	Less-than-load (LTL) shipments have a food security system in place to include ingredients, maintenance, sanitation, pest control, laboratory, and other received items		1		Procedure is not specific enough on LTL shipments	Review SOP
4.15	Written procedures to address quarantine and release, irregularities in amounts outside a predetermined range, evidence of tampering, or counterfeiting of goods received		1		Procedure is not specific enough on evidence of tampering	Review SOP
4.16	Tamper-resistant/tamper-evident packaging required for received materials, when feasible	1			Where feasible, tamper-resistant packaging utilized	Specify where used/not used in SOP
5.0	**Facility operations**	*Compliant*	*Improvement*	*Weakness*	**Comments**	**Actions required**
5.1	Assessment conducted to indicate sensitive areas, such as materials storage, water supply, steam, compressed air, ice system, air supply, mixing, batching, production, etc. (Documented)		1		Risk assessment conducted 2011	Update risk assessment
5.2	Access restricted to authorized individuals in sensitive areas identified in assessment	1			Access is documented in risk assessment	None

continued

WS E (Continued)						
5.3	Water supply and related critical components (storage tanks, backflow preventers, filters, etc.) are secured		1		Full assessment needed to confirm	Update risk assessment
5.4	Water portability testing conducted on a regular and random basis (Documented)		1		Water testing conducted by external contractor at regular intervals—SOP needs to be reviewed	Review SOP
5.5	Water treatment and/or filter systems monitored on a regular basis (Documented)	1			Water chlorination monitored continuously and checked manually and recorded on a weekly basis	Review SOP
5.6	Formal plan to address and react to a possible water safety issue (Documented)			1	SOP not in place/ not referred to in security procedures	SOP required
5.7	Physical barriers in place and/or access restricted to hazardous compounds, such as nitrite, cleaning, and sanitizing chemicals, maintenance chemicals, pesticides, etc.		1		Updated assessment required	Conduct assessment and confirm
5.8	Controls in place to prevent intentional contamination by contractors of maintenance, pest control, or sanitation crews		1		Updated assessment required	Conduct assessment and confirm
5.9	Program to identify any sampled or opened ingredient containers. Employees aware of program and understand procedures to follow if not properly identified (Documented)			1	SOP required	SOP required

continued

WS E TACCP Site Self-Assessment *(Continued)*

5.0	Facility operations *(Continued)*	*Compliant*	*Improvement*	*Weakness*	**Comments**	**Actions required**
5.10	Traceability provided for all ingredients, direct contact packaging, and rework (Documented)	1			Tracebility procedure in place	None
5.11	Access to food safety manufacturing components limited and controlled (retort controls, pasteurizer controls, heat control components, etc.)	1			Controls in place (password secured)	None
5.12	Unprocessed goods segregated from processed goods and a program to prevent deliberate mixing of these goods	1			Store segregation procedures	None
5.13	Tamper-resistant/tamper-evident packaging and/or seals provided for finished goods	1			In place	None
5.14	All finished goods have appropriate lot identification	1			In place	None
5.15	Labels held in a secure area		1		Labels printed and held in access controlled area, labels not secured within area	Discuss with TACCP team if securing of labels within bagging lines required
5.16	Labels provided on containers are verified			1	Labels verified by online system and store operators, SOP needs to be more specific	Review SOP
6.0	**Finished goods storage/shipping**	*Compliant*	*Improvement*	*Weakness*	**Comments**	**Actions required**
6.1	Finished goods appropriately segregated from raw materials or hazardous chemicals	1			In place	None

continued

CHAPTER 3

WS E (Continued)						
6.2	Quantities of finished goods are tracked and program in place to investigate missing or extra stock		1		Stock control system in place, SOP needs to specify procedure for extra/missing stock	Review and update procedure
6.3	List of all third-party storage and shipping companies/haulers used by the company is available and relevant third-party audit certification on file (Documented)		1		Confirmation needed from stores	Confirm with distribution manager
6.4	Third-party storage warehousing and shipping companies utilized by the facility practice food security (Documented)		1		Needs to be documented	Discuss with TACCP team
6.5	Written procedures for inspection of all vehicles prior to loading (bulk and nonbulk)		1		Checks conducted	SOP may require review
6.6	Inspection conducted of all outbound vehicles prior to loading (Documented)		1		In place, requires review	Review SOP
6.7	Amounts and lot numbers of materials verified during loading	1			In place	None
6.8	Driver identification verified (Documented)		1		Confirmation needed from stores	Confirm with distribution manager
6.9	Security of trucks and trailers maintained during transport to include multiple stops or deliveries		1		Confirmation needed from third-party haulers	Confirm with distribution manager
	Total	**31**	**25**	**9**		

WS F Site Threat Identification

Site name: Joe Bloggs LLC		Date: March 17, 2019		Completed by: G Moran	
Criteria/process step	**Further breakdown**	**Risk number**	**Threat**	**Vulnerability (motivation/ opportunity)**	**Preventive actions/ controls currently in place**
Food security programs	Documentation/ procedures	1.1	No operational risk management in place	Individual/group exploits gaps in management of operational risks	Operational risk management implemented, Doc ID RA-SOP-001 V1.0
		1.2	Off-site warehousing not controlled	Individual/group exploits lack of control of off-site warehouse in order to compromise safety of product	Approval of off-site/third-party warehousing
Outside grounds and roof	External	2.1	Site security not checked routinely	Individual/group exploits lack of security to compromise safety of product	24-hour security presence on-site, on-site CCTV monitoring of critical locations, plus physical security measures in place (fencing, barriers, etc.)
		2.2	Access to roof, silos, outbuildings, bulk storage not locked	Individual/group exploits lack of security to compromise safety of product	Outdoor buildings/silos secured, full assessment by third-party contractor anually
		2.3	Lack of exterior lighting which may allow access for unauthorized people	Individual/group exploits lack of lighting and hide until access can be gained	Adequate lighting in all areas
Employee and visitor programs	People	3.1	People/contractors are not screened prior to employment, people may cause malicious contamination to the site/products/other people or have reasons for extortion	Individual/group exploits lack of restrictions to gain access and compromise safety of product	People/contractors are screened prior to employment
		3.2	Inadequate storage for personnel items on-site which may allow malicious contamination of products/ equipment/harm to people	Disgruntled employee/ contractor exploits lack of storage for personal items to compromise safety of product	Adequate storage on-site in locker rooms for storage of personal items
Material receiving	Receipt	4.1	No documented specification for raw materials and packaging received on-site and what should be done if the materials/packaging are damaged or rejected	Individual/group exploits lack of documented specifications and lack of procedures to maliciously contaminate product through materials/ packaging	Specifications on-site for raw materials/packaging. PKG.SOP-003 V1.0

Risk assessment scoring																	
Likelihood of threat happening/detection					**Impact of threat**					**Overall score**	**Threat decision tree**						
Almost certain (5)	**Likely (4)**	**Possible (3)**	**Unlikely (2)**	**Rare (1)**	**Severe (5)**	**Major (4)**	**Moderate (3)**	**Minor (2)**	**Trivial (1)**	**If ≥6, proceed to decision tree**	**Q1**	**Q2**	**Q2a**	**Q3**	**Q4**	**Q5**	**PRP/ CCP/ VTP**
			2			4				8	Y						N/A
				1		4				4							N/A
				1		4				4							
				1		4				4	Y						N/A
				1			3			3	Y						PRP
			2			4				8	Y						N/A
			2			4				8	N	N	Y				PRP

continued

WS F Site Threat Identification *(Continued)*

Site name: Joe Bloggs LLC		Date: March 17, 2019		Completed by: G Moran	
Criteria/process step	**Further breakdown**	**Risk number**	**Threat**	**Vulnerability (motivation/ opportunity)**	**Preventive actions/ controls currently in place**
Material receiving *(continued)*	Receipt *(continued)*	4.2	Delivery driver not identified and access not restricted	Individual/group exploits lack of restrictions in order to compromise safety of the product	Delivery driver identified and access is restricted
		4.3	Contaminated raw materials received	Deliberately contaminated raw materials compromise safety of the product	Critical raw materials tested for key parameters. GQMS 0810.000.1004
Facility operations	Facilities	5.1	Unsecured water supply, storage tanks, back flow/ filters)	Individual/group exploits lack of security in order to compromise safety of the product	Water supply, storage tanks, back flow/filters secured
		5.2	Water contaminated and not monitored	Individual/group exploits lack of monitoring in order to compromise safety of the product	Water quality is monitored for relevant parameters
		5.3			
	Storage	5.4	Unauthorized access to bulk ingredients, chemical storage, hatches, voids, ceilings, vents, etc.	Individual/group exploits lack of access control in order to compromise safety of the product	Access control by keypad/ swipe card in place in all areas, full assessment required to confirm
		5.5	Pesticides (bait boxes) used on-site not restricted or controlled	Individual/group exploits lack of control in order to compromise safety of the product	Toxic baits used only in external (nonproduction, nonstorage) areas and are secured in place and checked regularly by pest control service provider
		5.6	Open raw materials not challenged	Individual/group exploits lack of control in order to compromise safety of the product through raw material tampering	Open raw materials not accepted
	Preparation	5.10	Unauthorized access to bulk ingredients, chemical storage, hatches, voids, ceilings, vents, etc.	Individual/group exploits lack of access control in order to compromise safety of the product	Access control by keypad/ swipe card in place in all areas, full assessment required to confirm
		5.11	Pesticides (bait boxes) used on-site not restricted or controlled	Individual/group exploits lack of control in order to compromise safety of the product	Toxic baits used only in external (nonproduction, nonstorage) areas and are secured in place and checked regularly by pest control service provider

Risk assessment scoring																	
Likelihood of threat happening/detection					Impact of threat					Overall score	Threat decision tree						
Almost certain (5)	Likely (4)	Possi-ble (3)	Unlikely (2)	Rare (1)	Severe (5)	Major (4)	Moder-ate (3)	Minor (2)	Triv-ial (1)	If ≥6, proceed to deci-sion tree	Q1	Q2	Q2a	Q3	Q4	Q5	PRP/ CCP/ VTP
				1		4				4							N/A
		3					3			9	N	N	Y				PRP
		3					3			9	N	N	Y				PRP
				1			3			3							PRP
		3				4				12	N	N	Y				N/A
				1				2		2							PRP
				1			3			3							PRP
		3				4				12	N	N	Y				N/A
				1				2		2							PRP

continued

CHAPTER 3

CHAPTER 3

WS F Site Threat Identification *(Continued)*

Site name: Joe Bloggs LLC		Date: March 17, 2019		Completed by: G Moran	
Criteria/process step	**Further breakdown**	**Risk number**	**Threat**	**Vulnerability (motivation/ opportunity)**	**Preventive actions/ controls currently in place**
Facility operations *(continued)*	Process Step	5.12	Contamination at point of manufacture (e.g., allergens, chemical)	Individual exploits lack of control at point of manufacture in order to compromise safety of the product	System interlocked to shut down/alarm if opened during production, proper supervision of staff in critical areas
	Packaging	5.14	Contamination at point of cutting or packing (e.g., allergens, chemical, physical), packaging not tamper resistant	Individual exploits lack of control at point of manufacture in order to compromise safety of the product	Tamper-evident packaging, enclosed filling systems in packaging lines
		5.15	Labels for primary packaging not restricted and obsolete labels available for use	Wrong labels are intentionally placed on finished product in order to damage the reputation of the company/ compromise safety of the product	Labels printed via SAP system and cannot be done retrospectively. Morning hygiene walks inspect line clearance of labels.
	Equipment	5.17	Contamination from equipment	Equipment is deliberately contaminated by and individual/group in order to compromise safety of the product	System interlocked to shut down/alarm if opened during production, proper supervision of staff in critical areas, approved suppliers used for purchase of equipment, CIP prerequisite program
Finished goods storage/shipping	Goods out	6.1	Sabotage at point of loading or during transit (e.g., vehicle hijack or damage)	Finished product safety is deliberately compromised at point of loading or during transit by individual/group	Vehicles loaded in secure areas within the perimeter, drivers identified
		6.2	Goods out procedures not followed and stock not accounted for	Individual exploits lack of control at goods out in order to compromise safety of the product or introduce counterfeit product	Stock control procedures in place, SAP system in place does not allow for unaccounted stock movement or usage
	Transport	6.3	Trucks not secured and product contaminated	Individual exploits lack of control in order to compromise safety of the product or introduce counterfeit product	Trucks are secured at loading and during transit
	Finished product storage	6.4	Finished product is not secured/no tamper-evident packaging used	Individual exploits lack of control in order to compromise safety of the product or introduce counterfeit product	Stores are access-controlled and tamper-evident packaging used

Note: CCP = critical control point; PRP = prerequisite program; VTP = vulnerable threat point.

Risk assessment scoring																	
Likelihood of threat happening/detection					Impact of threat					Overall score	Threat decision tree						
Almost certain (5)	Likely (4)	Possi-ble (3)	Unlikely (2)	Rare (1)	Severe (5)	Major (4)	Moder-ate (3)	Minor (2)	Triv-ial (1)	If ≥6, proceed to deci-sion tree	Q1	Q2	Q2a	Q3	Q4	Q5	PRP/ CCP/ VTP
				1		4				5							N/A
				1		4				4							N/A
	4						3			12	N	N	Y				PRP
				1			3			3							N/A
				1			3			3							N/A
			2			4				8	N	N	Y				N/A
				1		4				4							N/A
				1		4				4							N/A

CHAPTER 3

WS G Risk Register

			Mitigation			Corrective Action		
Criteria/ process step	Threat	Rationale	Short term	Medium term	Long term	Responsibility	Update	Date
Raw material receiving	Individual/group exploits gaps in management of operational risks	Financial self-interests	Operational risk management in place	Enhance detection	Appoint a full-time risk manager	C Flack	Implement risk management SOP	May 3, 2019

WS H Threat Decision Tree

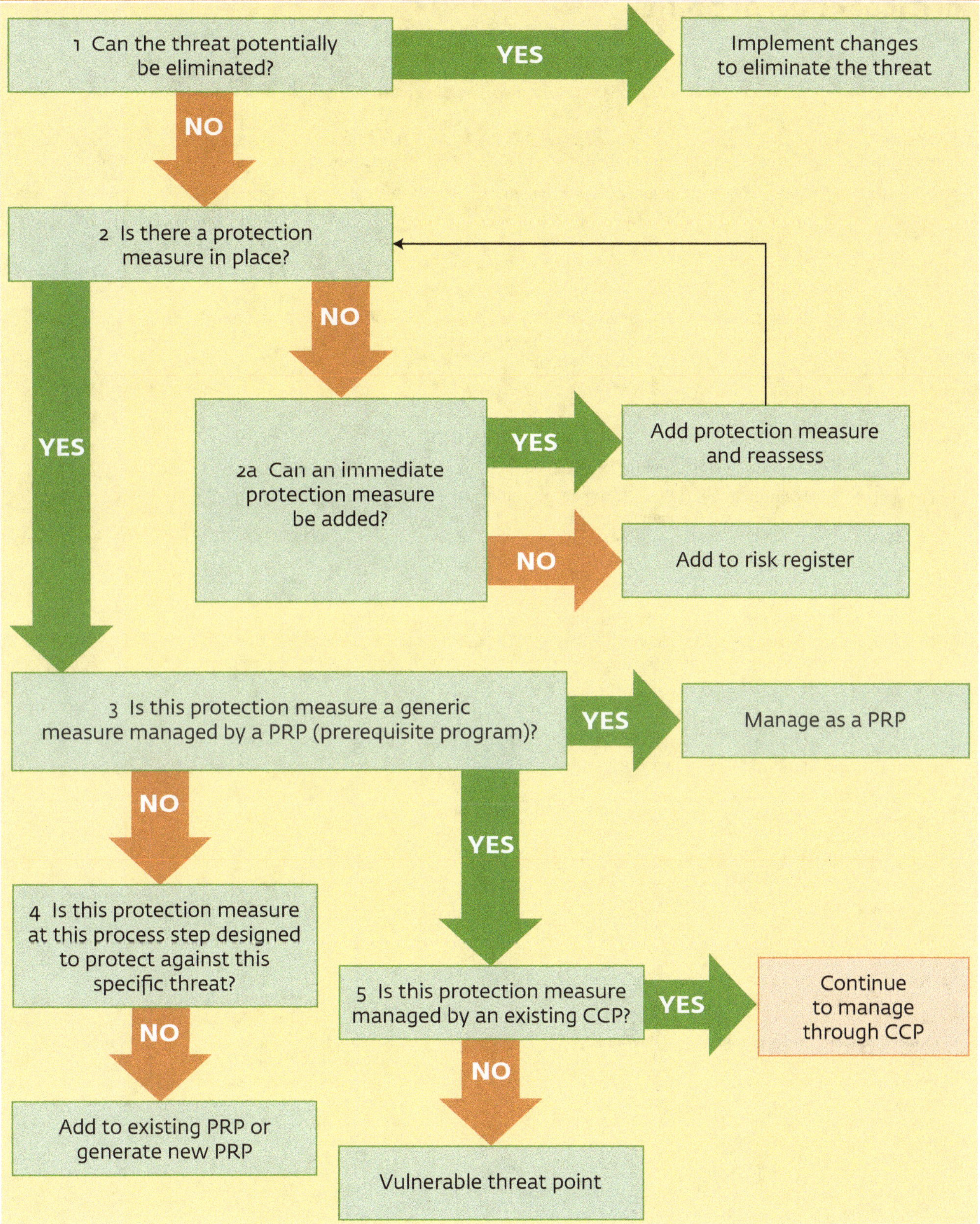

Annex 3B. Guidelines: Vulnerability assessment and critical control point

Table 3B.1 Guidelines and Definitions: The VACCP System					
Product/ product group	Process step	Vulnerability	Likelihood of occurrence	Impact of threat	Priority score
Whole milk	Milk Intake	Adulteration of raw milk to enhance volume	Possible	Trivial	3
	Purchase of lactose	Adulteration of lactose (see lactose vulnerability assessment) by addition of unapproved enhancers	Possible	Severe	15
Batching	Milk intake	Misbranding of non-Halal/Kosher ingredients as Halal/Kosher	Possible	Trivial	3
All	Maintenance	Substitution of food grade lubricants and/or equipment with cheaper alternatives	Unlikely	Major	8
	Cleaning	Substitution of approved cleaning agents with cheaper alternatives	Unlikely	Major	8
	Picking product for sale	Sale of flush/downgrade material as good product	Unlikely	Severe	10
Enter document ID					
Rev 0					
3-17-2019					
Instructions					
Product/product group Defines the product or product group under the scope of the VACCP study	**Process step** Defines and documents the relevant production steps concerning the product and/or a group of similar products	**Vulnerability** Outlines the vulnerable elements of the supply chain by the FBO TACCP/ VACCP team	**Likelihood of occurrence** Outlines the likelihood of the threat happening	**Impact of threat** Outlines the consequences of the threat happening	**Priority score** Outlines the risk profile number (RPN) assigned to the individual threat on the FBO based upon the food fraud and vulnerability SOP risk assessment scoring system

CHAPTER 3

Motivation/opportunity	Control measures	Mitigation strategy (applicable to scores of <10)	Verification
There is an economic motivation for the farmer to enhance the volume of milk by dilution. Opportunity exists at farm level to add the diluent before collection.	Add water test, document ID, Test. SOP.001 V1.2	Not applicable	Not applicable
Increasing perceived protein content with an approved enhancer to increase value is an economic motivation (horizon scanning examples: melamine, nondairy proteins)	Incoming material procedure, document ID, SOP-001 V1.1	FBO supplier management and verification standard, document ID, Stf-001, V1.4	Auditing of suppliers/vendors where risk is high, testing of finished product
Economic motivation is low. No food safety risk presented.	Certification maintained for all certified suppliers	Not applicable	Not applicable
There is an economic motivation to purchase cheaper maintenance equipment. Few people carry out purchasing so there is little opportunity to accomplish this without detection.	All lubricants and product-contact equipment is certified as food grade. If lubricants, a register is maintained.	Not applicable	Not applicable
There is an economic motivation to purchase cheaper cleaning materials. Few people carry out purchasing so there is little opportunity to accomplish this without detection.	A register of cleaning materials is maintained. Cleaning agents approved for use on product contact surfaces must be registered with the Department of Agriculture and are audited regularly.	Not applicable	Not applicable
Economic motivation exists to fraudulently market downgrade as good grade A product, however, opportunity is limited due to control offered by SAP System	All good grade A product is positively released to the customer. Flush and downgrade product is blocked on SAP system, preventing its movement. Flush and downgrade materials are disposed of by approved disposal company.	Not applicable	Not applicable

Instructions

Motivation/opportunity	**Control measures**	**Mitigation strategy**	**Verification**
Outlines the motivation/opportunity for the attacker if the threat is fully realized	Outlines the current controls within the FBO aimed at preventing the vulnerability from occurring	Outlines the mitigation approach to be taken by the FBO in the event of detection of an attack/incident	Outlines the action to verify effectiveness of the mitigation strategy actions, where relevant

Notes

1. See "ISO 22000:2018(en), Food Safety Management Systems: Requirements for Any Organization in the Food Chain," International Organization for Standardization, Geneva, https://www.iso.org/obp/ui/#iso:std:iso:22000:ed-2:v1:en. For any other ISO standards mentioned in this chapter, see OBP (Online Browsing Platform) (database), International Organization for Standardization, Geneva, https://www.iso.org/obp/ui/#home.
2. For any CAC guidelines mentioned in this chapter, see Guidelines (database), Codex Alimentarius, International Food Standards, Codex Alimentarius Commission Secretariat, Food and Agriculture Organization of the United Nations, Rome, http://www.fao.org/fao-who-codexalimentarius/codex-texts/guidelines/en/.
3. For any Codex Alimentarius standards mentioned in this chapter, see Standards (database), Codex Alimentarius, International Food Standards, Codex Alimentarius Commission Secretariat, Food and Agriculture Organization of the United Nations, Rome, http://www.fao.org/fao-who-codexalimentarius/codex-texts/list-standards/en/.
4. See "Codex Alimentarius: International Food Standards," Joint Food and Agriculture Organization of the United Nations–World Health Organization Food Standards Programme, Rome, http://www.fao.org/fao-who-codexalimentarius/en/.

References

CAC (Codex Alimentarius Commission). 2003. "General Principles of Food Hygiene." CAC/RCP 1-1969, Rev. 4-2003, CAC Secretariat, Food and Agriculture Organization of the United Nations, Rome.

Ministry of Health, Russian Federation. 2009. "Hygienic Requirements for the Quality and Safety of Food Raw Materials and Food Products." SanPin 63 (June 9), Sanitary Norms, Rules, and Hygienic Standards, Ministry of Health, Moscow.

Pennington, Thomas Hugh. 1997. "The Pennington Group: Report on the Circumstances Leading to the 1996 Outbreak of Infection with E.coli 0157 in Central Scotland, the Implications for Food Safety and the Lessons to be Learned." April 8, Stationery Office, Edinburgh.

WHO (World Health Organization). 1999. "Strategies for Implementing HACCP in Small and/or Less Developed Businesses." Report WHO/SDE/PHE/FOS/99.7, WHO, Geneva.

CHAPTER 4

FSMS Procedures and Documentation

Documentation overview

As noted in chapter 3, seventh and final principle of a hazard analysis critical control point (HACCP) system is to establish effective recordkeeping procedures to document the food safety management system (FSMS). Maintaining complete and accurate records is essential to ensuring the effective monitoring of an FSMS and demonstrating compliance with food safety requirements.

The structure of the documentation used in the FSMS may be described as a hierarchy. In International Organization for Standardization (ISO) 22000:2018, it is labeled "documented information."[1] This term was introduced as part of the typical high-level structure and common terms across management system standards. The definition of documented information can be found in ISO 22000:2018, clause 3.13, as follows: the "information required to be controlled and maintained by an organization and the medium on which it is contained." Documented information can be used to communicate a message, provide evidence of what was planned and what has actually been done, or share knowledge.

All the documented information that forms part of an FSMS must be controlled in accordance with ISO 9001:2015, clause 7.5. According to ISO 9001:2015, a document is information (meaningful data) and the medium on which it is contained, whereas document information is information (meaningful data) that is required to be controlled and maintained by an organization. HACCP and FSMS documents may be in any form or type of medium, such as paper, magnetic, electronic, or optical computer disc, photograph, or master sample.

According to ISO 9001:2015, clause 7.5.1, quality management system documentation shall include the documented information required by the international standard and determined by the organization as necessary for the effectiveness of the FSMS. The note at the end of the clause makes clear that the extent of FSMS-documented information can differ across organizations because of (1) the size of the organization and the type of activities, processes, products, and services it carries out or realizes; (2) the complexity of these processes and the interactions that result; and (3) the competence of the individuals involved. This facilitates the distribution, maintenance, and understanding of the documentation.

Figure 4.1 illustrates a typical hierarchy of HACCP or FSMS documentation. The development of the hierarchy depends on the circumstances of the organization.

Figure 4.1 Hierarchy of Food Safety Management System Documentation

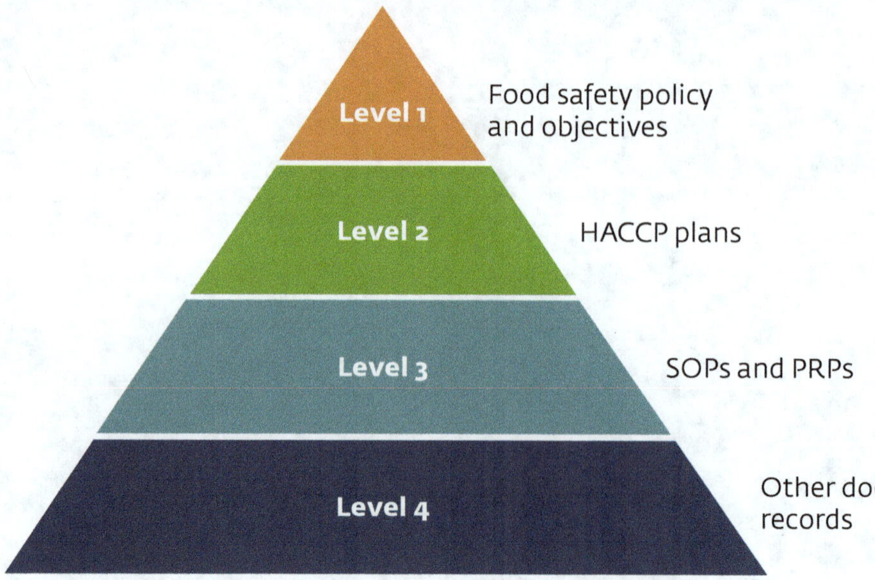

Note: PRP = prerequisite program; SOP = standard operating procedure.

Purpose and benefits

Among the organizational purposes and benefits of FSMS documentation are the following: (1) describing the FSMS of the organization; (2) affording information for cross-functional groups so that they may better understand interrelationships; (3) communicating management's commitment to food safety to employees; (4) helping employees understand their role within the organization, thus offering them an increased sense of purpose and of the importance of their work; (5) supporting mutual understanding between employees and management; (6) furnishing the groundwork for expectations around work performance; (7) stating how tasks should be carried out to achieve specified requirements; (8) supplying objective evidence that specified requirements have been achieved; (9) presenting a clear, efficient framework for operations; (10) establishing a platform for training new employees and the periodic retraining of current employees; (11) producing a footing for order and balance within the organization; (12) contributing to consistency in operations based on documented processes; (13) shaping an understructure for continual improvement; (14) maintaining customer confidence through documented systems; (15) demonstrating the capabilities of the organization to interested parties; (16) delivering a clear set of requirements for suppliers; (17) creating the authority for auditing the FSMS; and (18) building a foundation for evaluating the effectiveness and ongoing suitability of the FSMS.

Food safety policy and the associated objectives

The food safety policy and the associated objectives should be documented. The documentation may be independent or may be included in the FSMS. The food safety policy should contain the relevant defined requirements specified by the food safety scheme of the Global Food Safety Initiative (GFSI).

Food safety objectives should be SMART: specific, measurable, attainable (or achievable), realistic, and time-bound. They should be consistent with the food safety policy and with the primary aim of the GFSI food safety scheme, that is, to eliminate or reduce relevant food safety hazards.

Documented procedures

Structure and format

Organizations may select the form of the documentation on standard operating procedures (SOPs) that most closely fits their requirements. The structure and format of the documented SOPs, in hard copy or electronic media, might thus include text, flowcharts, tables, a combination of these, or any other suitable contents in accordance with the needs of the organization. The documented SOPs should contain all necessary information and be labeled according to a unique identification system. They may make reference to work instructions that define how an activity is performed. SOPs generally describe activities across various functions, while work instructions typically apply to tasks within a single function.

Sample templates of many possible documented SOPs follow in this chapter. They should be considered illustrative. They are each identified by a unique SOP number and are ordered in the sequence of these SOP numbers. There are gaps in the numbering, however. This is because procedures judged to be less relevant for the purposes of this handbook have not been included. The first procedure, Control of Documents (SOP-001), may be considered a model. The next subsections offer additional guidance.

Contents

The following structure is suggested for a documented procedure based on ISO 10013. However, organizations are free to select a document structure that meets their needs.

TITLE

The title should clearly identify the documented procedure.

PURPOSE

The purpose of the documented procedure should be defined.

SCOPE

The scope of the documented procedure, including the areas to be covered and areas not to be covered, should be described.

RESPONSIBILITY AND AUTHORITY

The responsibility and authority of individuals or organizational units, as well as the interrelationships between these individuals or units and the processes and activities described in the procedure, should be identified. These interrelationships should be described in the procedure through flowcharts and descriptive text as appropriate for clarity.

DESCRIPTION OF ACTIVITIES

The level of detail may vary depending on the complexity of the activities, the methods used, and the skills and training necessary for the activities to be accomplished. Irrespective of the level of detail, the description of activities should cover the following: (1) the needs of the organization and of the organization's customers and suppliers; (2) descriptions of the procedures through text and flowcharts on the required activities; (3) what is to be done, by whom or by which organizational unit, why, when, where, and how; (4) the process controls and controls on the identified activities; (5) the resources in personnel, training, equipment, and materials necessary for the accomplishment of the activities; (6) the appropriate documentation on the required activities; (7) the process inputs and outputs; and (8) the indicators and measurements to be developed. The organization may decide that work instructions are more appropriate in conveying some of the above information.

RECORDS

The records on the activities of the documented procedure should be described in this section of the documented procedure or in a related section. The forms to be filled out as part of the recordkeeping should be identified. The methods required to complete, file, and maintain the records should be outlined.

APPENDIXES

Appendixes containing supporting information, such as tables, graphs, flowcharts, and copies of forms on the documented procedure, may be included.

REVIEW, APPROVAL, AND REVISIONS

Evidence about reviews, approvals, status, and date of revision of the documented procedure should be indicated.

IDENTIFICATION OF CHANGES

Descriptions of any changes in the procedure should be identified in the document or in the appropriate attachments.

Work instructions

Structure and format

Work instructions should be developed and maintained to describe the steps in performing any work that might be adversely affected if such instructions were not included. The work instructions should include a title and a unique identification label. The structure, format, and level of detail of the work instructions should be tailored to the needs of the organization's personnel. They depend on the complexity of the work, the methods used, the training undertaken, and the skills and qualifications of the personnel. The structure of the work instructions may vary from that of documented procedures. The work instructions may be included in the documented procedures or referenced in them.

Contents

Work instructions should describe critical activities. Details that do not support more control over the activities should be avoided. Training can reduce the need for detailed instructions, provided the individuals obtain the information they need to do their jobs.

TYPES OF WORK INSTRUCTIONS

In general, although work instructions have no required structure or format, they should convey the purpose, scope, and objectives of the work with reference to the pertinent documented procedures. Whichever format or combination is chosen, the work instructions should follow the order or sequence of the operations and accurately reflect the requirements of the relevant activities. To reduce confusion and uncertainty, a consistent format or structure should be established and maintained.

REVIEW, APPROVAL, AND REVISIONS

The organization should provide clear evidence on reviews and approvals of the work instructions, any revisions, and the dates of revisions.

RECORDS

Where applicable, the records specified in the work instructions should be defined in this section or in related sections. The minimum food safety records required are identified in the relevant GFSI food scheme. The methods required to complete, file, and keep the records should be described. The forms to be used for these records should be identified.

IDENTIFICATION OF CHANGES

The nature of any changes should be identified in the document or in appropriate attachments.

Hazard control plan

A hazard control plan or HACCP plan is a set of written procedures that help minimize the potential of causing an adverse health effect. The contents of the hazard control plan are defined by the Codex Alimentarius Commission (CAC) (CAC 2003; see chapter 3).

Specifications

Raw materials, ingredients, and product-contact materials

All raw materials, ingredients, and materials that come into contact with food and food products, including packaging, are to be described in documents to the extent necessary to conduct the hazard analysis. This covers the following, as appropriate: (1) biological, chemical, and physical characteristics; (2) the composition of formulated ingredients, including additives and processing aids; (3) sources (animal, mineral, or vegetable); (4) place of origin (provenance); (5) method of production; (6) method of packaging and delivery; (7) storage conditions and shelf life; (8) preparation and handling before use or processing; and (9) acceptance criteria related to food safety or the specifications of purchased materials and ingredients appropriate to the intended uses. The organization is to identify and follow statutory and regulatory food safety requirements relevant to the above or, if they are more strict than the statutory and regulatory food safety requirements, the organization's food safety requirements. The descriptions are to be kept up-to-date as required.

Finished product specification

The characteristics of end products are to be described in documents to the extent needed to conduct the hazard analysis, including information on the following, as appropriate: (1) product name or similar identification; (2) composition; (3) biological, chemical, and physical characteristics relevant to food safety; (4) intended shelf life and storage conditions; (5) packaging; (6) labeling related to food safety and instructions for handling, preparation, and intended use; and (7) methods of distribution and delivery. The organization is to identify statutory and regulatory food safety requirements related to these characteristics. The descriptions are to be kept up-to-date as required.

Forms

Forms are developed and maintained to record data demonstrating compliance with the requirements of the FSMS. The forms should include a title, an identification number, and information on revisions, including dates of revision. Forms should be referenced in or attached to the quality manual, documented procedures, and work instructions.

Records

FSMS records supply information on results achieved or provide evidence that the activities described in the documented procedures and work instructions have been performed. (Records are not generally the subject of revision control because they typically do not change.) Records should demonstrate compliance with the requirements of the FSMS and the specified requirements on food safety. The responsibility for the preparation of records should be addressed in the FSMS documentation. Records furnish the only reference information available to trace the production history of a finished product. They can be used as a tool to alert the food business operator (FBO) to potential problems before these lead to the violation of a critical limit. They can serve as evidence that proper procedures are being followed.

The approval, issue, and control of FSMS documents

Review and approval

Prior to publication, documents should be reviewed by authorized individuals to ensure clarity, accuracy, adequacy, and proper structure. Intended users should also have the opportunity to assess and comment on the usefulness of the documents and the extent to which documents reflect practice. The release of documents should be approved by the management authorities responsible for the implementation of the documents. Each copy of a document should show evidence of the release authorization. Likewise, evidence of the approval of a document should be retained.

Distribution

The method of distribution of documents by authorized personnel should ensure that pertinent documents are available to all personnel who need the information included in the documents. Proper distribution and control may be supported, for example, through the use of serial numbers on individual copies of the documents. Documents, such as an HACCP manual, that is, a set of documents used to establish and support the development and operation of an HACCP system, may be distributed to external groups and entities, for instance, customers, certification bodies, and regulatory authorities.

The incorporation of changes

A process for the initiation, development, review, control, and incorporation of changes to documents should be provided. The same review and approval process used in developing the original documents should apply in processing changes.

ISSUE AND CHANGE CONTROL

Control over the publication of and changes to documents is essential to ensuring that the contents of documents are properly approved by authorized personnel and that the approval is clearly demonstrated. Various methods may be considered for facilitating the process of making changes. Responsible authorities within organizations should consider maintaining a record of the changes to documents for legal purposes and to preserve knowledge.

A process should be established to ensure that only appropriate documents are being used. Under certain circumstances, the appropriate document may not be the latest revision.

Documents that have been revised should be replaced by the latest versions. A document master list with revision status may be produced to help assure users that they possess the correct versions of authorized documents.

UNCONTROLLED COPIES

For the purpose of tenders, customer off-site usage and other, special distributions of documents on which the control over changes is not intended to be exercised, the documents so distributed should be clearly identified as uncontrolled copies. Failure to provide such identification may lead to the use of obsolete documents.

Record retention

Storing records

Records can be stored as case files, logbooks, data in databases, and so on. FBOs should take reasonable steps to ensure that training records are stored in a secure location and are not available to individuals who are not authorized to have access. FBOs should also adopt a policy on backing up data, access rights, and security. Precautions should be taken to protect soft-copy records from electronic viruses or technical failures, and written records should be protected from damage by fire, water, or even rodents, termites, and other pests.

Privacy protection and access to records

FBOs should develop a policy to maintain the confidentiality of written and electronic records, including sensitive information on trainees and employees. All FBO personnel should be required to abide by the policy. FBOs should seek to balance the individual's right to privacy with the needs of services provision. In providing auditors with access to training records, for instance, FBOs should take steps to protect the privacy of employees and other individuals identified or discussed in the records. Both auditor requests and the reasons for withholding records should be documented in client files. Sensitive and confidential information must be released only to authorized parties with the consent of the individuals identified in the records.

Record maintenance and destruction

FBOs should ensure that their recordkeeping practices comply with all contractual, regulatory, and legal requirements. FBOs should store training records for at least six years where practical; this is a general GFSI scheme requirement. The transfer or disposal of FBO training records should be conducted in a manner that protects employee confidentiality. FBOs should develop an internal policy on the time frame for updating records.

Electronic or hard-copy records?

FBOs could consider all factors involved in maintaining electronic or written records. They should choose the system that meets their needs and more clearly benefits the FBO, employees, and auditors.

ELECTRONIC RECORDS

Maintaining training records as soft copies allows for easy access, transfer, and storage. However, keeping records using an electronic tool, such as the personal digital assistant or a smartphone, while conducting intake assessments of new clients, for instance, may appear impersonal and inappropriate. If documentation and records are stored electronically, the FBO should develop relevant policies and procedures for information management and the use of information technology, including system maintenance, access monitoring, and staff training.

WRITTEN RECORDS

Written records are common and may be more user-friendly among employees and auditors. However, they may sometimes become difficult to read because of variations in handwriting. In addition, duplicate copies have to be made for transmission to additional individuals or agencies.

Record review

It is good practice to review records so that improvements in training design and delivery can be undertaken. Records should be reviewed periodically to establish the following:

- The thoroughness, completeness, and timeliness of assessments
- The active involvement of clients in making informed choices among services
- The ability of the services provided to clients to achieve appropriate client outcomes
- The identification of the need for improvement in client outcomes

Note

1. See "ISO 22000:2018(en), Food Safety Management Systems: Requirements for Any Organization in the Food Chain," International Organization for Standardization, Geneva, https://www.iso.org/obp/ui/#iso:std:iso:22000:ed-2:v1:en. For any other ISO standards mentioned in this chapter, see OBP (Online Browsing Platform) (database), International Organization for Standardization, Geneva, https://www.iso.org/obp/ui/#home.

Reference

CAC (Codex Alimentarius Commission). 2003. "General Principles of Food Hygiene." CAC/RCP 1-1969, Rev. 4-2003, CAC Secretariat, Food and Agriculture Organization of the United Nations, Rome.

Food safety management system procedures: Templates and instructions

CONTROL OF DOCUMENTS	
An FBO Procedure	
Document No.	Standard operating procedure SOP-001
Created	April 20, 2018
Updated	January 13, 2019
Controller	Document Controller
Owner	Food Safety Manager
Confidentiality Statement Information in this document must be kept confidential as per the document's classification below and the rules of disclosure.	
All FBO documents are classified in the following way. PUBLIC documents are intended for anyone. COMMERCIAL IN CONFIDENCE documents are to be kept confidential among restricted individuals within the FBO and partner organizations. COMPANY CONFIDENTIAL documents are to be kept confidential within the FBO and used for normal business activities by the general office population. HIGHLY CONFIDENTIAL documents are to be kept confidential among restricted individuals within the FBO.	
© Copyright FBO. All rights reserved. No part of this publication may be reproduced, stored in a retrieval system, or transmitted in any form or by any means, electronic, mechanical, photocopying, recording, or otherwise, without the written permission of the FBO.	
Classification	Company Confidential

Revision History

Date	Version	Author	Comments (including review history)
April 20, 2018	Draft 01	Joe Bloggs	Initial document for review and discussion
April 24, 2018	V1.0	Joe Bloggs	Approved and released by process owner
January 13, 2019	V1.1	Joe Bloggs	Updated definitions section

Contents

CONTROL OF DOCUMENTS

1 Summary

Purpose	The purpose of this procedure is to describe the following: • The methodology in place to control all documentation relevant to the food safety management system (FSMS)
Scope	This procedure applies to the following: • The creation, review, approval, obsolescence, archiving, disposal/destruction of FSMS documentation • The control of documents of external origin determined to be necessary for the planning and operation of the FSMS • The control of the company portal, website, and marketing materials
Functional responsibility	The functional responsibility for this procedure lies with the food safety manager, who is responsible for the effective implementation and maintenance of this procedure.

2 Related Documents

Policies	Food Safety Policy, POL-001
Processes	Not applicable
Procedures	Control of Records, SOP-002
Work instructions	Not applicable
Forms	Document Request Form Disposal/Archival Request Form
Other	Not applicable

3 Definitions

Term or acronym	Description
DMS	document management system
Document controller	The person responsible for the control of documentation; this is the document controller
Documented information	Information required to be controlled and maintained by the organization and the medium upon which it is contained (clause 3.13 of International Organization for Standardization [ISO] 22000:2018)
Document template	The template used to create documentation
FBO	food business operator
FSMS	food safety management system
HACCP	hazard analysis critical control point, a system that identifies, evaluates, and controls hazards that are significant for food safety
Hazard control plan	A document prepared in accordance with the principles of HACCP to ensure control of hazards that are significant for food safety in the segment of the food chain under consideration
OPRP	operational prerequisite program
PRP	prerequisite program

CHAPTER 4

CONTROL OF DOCUMENTS

4 Introduction

4.1 General

Documentation is used by an organization to ensure communication and consistency of action. The effective use of documentation enables the following:

- Achievement of conformity to customer requirements and quality improvement
- Provision of appropriate training
- Repeatability and traceability
- Provision of objective evidence
- Evaluation of the effectiveness and continuing suitability of the management system

In a food safety management system (FSMS), the following documentation may typically occur:

- Documents that provide consistent information, both internally and externally, about the organization's management system, referred to as *management system manuals*, for example, food safety manual or FSMS manual
- Documents that describe how the FSMS is applied to a specific product, referred to as *prerequisite programs* (PRPs), *operational prerequisite programs* (OPRPs), *hazard control plans*, and so on
- Documents stating requirements, referred to as *specifications*
- Documents stating recommendations or suggestions, referred to as *guidelines*
- Documents that provide information about how to perform activities and processes consistently, referred to as *documented procedures*, *work instructions and drawings*, *forms*, *document templates*, and other documentation
- Documents that provide objective evidence of activities performed or results achieved, referred to as *records*

4.2 Document Control Policy

An electronic document management system (DMS) is implemented to control all documents falling under the scope of the FSMS. This system allows documentation, in electronic format, to be available, accessible, and controlled.

The controlled master documents are held in the DMS. Any printed copies are valid only on the day of printing and are deemed uncontrolled thereafter.

Employees are not permitted to hold any versions of FSMS documentation on their personal hard drives and must review/obtain all copies of required documents from the DMS.

Records are a special type of document and are controlled as per standard operating procedure SOP-002 Control of Records.

CONTROL OF DOCUMENTS

4.3 Content of Documents

As part of the standardization process, all FSMS documentation will follow the same format. In general, all company documentation must accomplish the following:

- Clearly display the company logo in the header
- Identify, in the footer, the number of the current page and the total number of pages
- Show the control number
- Display the document name
- Show the revision number

For procedures and work instructions, the following sections are required:

- Summary, including purpose, scope, and functional responsibility
- Related documents table, including policies, processes, procedures, work instructions, forms, and others
- Definitions table
- Introduction to the procedure
- Procedure flowchart
- Procedure notes
- Records table

Subsections may be added as necessary. The layout of this procedure—the document control procedure—should be used as a model.

The format of the header and footer in this procedure—the document control procedure—must be used and edited appropriately in all other procedures.

4.4 Documents of External Origin

Where deemed necessary for the planning and operation of its processes and activities, the organization may obtain documents from external sources. These documents may be in any medium, for instance, DVD, compact disc, Internet, or a supplier or client portal. They must be controlled if a library is maintained by the food business operator (FBO). Most food safety schemes require access to such documents, but not necessarily the physical or electronic storage of documents of external origin.

Examples of external documents include the following:

- Equipment manuals in hard copy, compact disc, or DVD
- Building blueprints
- Customer specifications
- Other legislative or regulatory requirements
- International standards (for example, ISO 22000:2018, FSSC 22000 V4.1)

CONTROL OF DOCUMENTS

On receipt of or notification of an external document of relevance, the relevant department must inform the document controller so the document can be recorded in and controlled through the DMS.

This control, through the DMS, will extend to the following:

- Assigning a control number if one does not already exist
- Assigning a receipt date, that is, the receipt of the document by the company
- Assigning a revision number if one does not already exist
- Recording the distribution of the document within the company

Documents of external origin requiring a control number and a revision number should take the following format:

EXT xxx yyy Name of Document Revision zzz,

where *EXT* signifies that the document is of external origin; *xxx* identifies the applicable company department, for example, compliance; *yyy* = the next control number available; and *zzz* = the revision control number.

The document controller will be responsible for naming and numbering all documents of external origin. The receipt date will be noted in the DMS as a note to the document.

If a document of external origin of relevance is referenced in the DMS, but not stored, the link to the online location should be recorded and maintained by the document controller.

Any updates to documents of external origin will be reviewed by the relevant department. An assessment of applicability will be carried out, and the appropriate actions will be taken. The newer version of the external document will be controlled as outlined above, and the previous version will be obsoleted.

CONTROL OF DOCUMENTS

5 Procedure Flowchart

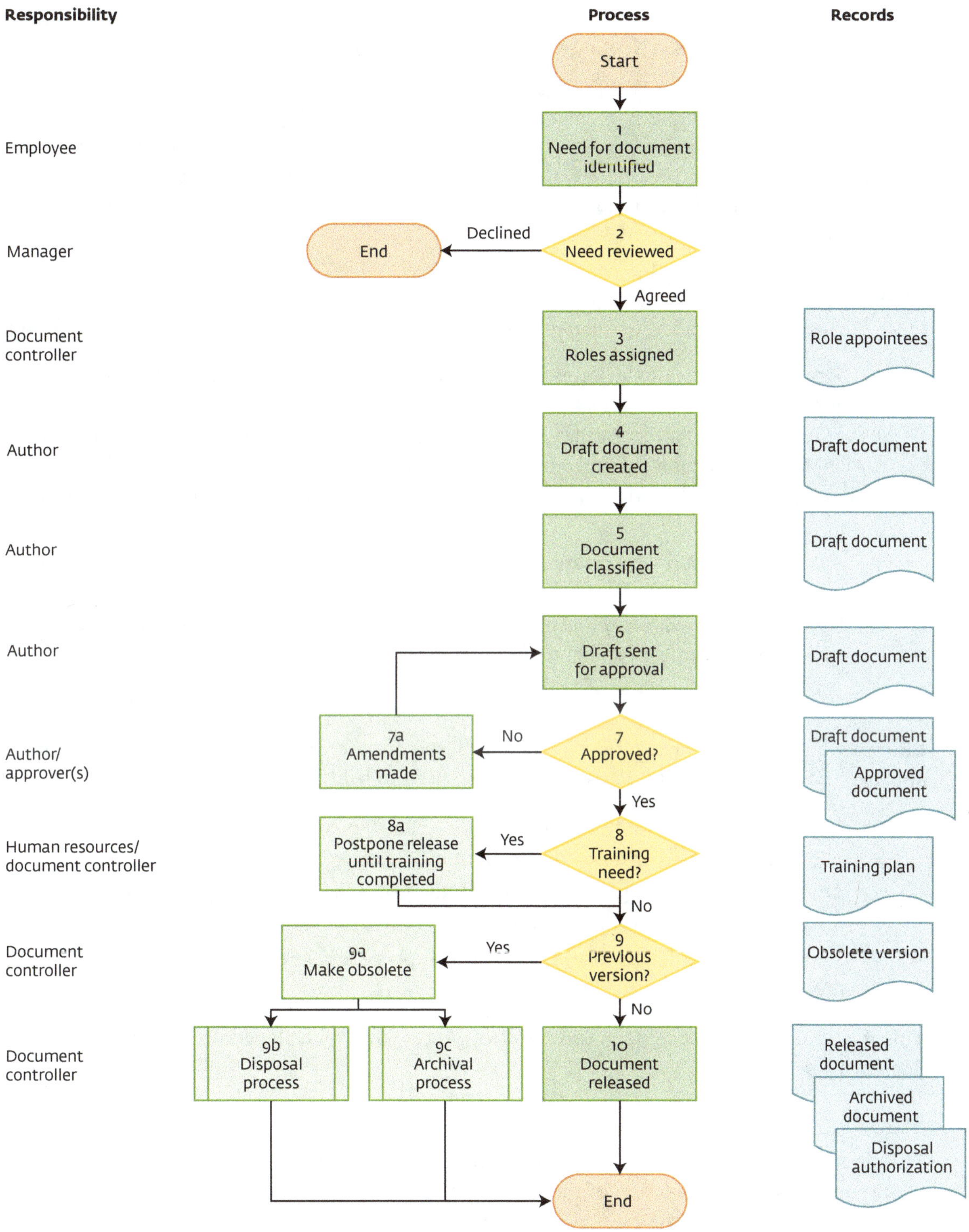

6 Procedure Notes

Step 1

The potential need for a new document can be raised by any employee.

Step 2

The need for a new document must be reviewed by the process owner or the head of the department to ensure that the need represents a valid requirement and that no other existing document already covers the need identified or could be amended to meet the need. If the need is declined, the process ends at this point.

Step 3

Once the need has been accepted, the relevant actors in documentation are appointed, as follows:

- *Sponsor*, the person who determines whether the need is valid; the sponsor may also be the process owner
- *Manager*, the person with the responsibility and authority to undertake the flawless implementation and management of the procedure
- *Approver(s)*, those with review and approval responsibility and authority related to the document
- *Author*, the person who creates/writes the document utilizing the approved document template
- *User,* the person with responsibility to ensure conformity with the procedure and to advise on any changes, if required

Step 4

Either within the DMS or external to the DMS, the author will create the document utilizing the approved document template. The following also need to be defined at this stage:

- The effective date of the procedure
- The review period, for instance, 12 months or sooner
- Any verification (testing) associated with the procedure, for example, a quiz
- Identifying the relevant interested parties
- Identifying other documents affected by this procedure and notifying the relevant process owners

This is also the point in the document control process at which the control of changes to existing documents begins.

Step 5

The author, in association with the sponsor and owner, will classify the document in accordance with the proposed use and circulation of the document. Classifications include the following:

- *Public* documents are intended for anyone.
- *Commercial in confidence* documents are to be kept confidential between restricted individuals within the FBO and partner organizations.

CONTROL OF DOCUMENTS

- *Company confidential* documents are to be kept confidential within the FBO and used for normal business activities by the general population.
- *Highly confidential* documents are to be kept confidential among restricted individuals within the FBO.

Steps 6 and 7

Once authors are satisfied with the level of detail in the procedure and so on, they will verify that the procedure matches current operational practices and relevant statutory, regulatory, and conformity requirements prior to submitting the document for approval. This can be achieved through discussions with relevant departments.

The document is then sent for approval through the DMS or manually to each of the specified approvers. All approvers are required to read and evaluate the document and specify their approval or disapproval of the contents. If approved, the document moves to the next stage of the process. However, if one or more approvers reject the document, it returns to the author for the appropriate amendments to be made. It will then be resubmitted for approval.

This approvals loop may pass through several iterations.

Steps 8 and 8a

Once the document is approved, the impact of the document will be assessed, and any training needs identified. The effective date of this procedure will be postponed until any required training has been successfully completed. The DMS has the ability to test persons on their understanding of a process or procedure through a quiz.

Step 9

The DMS will automatically remove obsoleted versions of a document.

Steps 9a, 9b, and 9c

The DMS will automatically archive/dispose of obsoleted documents.

Step 10

The approved document is released on the DMS, and the relevant personnel are informed of the release.

7 The Document Management System

7.1 General

A DMS has been implemented within the company to ensure the necessary control of all documentation that falls under the scope of the FSMS. This DMS covers and provides evidence of the control of documentation in line with the flowchart outlined in section 5 of this procedure and the notes outlined in section 6 of this procedure.

CHAPTER 4

CONTROL OF DOCUMENTS

7.2 Access Rights

Access rights to the DMS have been assigned as follows:

- Full access:
  - The food safety manager
  - The document controller
- Edit/amendments:
  - Document owners
  - Document approvers
  - Document authors
- Read only:
  - Authorized employees

Only the document controller and the food safety manager may release a document in the DMS, subject to the completion of a successful approval process.

7.3 Document Review

At placement of a document within the DMS structure, the definition of a review time frame is required. The document controller will monitor and ensure that the time frame is respected.

7.4 Obsolete Documents

The DMS will automatically remove obsolete documents from view. If obsolete, hard-copy documents held for legal, knowledge retention, or other purposes will be clearly marked as obsolete to prevent unintended use. Obsolete documentation held on the DMS may be accessed only by the document controller and the food safety manager.

7.5 Document Disposal

Authorization through the DMS must be granted before the disposal of a document. Documents may be disposed of through deletion from the DMS or the shredding of physical documents.

7.6 Document Archiving

The archiving of documents is managed automatically within the DMS.

CONTROL OF DOCUMENTS

7.7 Document Numbering

All documents within the scope of the management system shall follow the naming structure outlined below:

Document type	Numbering structure	Example
Policy	POL. xxx yyy Name Revision number	POL FSMS 001 Food Safety Policy Rev 01
Procedure	PRO xxx yyy Name Revision number	PRO QMS 001 Document Control Rev 01
Work instruction	WI xxx yyy Name Revision number	WI QMS 001 Writing a Job Description Rev 01
Specification	SPEC xxx yyy Name Revision number	SPEC QMS 001 Specification Rev 01
Form/document template	FRM xxx yyy Name Revision number	FRM QMS 001 Master Document Register

Where:

xxx = department identification and *yyy* = document number.

Document numbers will be assigned by the document controller based on the documentation master list. Only the document controller is authorized to change the naming structure.

External document naming criteria are outlined in section 4.4 of this procedure.

8 Records

Document	Location	Duration of record	Responsibility
Documentation master list	DMS	Indefinitely	Document controller
Documentation review report	DMS	Indefinitely	Document controller
Disposal/archival request form	DMS	Indefinitely	Document controller

CONTROL OF RECORDS	
An FBO Procedure	
Document No.	Standard operating procedure SOP-002
Created	April 20, 2018
Updated	January 13, 2019
Controller	Document Controller
Owner	Food Safety Manager
Confidentiality Statement	
Information in this document must be kept confidential as per the document's classification below and the rules of disclosure.	
All FBO documents are classified in the following way. PUBLIC documents are intended for anyone. COMMERCIAL IN CONFIDENCE documents are to be kept confidential among restricted individuals within the FBO and partner organizations. COMPANY CONFIDENTIAL documents are to be kept confidential within the FBO and used for normal business activities by the general office population. HIGHLY CONFIDENTIAL documents are to be kept confidential among restricted individuals within the FBO.	
© Copyright FBO. All rights reserved. No part of this publication may be reproduced, stored in a retrieval system, or transmitted in any form or by any means, electronic, mechanical, photocopying, recording, or otherwise, without the written permission of the FBO.	
Classification	Company Confidential

Revision History

Date	Version	Author	Comments (including review history)
April 20, 2018	Draft 01	Joe Bloggs	Initial document for review and discussion
April 24, 2018	V1.0	Joe Bloggs	Approved for release by process owner
January 13, 2019	V1.1	Joe Bloggs	Annual review, no changes

Contents

CONTROL OF RECORDS

1 Summary

Purpose	The purpose of this procedure is to describe the methodology used to control records developed as part of the food safety management system (FSMS).
Scope	This procedure applies to the distribution, storage, preservation, legibility, retention, disposition, and access to and retrieval of records.
Functional responsibility	The functional responsibility for this procedure lies with the food safety manager, who is responsible for the effective implementation and maintenance of this procedure. Departmental managers are responsible for ensuring that the records under their control are managed in accordance with this documented procedure.

2 Related Documents

Policies	Food Safety Policy, POL-001
Procedures	Control of Documents, SOP-001
Work instructions	Not applicable
Forms	Master Document Register
Other	Document Management System (DMS)

3 Definitions

Term or acronym	Description
DMS	document management system
FBO	food business operator
FSMS	food safety management system

4 Introduction

Records are documents stating the results achieved or providing evidence of the activities performed. Records can be in either hard copy or soft copy (respectively, for example, paper or electronic) and must be managed. The management of records is a critical factor in a food safety management system (FSMS) because, without the availability of records, the company is unable to verify that required activities have taken place or that results have been achieved.

CONTROL OF RECORDS

5 Procedure Flowchart

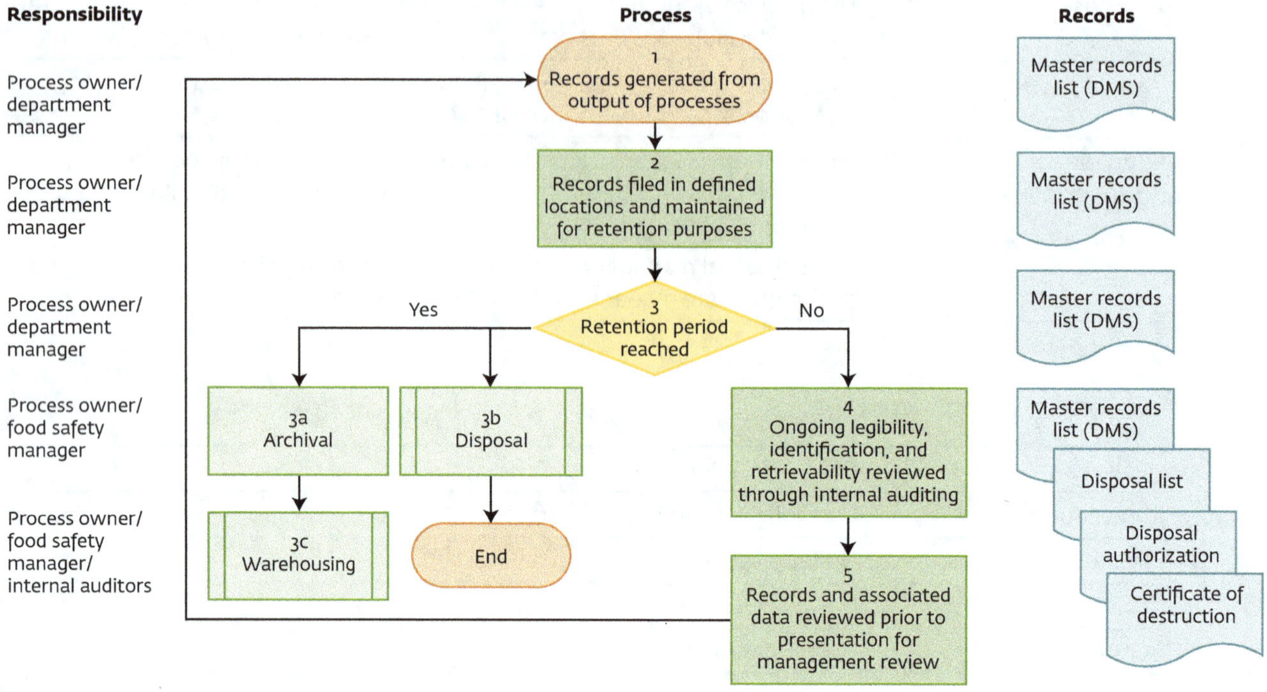

6 Procedure Notes

Steps 1 and 2

Through daily activities, food safety records are generated that provide evidence of the completion of activities and the achievement of results. These records are held in accordance with defined retention times and to ensure the preservation of the contents, their identification, and their legibility. Retention times vary across regions based on local law. The food business operator (FBO) should include an annex in this procedure showing the retention times per controlled record. The storage of records should ensure that the original records are maintained despite subsequent amendments. These requirements are listed on the records master list.

Step 3

Once the retention period relevant to the records has been reached, a decision must be made about what is going to happen to the records. This decision is made by the process owner and the food safety manager. The food safety team will be queried to determine if any compliance issues related to the specific records exist and need to be met.

Steps 3a and 3b

If a decision to archive the records is made, these records must be suitably boxed to preserve their integrity. The contents of the box must be labeled clearly (date, type of record, origin of record, and so on). An e-mail is then sent by the process owner or the process owner's delegate to the warehouse informing the warehouse team to expect the delivery of the box. The process owner or the process owner's delegate will arrange for the delivery of the box to the warehouse.

CONTROL OF RECORDS

Step 3c

If the decision is to dispose of the records, the process owner and the food safety manager must authorize this disposal. A list of all documents to be disposed of must be created and signed off by the above to signify their approval to dispose of the records. It is the responsibility of the process owner to create the disposal list and obtain the necessary approvals for disposal. If such records are held in the warehouse, a written instruction must be sent, following authorization, to the warehouse instructing the warehouse team to dispose of the records. Records must be disposed of in a fully traceable and confidential manner using an approved disposal company. Shredding is the preferred method for the disposal of records. It is the responsibility of the food safety manager to obtain a certificate of destruction from the disposal company. The certificate of destruction must be attached to the disposal list and maintained by the food safety manager.

Step 4

If the retention times are not reached, the control of records is monitored as part of the internal auditing process, that is, FSMS internal auditing.

Step 5

The management and control of records will be reviewed as part of the management review process, under the agenda heading of documentation.

7 Records

Document	Location	Duration of record	Responsibility
Records master list	Food Safety Office	Indefinitely	Food safety manager
Disposal list	Food Safety Office	Three years	Food safety manager
Disposal authorization	Food Safety Office	Three years	Food safety manager
Certificate of destruction	Food Safety Office	Three years	Food safety manager
Warehouse storage location list	Warehouse manager	Indefinitely	Warehouse manager

CONTROL OF NONCONFORMING PRODUCT	
An FBO Procedure	
Document No.	Standard operating procedure SOP-003
Created	April 20, 2018
Updated	January 13, 2019
Controller	Document Controller
Owner	Food Safety Manager
Confidentiality Statement	
Information in this document must be kept confidential as per the document's classification below and the rules of disclosure.	
All FBO documents are classified in the following way. PUBLIC documents are intended for anyone. COMMERCIAL IN CONFIDENCE documents are to be kept confidential among restricted individuals within the FBO and partner organizations. COMPANY CONFIDENTIAL documents are to be kept confidential within the FBO and used for normal business activities by the general office population. HIGHLY CONFIDENTIAL documents are to be kept confidential among restricted individuals within the FBO.	
© Copyright FBO. All rights reserved. No part of this publication may be reproduced, stored in a retrieval system, or transmitted in any form or by any means, electronic, mechanical, photocopying, recording, or otherwise, without the written permission of the FBO.	
Classification	Company Confidential

Revision History

Date	Version	Author	Comments (including review history)
April 20, 2018	Draft 01	Joe Bloggs	Initial document for review and discussion
April 24, 2018	V1.0	Joe Bloggs	Approved and released by the process owner
January 13, 2019	V1.1	Joe Bloggs	The related documents have been updated, and the sections on effectiveness have been verified

Contents

CONTROL OF NONCONFORMING PRODUCT

1 Summary

Purpose	The purpose of this procedure is to describe the methodology utilized by the food business operator (FBO) to control instances where expected outputs have not met requirements.
Scope	This procedure applies to product delivery, that is, ingredients, raw materials, and intermediate or finished products at all levels in the organization.
Functional responsibility	The functional responsibility for this procedure lies with the food safety manager. The food safety manager is responsible for the effective implementation and maintenance of this procedure.

2 Related Documents

Policies	Food Safety Policy, POL-001 Customer/Consumer Complaints Policy, POL-002
Processes	Departmental process descriptions
Procedures	Correction and Corrective Action, SOP-009
Work instructions	Not applicable
Forms	Nonconformance log (document management system [DMS])
Other	Not applicable

3 Definitions

Term or acronym	Description
Characteristic	Distinguishing feature, inherent or assigned, qualitative or quantitative
Complaint	Expression of dissatisfaction made to an organization related to the organization's products or services, or the complaints-handling process itself where a response or resolution is explicitly or implicitly expected
Concession	Permission to release a product or a service that does not conform to specified requirements
Correction	Action to eliminate a detected nonconformity
Corrective action	Action to eliminate the cause of a nonconformity and to prevent recurrence
Customer	Person or organization that could not or does not receive a product or a service that is intended for or required by this person or organization
Customer satisfaction	Customer's perception of the degree to which the customer's expectations have been fulfilled
Defect	Nonconformity related to an intended or specified use
FBO	food business operator
Feedback	Opinions, comments, and expressions of interest in a product, a service, or a complaints-handling process

CONTROL OF NONCONFORMING PRODUCT

Term or acronym	Description
FSMS	food safety management system
FSSC	Food Safety System Certification
Nonconformity	Nonfulfillment of a requirement
Product	Output that is a result of activities none of which are necessarily performed at the interface between the provider and the customer. For the FBO, this can be an ingredient, raw material, intermediate product, or finished product supplied to a customer or consumer.
Regulatory requirement	Obligatory requirement specified by an authority mandated by a legislative body
Risk	The effect of uncertainty on an expected result
Root cause	A cause that, once removed from the problem-fault sequence, prevents the final undesirable event from recurring
Root cause analysis	A method of problem solving that involves an attempt to identify the root cause of the fault or problem
Service	Intangible output that is the result of at least one activity necessarily performed at the interface between the provider and the customer
Statutory requirement	Obligatory requirement specified by a legislative body

4 Introduction

A nonconforming product results if a defined requirement is not being met. Examples of a nonconforming product include, but are not limited to, the following:

- Breach of statutory or regulatory compliance
- Failure to implement and maintain a requirement of Food Safety System Certification (FSSC) 22000, BRC Global Standards, Safe Quality Food Safety Code, or other
- Failure to meet a customer requirement, whether specified or implied
- Failure to deliver a required process output

Customer complaints are handled in accordance with POL-002 Customer/Consumer Complaints Policy.

All instances of nonconforming service must be identified, investigated, and resolved to ensure continual improvement of the food safety management system (FSMS) and the service provided by the organization.

CONTROL OF NONCONFORMING PRODUCT

5 Procedure Flowchart

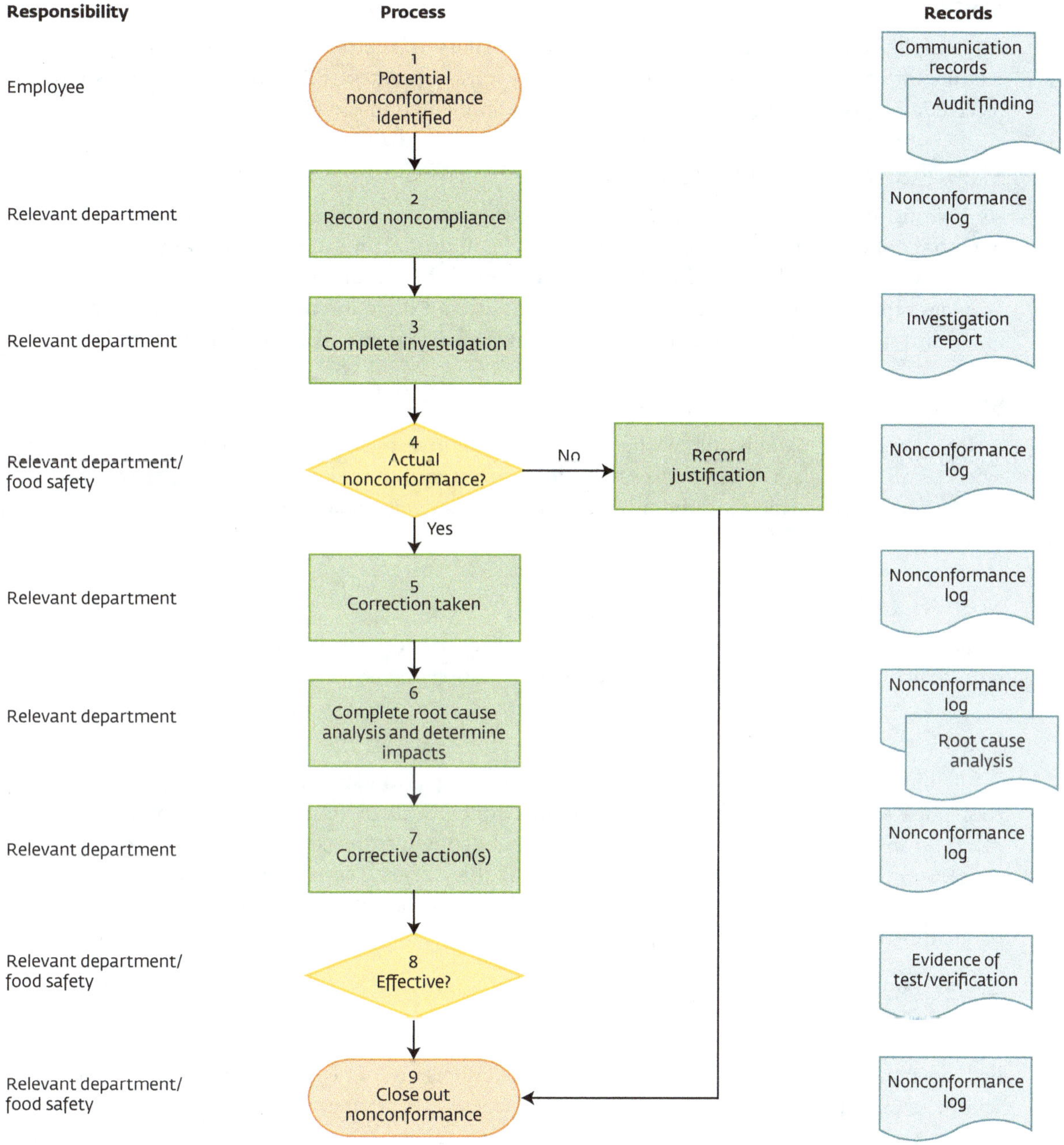

CONTROL OF NONCONFORMING PRODUCT

6 Procedure Notes

Steps 1 and 2 Identification and Recording of Nonconformance

Any employee can identify a potential nonconformity in relation to the provision of a service, or a nonconformity can be notified to the food business operator (FBO) by an external source. Once received, the potential nonconformity must be documented in the FSMS.

Step 3 Complete Investigation

An investigation must be conducted by the appropriate department to determine the validity of the potential nonconformity. This investigation will be in proportion to the potential risks that may arise based on the potential nonconformity. If there is a risk to compliance, the food safety manager/management representative will be notified immediately, and direction and assistance sought. The results of the investigation will be documented and forwarded to the food safety departments for review.

Step 4 Actual Nonconformance

A decision will be made based on the outcome of the investigation as to whether a nonconformance exists. If all parties agree (the food safety department and the relevant affected department) that no nonconformance exists, then the justification for this decision will be documented and the matter closed.

Step 5 Correction

If, based on the outcome of the investigation, it is found that a nonconformance does exist, the required correction will be taken immediately to resolve the issue.

Step 6 Root Cause Analysis

A full and thorough root cause analysis will be conducted to identify the root cause of the issue. This root cause analysis will be based on a recognized methodology (the 5 whys, a fishbone diagram, the 8Ds, and so on) and documented. Assistance may be sought from external parties if required. The root cause cannot be stated simply as human error. If this occurs, the root cause analysis must be rerun. If the root cause analysis identifies other potential risks, then the appropriate preventive action must be identified, documented, and implemented.

Step 7 Corrective Action

Based on the root cause identified in the previous step, the required corrective actions will be identified, documented, and implemented. The responsibilities and time frames for these corrective actions will be established and documented. If corrective action is planned to occur over a long time period, appropriate monitoring or measurement must be established to track the progress and effectiveness of the corrective actions.

Step 8 Verification of Effectiveness

After a suitable period of time has elapsed following the implementation of corrective action, the effectiveness of the corrective action must be determined. The corrective action is aimed at eliminating the cause of the nonconformity and preventing recurrence. Hence, the validation and verification of effectiveness must test the possibility that the nonconformity will recur. This step must be completed before the corrective action can be closed. If the test is satisfactory, the corrective action can be closed. The test that is performed or the data that are reviewed as part of this process must be documented.

Assistance from external sources may be utilized for the review of effectiveness. If the test shows that the corrective action has not been effective, then the root cause analysis must be revisited to ensure that the correct cause was identified, and the process must be repeated.

Step 9 Closure

If the verification of the effectiveness of the corrective action is successful in determining that the nonconformity has been rectified, then the matter is closed out and recorded as closed.

7 Records

Document	Location	Duration of record	Responsibility
Nonconformance log	Document management system (DMS)	Indefinitely	Food safety manager/ management representative
Root cause analysis	Relevant department	Indefinitely	Process owner/department manager
Investigation report	Relevant department	Indefinitely	Process owner/department manager
Evidence of the verification of effectiveness	Relevant department	Indefinitely	Process owner/department manager

HYGIENE PROCEDURE	
An FBO Procedure	
Document No.	Standard operating procedure SOP-005
Created	April 20, 2018
Updated	January 13, 2019
Controller	Document Controller
Owner	Food Safety Manager
Confidentiality Statement Information in this document must be kept confidential as per the document's classification below and the rules of disclosure.	
All FBO documents are classified in the following way. PUBLIC documents are intended for anyone. COMMERCIAL IN CONFIDENCE documents are to be kept confidential among restricted individuals within the FBO and partner organizations. COMPANY CONFIDENTIAL documents are to be kept confidential within the FBO and used for normal business activities by the general office population. HIGHLY CONFIDENTIAL documents are to be kept confidential among restricted individuals within the FBO.	
© Copyright FBO. All rights reserved. No part of this publication may be reproduced, stored in a retrieval system, or transmitted in any form or by any means, electronic, mechanical, photocopying, recording, or otherwise, without the written permission of the FBO.	
Classification	Company Confidential

Revision History

Date	Version	Author	Comments (including review history)
April 20, 2018	Draft 01	Joe Bloggs	Initial draft for review and discussion
April 24, 2018	V1.0	Joe Bloggs	Reviewed and approved by process owner
January 13, 2019	V1.1	Joe Bloggs	Updated the related documents section

Contents

HYGIENE PROCEDURE

1 Summary

Purpose	To comply with legal requirements, all workers in direct contact with food and food processing lines must maintain a high standard of personal hygiene and hygiene routines, which are outlined in this procedure.
Scope	This procedure is valid for the food business operator (FBO) and applies to all staff working at the FBO and to visitors, contractors, and part-time and temporary workers present on the premises.
Functional responsibility	The functional responsibility for this procedure lies with the food safety manager.

2 Related Documents

Policies	Food Safety Policy, POL-001
Processes	Food Safety, PRO-001
Procedures	Visitor Control, SOP-004 Internal Audit, SOP-006 Correction and Corrective Action, SOP-009 Change Management, SOP-011 Contractor Control, SOP-049
Work instructions	Not applicable
Forms	Not applicable
Other	Not applicable

3 Definitions

Term or acronym	Description
Basic hygiene area	The area of food tasting and handling for research and development; at the food business operator (FBO), includes development and sensory labs
Correction	Immediate action to correct a problem or potential problem
Corrective action	Action aimed at addressing the root cause of a problem and preventing recurrence
FBO	food business operator
High hygiene area	A critical hygienic area within the plant in which products and ingredients vulnerable to contamination or microbial growth are processed, treated, handled, or stored
Medium hygiene area	The area of food handling in which food is produced, processed, stored, and packaged; at the FBO, includes only the production plant

CHAPTER 4

HYGIENE PROCEDURE

4 Introduction

The great majority of people will experience a food- or waterborne disease at some point in their lives. This highlights the importance of ensuring that food is not contaminated with potentially harmful bacteria, parasites, viruses, toxins, or chemicals.

Over the past half century, the process by which food travels from the farm to the plate has changed drastically. Food contamination that occurs in one place may affect the health of consumers living on the other side of the planet. This means that everyone along the production chain, from producer to consumer, must observe safe food handling practices.

Good food hygiene is essential if the food business operator (FBO) is to make or sell food that is safe to eat. The FBO and staff must understand what good food hygiene is.

Good food hygiene helps the FBO to accomplish the following:

- Obey laws and regulations and maintain standards
- Reduce the risk of food poisoning among consumers
- Protect the business's reputation

5 Procedure Flowchart

Not applicable.

6 Procedure Notes

6.1 Hygiene Rules

PERSONAL HYGIENE RULES

- Nails must be clean and neatly trimmed, without nail polish or artificial nails.
- No strong perfumes or strongly scented personal care products or heavy makeup are to be worn, for example, false eyelashes.
- Cuts and lesions must be fully covered with approved (blue), waterproof, metal detectable Band-Aids, which can be obtained with any first aid kit. Any lost dressing must be reported to the supervisor immediately.
- All unhygienic practices, such as spitting, coughing or sneezing over food, or consuming food dropped on the floor, are unacceptable.
- Personnel and others must wash their hands before they enter the premises and after handling any dirty objects, including waste, floors, shoes, money, and so on.
- Gloves should be worn if only the aesthetic appeal of products is endangered or for personal safety reasons. They may never be used to replace handwashing.
- White workwear and protective clothing must be removed before entering toilet cubicles and should not be replaced until hands have been washed.
- The FBO is a nonsmoking site; smoking is allowed only in designated areas.
- Personal safety gear must be worn whenever necessary.

HYGIENE PROCEDURE

- Personal items, such as smoking materials and medicines, are allowed in designated areas only.
- Personal lockers should be maintained clean and tidy and kept free of rubbish and soiled clothing.

HEALTH STATUS CONTROL

- Subject to legal restrictions in the country of operation, employees must undergo a medical examination prior to employment in food contact operations, including site catering, unless documented hazards or medical assessment indicate otherwise.
- Additional medical examinations, where permitted, shall be carried out at intervals defined by the organization.

BASIC HYGIENE AREAS (DEVELOPMENT AND SENSORY LABS)

- Maintain a high level of personal hygiene according to the personal hygiene rules listed above.
- Wear workwear and protective clothing coats and hairnets while handling products that will be tasted.
- For bench tasting, workwear and protective clothing is a minimum requirement.
- Additional hygiene rules may be set by the tasting organizer if necessary.
- Employee's private foods should be stored separately, and the private foods should not be handled or consumed where FBO food products are handled or tasted.
- Clean and sanitize hands after handling private foods.

MEDIUM HYGIENE AREAS (PROCESSING PLANT)

- Maintain a high level of personal hygiene according to the personal hygiene rules listed above.
- Workwear and protective clothing should be changed daily.
- Wear clothing that is clean and in appropriate colors, if required, including workwear, protective clothing, hairnets, and safety shoes while working.
- Wash hands before entering work.
- Eating, drinking, or chewing is forbidden in the medium hygiene area.
- Remove all jewelry before entering work except any plain solid band wedding ring.
- Placing writing implements behind the ears is prohibited.
- Product contact tools and equipment should not be stored in personal lockers.
- Fully enclosed shoes must be worn when entering or working in the processing plant.

HIGH HYGIENE AREAS (PROCESSING AREA: FILING)

- Access to high hygiene areas is allowed only to those wearing clothing that is clean and in appropriate colors, if required, including workwear and protective clothing. Hair must be covered. Hands must be washed and, if necessary, disinfected before each entry. Disinfectant boot dip mats may be required.
- Prior to the start of a new process, stringent controls must be run on cleanliness and disinfection.

HYGIENE PROCEDURE

- For a new process, access is only allowed to specially trained personnel.
- No wooden pallets, cardboard, or other unhygienic materials are permitted.
- Unimpeded air flows and proper ventilation out of the area are required to maintain the higher pressure within the area.

VISITORS AND CONTRACTORS

- It is the responsibility of FBO employees to ensure that all visitors and contractors understand the hygiene and safety rules and to check that visitors and contractors follow the rules when on site.
- The contact person will give the visitor control form to visitors and contractors when they arrive. Visitors and contractors are to read the form carefully, understand it, and then sign it in the place provided at the bottom.
- The contact person should keep the signed form and is responsible for ensuring that the visitors and contractors follow the rules listed on the form.
- White coats for visitors and contractors must be available, and they will be given to the visitors and contractors by the contact person.

6.2 Cleaning and Housekeeping

ALL HYGIENE AREAS

- Working areas must be clean and tidy at all times.
- A clean-as-you-go approach should be adopted in cleaning; regularly inspect for residues.

MEDIUM AND HIGH HYGIENE AREAS

- Follow the cleaning procedure and schedule in the processing plant master plans. Equipment must always be cleaned after each use to prevent hygiene issues, such as pest infestation and microbiological contamination.
- Post signs, as follows, in medium hygiene areas. The signs should be color-coded to indicate the types of tools that may be used or the places where the tools may be used. Tools associated with separate colors must be stored separately from each other.

White–food contact surfaces only

Yellow–the surfaces of food equipment or packaging only (drums, boxes, bags, and so on)

Red–warehouse and maintenance shop only

Black–floors, walls, pipes, and ceiling surfaces only

Black oval–drain surfaces only

Classification	Company Confidential	Hygiene Procedure This document is uncontrolled if printed.	
Doc ID: SOP-005 Created: April 20, 2018	Printed: Updated: January 13, 2019	Controller: Document Controller Owner: Food Safety Manager	Page 5 of 7

HYGIENE PROCEDURE

6.3 Hygienic Maintenance in Medium and High Hygiene Areas

- Equipment sent for maintenance should be cleaned before reinstallation in the processing plant. Particular attention should be paid to food contact surfaces, which require thorough cleaning and sanitizing.
- Working tools must be stored in assigned containers and must not be placed on or above food contact surfaces. The tools should be removed from the processing plant immediately after work.
- The use of food-grade lubricants is mandatory unless technological reasons prevent their use. All exceptions should be approved by the manager. Lubricants should be applied in appropriate quantities to avoid excess lubricant falling onto or into products.
- Material that could taint any food product or ingredient (such as paint, glue, and so on) must not be brought onto the site. Contractors need to acquire written permission from the FBO food safety department to use such materials.
- Obsolete or unused equipment should be removed on a regular basis.
- Apply the change management procedure for any equipment change.

6.4 Waste in Medium and High Hygiene Areas

Food contact waste and other garbage should be stored and eliminated separately. Orange bags should be used for food scraps and animal feed.

6.5 Hygiene Training

- New personnel will undergo an initial hygiene induction training session
- Once a year, all staff working in hygiene areas must be retrained by the food safety department
- Training may be required of contractor staff working in medium hygiene areas over a period of time or on a regular basis

6.6 Reporting Illness and Injury

If an employee or a member of an employee's household suffers from one of the following conditions, then the employee must report this immediately upon returning to work to the line manager. It is the manager's responsibility to discuss the symptoms with the employee.

- Jaundice
- Diarrhea
- Vomiting
- Fever
- Sore throat with fever
- Visibly infected skin (boils, cuts)
- Discharge from ear, eye, or nose

HYGIENE PROCEDURE

No person with such a health problem shall be permitted to work in medium hygiene areas. Each such person must avoid handling food for at least 48 hours after the last episode of vomiting or diarrhea has occurred to prevent contamination of the food produced at the FBO.

If an employee contracts an ailment or disease while traveling abroad for business or personal reasons, it is the employee's responsibility to contact a doctor upon return to obtain information and advice about the ailment or the disease and report the health problem to the line manager on the first day back to work.

7 Records

Document	Location	Duration of record	Responsibility
Signed training participant lists (hard copies)	Food Safety Office	Seven years	Food safety manager
Visitor control form	Food Safety Office	Seven years	Food safety manager

INTERNAL AUDITING	
An FBO Procedure	
Document No.	Standard operating procedure SOP-006
Created	April 20, 2018
Updated	January 13, 2019
Controller	Document Controller
Owner	Food Safety Manager
Confidentiality Statement Information in this document must be kept confidential as per the document's classification below and the rules of disclosure.	
All FBO documents are classified in the following way. PUBLIC documents are intended for anyone. COMMERCIAL IN CONFIDENCE documents are to be kept confidential among restricted individuals within the FBO and partner organizations. COMPANY CONFIDENTIAL documents are to be kept confidential within the FBO and used for normal business activities by the general office population. HIGHLY CONFIDENTIAL documents are to be kept confidential among restricted individuals within the FBO.	
Classification	Company Confidential

Revision History

Date	Version	Author	Comments (including review history)
April 20, 2018	Draft 01	Joe Bloggs	Initial document for review and discussion
April 24, 2018	V1.0	Joe Bloggs	Approved for release by process owner
January 13, 2019	V1.1	Joe Bloggs	Updated to reflect changes in International Organization for Standardization (ISO) 19011:2018

Contents

CHAPTER 4

INTERNAL AUDITING

1 Summary

Purpose	The purpose of this procedure is to describe the following: • The internal audit methodology employed based on International Organization for Standardization (ISO) 19011:2018 to ensure that the food safety management system (FSMS) remains suitable, adequate, and effective in meeting business and customer compliance requirements and the requirements of ISO 22000:2018 and Food Safety System Certification (FSSC) 22000 and that the FSMS is effectively implemented and maintained
Scope	This procedure applies to the following: • Audit program planning, performance, and follow-up, including audit initiation, audit preparation, conducting the audit, preparing and distributing the audit report, completing the audit, and audit follow-up if required • Compliance and conformance auditing
Functional responsibility	The functional responsibility for this procedure lies with the food safety manager, who is responsible for the effective implementation and maintenance of the procedure.

2 Related Documents

Policies	Food Safety Policy, POL-001
Processes	Departmental process descriptions
Procedures	Correction and Corrective Action Procedure, SOP-009 Management Review Procedure, SOP-021
Work instructions	Not applicable
Forms	FSMS auditing checklist
Other	Statutory and regulatory requirements: FSSC 22000 ISO 22000:2018 ISO 19011:2018

3 Definitions

Term or acronym	Description
Audit conclusion	The outcome of the audit after consideration of the audit objectives and all audit findings
Audit criteria	Set of requirements used as a reference against which objective evidence is compared
Audit evidence	Records, statements of fact, or other information that is relevant to the audit criteria and verifiable

INTERNAL AUDITING

Term or acronym	Description
Audit finding	Results of the evaluation of collected audit evidence against the audit criteria
Audit plan	Description of the activities and arrangements in an audit
Audit program	Arrangements for a set of one or more audits planned for a specific time frame and directed toward a specific purpose
Audit scope	The extent and boundaries of the audit
Auditor	An individual with the demonstrated personal attributes and competence to conduct an audit
Combined audit	An audit carried out together at a single auditee on two or more systems, for example, FSSC 22000, ISO 9001:2015, and so on
Compliance auditing	Determination of the compliance with defined statutory, regulatory, and customer legal obligation requirements
Conformance auditing	Determination of conformity with defined international standards, such as FSSC 22000, ISO 22000:2018, and so on
Correction	Action taken to eliminate a detected nonconformity
Corrective action	Action taken to eliminate the cause of a nonconformity and prevent recurrence
FBO	food business operator
FSMS	food safety management system
FSSC	Food Safety System Certification
High-risk finding	A significant weakness in the system or process requires immediate rectification
Internal audit	A systematic and independent process for obtaining audit evidence and evaluating it objectively to determine the extent to which audit criteria are fulfilled
ISO	International Organization for Standardization
Low-risk finding	A general weakness in the system or process that, if rectified immediately, could improve efficiency
Major finding	A total breakdown or absence of objective evidence to satisfy one or more FSMS requirements or a situation that would, on the basis of available objective evidence, raise significant doubt as to the quality of the product that the organization is supplying
Medium-risk finding	A potentially significant weakness in the system or process that, if not rectified immediately, may lead to high risk
Minor finding	A finding, in a defined and documented system that generally satisfies one or more FSMS requirements, that a situation, on the basis of available objective evidence, raises a concern about the potential quality of what the organization is supplying, for example, the system or one or more processes have not reached an acceptable level of maturity
Nonconformity	The nonfulfillment of a requirement

INTERNAL AUDITING

Term or acronym	Description
Objective evidence	Data supporting the existence or verity of an event or item
Risk	The effect of uncertainty
Root cause analysis	A method of problem solving that involves the attempt to identify the root cause of faults or problems
SWOT analysis	A section of the audit report in which the audit team categorizes the audit findings on strengths, weaknesses, opportunities, and threats (SWOT)

4 Introduction

Internal auditing is undertaken to monitor and measure the company's compliance against statutory and regulatory requirements and the company's conformity with the requirements of the food safety management system (FSMS).

Internal audits are scheduled on a planned basis and conducted by trained internal auditors, whose findings are reported to management for review and action. If the audit findings highlight problems, the auditee is required to provide a commitment to addressing and resolving the issues. The internal auditor seeks evidence of the effective implementation of the subsequent actions of the auditee. The results of the internal audits and the overall effectiveness of the internal audit program are reported at the management review meeting.

INTERNAL AUDITING

5 Procedure Flowchart

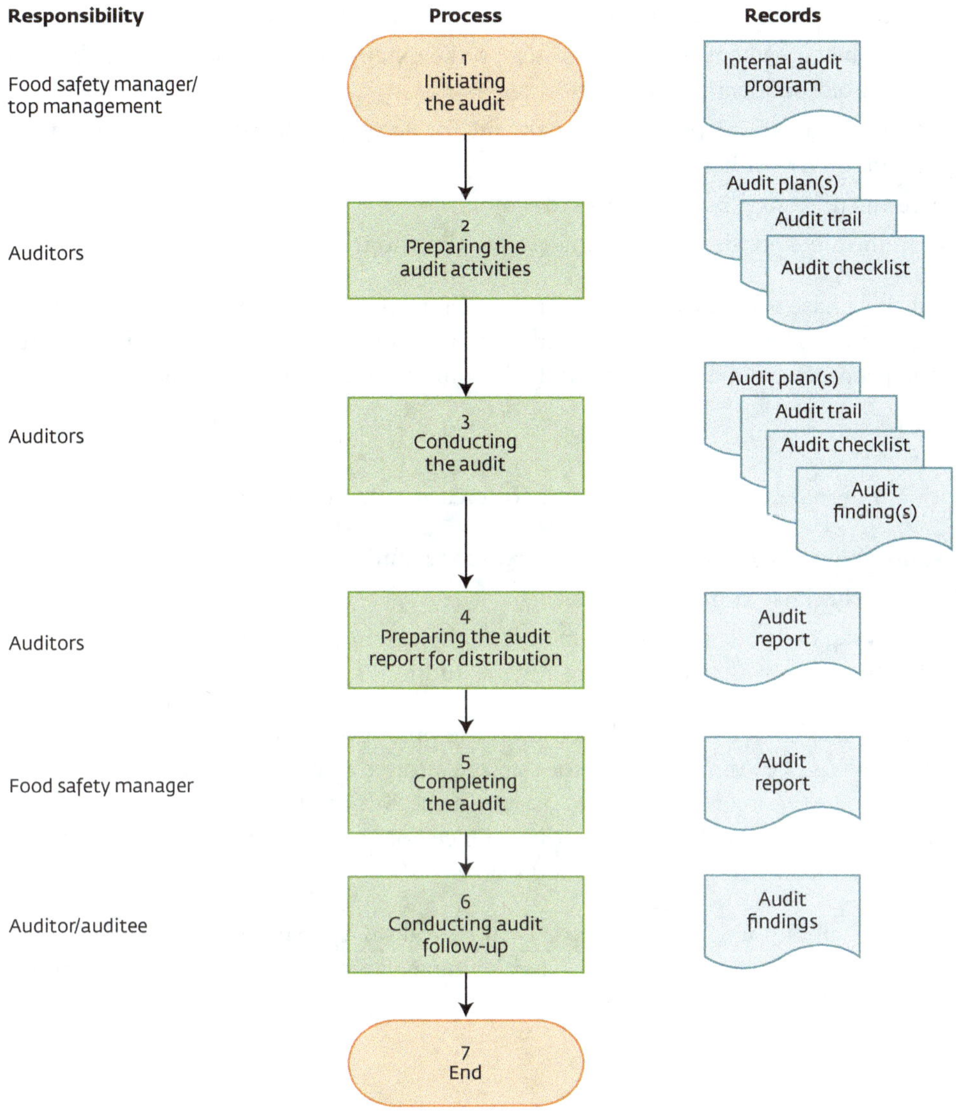

6 Procedure Notes

Step 1

The food safety manager has the responsibility to create and manage the internal audit process. This involves establishing initial contact with the auditee(s) and reaching agreement on the following:

- Audit objectives, scope, criteria, methods, and audit team composition, including any technical experts
- Provide relevant information for planning, including information on the risks and opportunities the organization has identified and how they are being addressed
- Agree on the dates of the audit

INTERNAL AUDITING

- Identify the resources needed to complete the audit, including access to the required people, processes, activities, and documentation
- Assess the statutory and regulatory requirements during the audit
- Confirm the agreement with the auditee on the extent of the disclosure and the treatment of confidential information
- Confirm any location-specific arrangements for access, health and safety, security, confidentiality, or other
- Gauge the need for observers or guides
- Determine any specific areas of concern for the auditee

The output of this phase is the development of an audit program outlining the audits to be completed over a defined period. The process may also identify the internal auditor assigned to the audit. Once completed, the program will be published and communicated across the company.

Step 2

Each individual internal auditor is responsible for creating the following:

- An audit plan, including audit objectives, scope, and criteria
- An audit checklist or audit protocol
- The auditing methods to be used, including the extent to which audit sampling is needed to obtain sufficient evidence for the audit

Audit plans, checklists, and trails will be based on templates to ensure consistency. Audit planning should consider the risks of the audit activities on the auditee's processes and provide the foundation for agreement among the interested parties based on the information in the audit program and the documented information provided by the auditee. Once documented by the internal auditor, the audit plan will be communicated to the relevant auditee(s).

Some audits will be unannounced, as directed by the food safety manager. If this is the case, no audit plan may be produced. However, the food safety manager will fully brief the internal auditor on the objectives, scope, and criteria of the audit.

Step 3

The internal auditor will conduct the audit in accordance with the plan. Audit checklists or audit trails will be used by the auditor to record audit evidence. Audits will be conducted using interview, observation, reviews of records and documents, and analysis of data. Trend analysis and tests may also be utilized to gather evidence as required. The details to be recorded on the checklist or audit trails include information on the requirement that is being checked, the evidence gathered, the conformance indication, and the identification of the auditee.

In the event that an internal auditor identifies a nonconformity based on objective evidence, the internal auditor will inform the process owner/head of department about the issue and explain the nature of the nonconformity, why it is a nonconformity, and the requirement that has not been fulfilled. The internal auditor will document the nonconformity in the checklist or audit trail and obtain the signature of the auditee signifying the auditee's acceptance of the issue and the commitment to rectify the issue. The internal auditor will classify the audit finding as major, minor, or an opportunity for improvement based on the risk.

INTERNAL AUDITING

It is solely the responsibility of the process owner/head of department, if audit findings highlight problems, to rectify the issues. Correction must be undertaken; a root cause analysis using a recognized root cause analysis methodology, for example, 5 whys, a fishbone diagram, and so on, must be completed, and corrective action identified and implemented. A response plan must be submitted to the internal auditor by the auditee within an agreed time frame of the audit. It must outline the correction, root cause analysis, and corrective action(s), including a risk assessment. The internal auditor will review the response plan and approve or reject it. Thus, if there is no root cause analysis, the root cause analysis is inadequate, and so on. If the plan is rejected, the auditee must correct the response plan and resubmit it for approval. All audit findings should be closed out within 12 weeks of the issuance of the findings. Exceptions may be granted, subject to the approval of the internal auditor and the food safety manager/management representative. As appropriate, the auditee should keep the individuals managing the audit program or the audit team informed of the status of these actions.

The outputs of this phase should be that the audit objective has been achieved, the audit plan has been carried out, the checklists/audit trails have been completed, if applicable, and audit findings and a response plan have been received from the process owner/head of department.

Step 4

The internal auditor will prepare an audit report outlining the audit conclusions. The conclusions are based on a comparison of all the audit findings against the audit objective. The report will be detailed and cover the following points at a minimum:

- Identification of the audit objective, scope, and criteria
- Identification of the auditor and process owner(s)/head of department
- The audit conclusions
- An executive summary
- The audit findings on strengths, weaknesses, opportunities, and threats (SWOT)
- A description of the process, critical process parameters, and process performance
- The number and classification of the audit findings
- The audit findings in detail
- Sample/confidentiality statement
- Audit follow-up
- Audit checklist or audit trail, as an attachment

The audit report will then be released to the food safety manager and the process owner/head of department.

Step 5

The audit is completed when all planned audit activities have been completed or otherwise agreed with the process owner. For instance, there may have been an unexpected event that prevented the audit plan from being completed.

The food safety manager will review the audit report to ensure that all technical aspects of the audit plan have been covered, the evidence gathered is objective and related to the audit criteria, and the audit conclusions reached are correct. The food safety manager will also manage any appeals raised by the process manager/head of department in relation to an audit finding. If agreement cannot be reached between the food safety manager and the process owner/head of department, the food safety manager will elevate the issue to the executive management team for resolution.

Step 6

Based on the response plan submitted by the process owner and the agreed closure time frame, the internal auditor will follow up to ensure that all audit findings have been effectively closed out. This will be achieved through the evaluation of the risk assessment and effectiveness checks.

The effectiveness checks must be completed before the corrective action risk assessment can be closed. The purpose of these follow-up checks is to ensure that the stated actions have been implemented and that they have been effective in solving the stated problem. If satisfied, the internal auditor will close the audit findings.

If the internal auditor does not agree to close the audit findings, agreement on the actions to be taken will be determined between the internal auditor and the auditee.

7 Audit Records

The following documentation will be maintained as evidence that the audits have been performed:

- Audit plan
- Audit checklist/audit trail
- Audit report
- Root cause analysis data/response plan

8 Records

Document	Location	Duration of record	Responsibility
Internal audit program	Food Safety Office	One year	Food safety manager
Internal audit plan	Food Safety Office	Indefinitely	Food safety manager
Audit checklist/audit trail	Food Safety Office	Indefinitely	Food safety manager
Audit report	Food Safety Office	Indefinitely	Food safety manager
Response plan	Food Safety Office	Indefinitely	Food safety manager

ALLERGEN CONTROL	
An FBO Procedure	
Document No.	Standard operating procedure SOP-007
Created	April 20, 2018
Updated	January 13, 2019
Controller	Document Controller
Owner	Food Safety Manager
Confidentiality Statement	
Information in this document must be kept confidential as per the document's classification below and the rules of disclosure.	
All FBO documents are classified in the following way. PUBLIC documents are intended for anyone. COMMERCIAL IN CONFIDENCE documents are to be kept confidential among restricted individuals within the FBO and partner organizations. COMPANY CONFIDENTIAL documents are to be kept confidential within the FBO and used for normal business activities by the general office population. HIGHLY CONFIDENTIAL documents are to be kept confidential among restricted individuals within the FBO.	
© Copyright FBO. All rights reserved. No part of this publication may be reproduced, stored in a retrieval system, or transmitted in any form or by any means, electronic, mechanical, photocopying, recording, or otherwise, without the written permission of the FBO.	
Classification	Company Confidential

Revision History

Date	Version	Author	Comments (including review history)
April 20, 2018	Draft 01	Joe Bloggs	Initial document for review and discussion
April 24, 2018	V1.0	Joe Bloggs	Approved for release by process owner
January 13, 2019	V1.1	Joe Bloggs	Updated formatting and consistency issues

Contents

CHAPTER 4

ALLERGEN CONTROL

1 Summary

Purpose	To ensure the effective use, storage, and labeling of allergens and food allergen management at the food business operator (FBO).
Scope	This procedure is applicable to products, processes, storage and production environments, and suppliers of raw materials at the FBO.
Functional responsibility	The functional responsibility for this procedure lies with the food safety manager, who is responsible for the effective implementation and maintenance of the procedure.

2 Related Documents

Policies	Food Safety Policy, POL-001
Processes	Food Safety, PRO-001
Procedures	Control of Documents, SOP-001 Traceability, SOP-012
Work instructions	Not applicable
Forms	Master Document Register
Other	Document management system (DMS)

3 Definitions

Term or acronym	Description
FBO	food business operator
Food allergy	Immunological-based reaction to chemical substances, usually proteins or protein fragments, by individuals who have previously been sensitized to the same substance and have formed antibodies. Allergic reactions can be initiated by small quantities of allergens. Reactions are usually mild and transitory, but, in a small share of the population, reactions can be severe and may, in some cases, lead to death.
Food Information for Consumers	A European Union (EU) initiative to provide consumers with information about food
Major food allergens at the FBO	Milk, soy, and gluten allergens

4 Introduction

Under the Food Information for Consumers Regulation of the European Union (EU), all food business operators (FBOs) should declare the presence—whether for use as an ingredient or a processing aid—of any of the 14 major allergens listed in the regulation. In accordance with the regulation, the mandatory information should be easily accessible, in a conspicuous place, readily visible, and legible. The display of the information should be indelible (permanent) where appropriate, for example, on food labels, where it needs to withstand handling. The information should not be hidden, obscured, detracted from, or interrupted by other written or pictorial matter or any other intervening material.

The 14 allergens listed in the regulation are recognized across Europe as the most common ingredients or processing aids that cause food allergies and intolerances. If a food product contains or uses an ingredient or processing aid (such as wheat flour used to roll out dough made from rye flour) derived from one of the substances or products listed in the regulation, this fact must be declared by the FBO to the consumer.

The information supplied in this procedure is not exhaustive and does not cover other labeling requirements, such as other general labeling, for example, country of origin, lactose content, quantities, additives, nutrition, and so on.

5 Procedure Flowchart

Not applicable.

6 Procedure Notes

6.1 Storage of Raw Materials Containing Allergens

- Allergen-containing raw materials should be stored separately from nonallergenic materials.
- Allergen-containing raw materials should not be stored over nonallergenic materials.
- Milk allergen pallets should not be stored over soy allergen pallets or vice versa.

See Raw Material Management Procedure, SOP-010 for details.

6.2 Labeling

All allergen-containing raw materials are initially received with orange labels from factories. Then milk and soy allergens are labeled with purple and green labels, respectively. The labels of all relevant finished food products shows the declaration "Contains allergens." The identity of the allergen is included on the product labeling in line with the relevant International Organization for Standardization/technical specification (ISO/TS) 22002 standard series.

6.3 External Panel and Consumer Screening

External panelists and consumers who participate in product tasting are screened for sensitivity to major allergens. Only panelists who are not allergic to foods are permitted to participate in consumer tests.

6.4 Internal Panel Screening

Internal panelists are alerted that samples consumed at the FBO may contain any one of the known allergens indicated in the definitions section above.

ALLERGEN CONTROL

6.5 Preventing Allergen Cross Contamination

- Use a dedicated scoop for each raw material during transfer.
- Wipe down all affected surfaces after weighing out an allergen.
- Change gloves or wash hands after an allergen is handled.
- Keep all containers with allergens sealed.
- As much as possible, store allergens on the lower section of storage racks. Some FBOs use dedicated production lines and equipment in processing allergen raw materials in products.

7 Records

Document	Location	Duration of record	Responsibility
Allergen file	Food Safety Office	Indefinitely	Food safety manager

MOCK RECALL	
An FBO Procedure	
Document No.	Standard operating procedure SOP-008
Created	April 20, 2018
Updated	January 13, 2019
Controller	Document Controller
Owner	Food Safety Manager

Confidentiality Statement

Information in this document must be kept confidential as per the document's classification below and the rules of disclosure.

All FBO documents are classified in the following way. PUBLIC documents are intended for anyone. COMMERCIAL IN CONFIDENCE documents are to be kept confidential among restricted individuals within the FBO and partner organizations. COMPANY CONFIDENTIAL documents are to be kept confidential within the FBO and used for normal business activities by the general office population. HIGHLY CONFIDENTIAL documents are to be kept confidential among restricted individuals within the FBO.

Classification	Company Confidential

Revision History

Date	Version	Author	Comments (including review history)
April 20, 2018	Draft 01	Joe Bloggs	Initial document for review and discussion
April 24, 2018	V1.0	Joe Bloggs	Approved and released by process owner
January 13, 2019	V1.1	Joe Bloggs	Updated related documents section and formatting

Contents

CHAPTER 4

MOCK RECALL

1 Summary

Purpose	The purpose of this procedure is to describe the process for effectively conducting a mock recall and potentially removing a product from the external supply chain/distribution.
Scope	This instruction covers all products manufactured or distributed by the food business operator (FBO). Local regulations and laws take precedence over this guideline.
Functional responsibility	The functional responsibility for this procedure lies with the food safety manager, who is responsible for the effective implementation and maintenance of this procedure.

2 Related Documents

Policies	Food Safety Policy, POL-001 Customer/Consumer Complaints Policy, POL-002
Processes	Departmental process descriptions
Procedures	Control of Nonconforming Product, SOP-003 Correction and Corrective Action, SOP-009 Communication, SOP-020 Product Recall and Withdrawal, SOP-023 Crisis Management, SOP-029
Work instructions	Not applicable
Forms	Recall/withdrawal log Communication log Root cause analysis/corrective action
Other	Not applicable

3 Definitions

Term or acronym	Description
Complaint	An expression of dissatisfaction communicated to an organization in relation to the organization's products or services or the complaints-handling process, during which a response or resolution is explicitly or implicitly expected
Correction	Action to eliminate a detected nonconformity
Corrective action	Action to eliminate the cause of a nonconformity and to prevent recurrence
FBO	food business operator
FSMS	food safety management system
Nonconformity	Nonfulfillment of a requirement

Classification	Company Confidential	Mock Recall Procedure	
Doc ID: SOP-008 Created: April 20, 2018	Printed: Updated: January 13, 2019	Controller: Document Controller Owner: Food Safety Manager	Page 2 of 8

MOCK RECALL

Term or acronym	Description
Product	Output that is a result of activities none of which is necessarily performed at the interface between the provider and the customer; for the food business operator (FBO), this may be an ingredient, raw material, intermediate product, or finished product supplied to a customer or consumer
Recall	The process by which a product is removed from the external supply chain/ distribution and consumers are publicly advised to take specific actions with the product, for example, do not consume the product, or return the product to the shop or manufacturer; this includes the U.S. Food and Drug Administration (FDA) class I and class II recalls
Regulatory requirement	Obligatory requirement specified by an authority mandated by a legislative body
Risk	The effect of uncertainty on an expected result
Root cause	A cause that, once removed from the problem fault sequence, prevents the final undesirable event from recurring
Root cause analysis	A method of problem solving that involves an attempt to identify the root cause of faults or problems
Statutory requirement	Obligatory requirement specified by a legislative body
Traceability	The ability to track a food through all stages of production, processing, and distribution, including importation and retail; traceability should mean that movements can be traced one step back and one step forward at any point in the supply chain

4 Introduction

A food recall is an action taken to remove food that is unsafe or potentially unsafe from distribution, sale, and consumption. An unsafe food is a food that may cause illness or other physical harm to a person consuming the food. The food industry recall procedure protocol provides information for food businesses operators (FBOs) on recalling food. It also offers guidance to FBOs in the development of a written mock recall or recall plan for food. A mock recall represents a method of verifying the effectiveness of an FBO's recall procedure. The primary objective of a mock recall is as follows:

- Effectively and efficiently verify whether an FBO's arrangements in the event of a recall are likely to be successful

The procedure protocol provides guidance only and is not legally binding; however, it outlines the legal requirements relating to mock recalls and recalls that are enforceable by applicable national, federal, or territorial governments. If there is no legal obligation to enforce a recall, customer or food safety scheme standards should be followed.

Recall systems should be tailored to the individual needs of the FBO. A business may seek independent advice, including legal advice, about the system it develops for mock recalls.

5 Procedure Flowchart

Not applicable.

6 Procedure Notes

6.1 Data Collection and Management

A food business may become informed of a problem with any of its food products, raw materials, ingredients, intermediate products, or finished products through any of the following:

- In-house testing indicating there may be a potential problem with a particular food product or batch
- Customer/consumer complaints/feedback, for example, a phone call or e-mail from a customer or wholesaler informing the business about a potential problem
- A supplier of a raw material that is used by the company to make its food products may inform the business that there is a problem with an ingredient
- Government entities, such as health departments, local councils, or the police, may indicate that there may be a problem with a particular food product

Such problems may include any of the following:

- The presence of pathogenic bacteria, such as *Salmonella*
- Chemical contamination, for instance, a chemical sanitizer
- Foreign matter contamination, for example, pieces of glass, metal, or plastic, that could cause physical harm to a person consuming the food
- Labeling errors, such as incorrect or insufficient cooking instructions
- Undeclared allergens, for instance, peanut, milk, or soy ingredients that are not properly declared on the label
- Packaging defects, for example, the integrity of a package has been compromised, and a piece of the packaging becomes a choking hazard
- The underprocessing of food, resulting in potentially unsafe food

All necessary information about the nature of the problem or hazard must be obtained to support an assessment to establish whether a food product is unsafe and a recall action is required. In assessing the risks, a sponsor needs to accomplish the following:

- Identify the hazard associated with the food, for example, is it a microbiological, physical, chemical, or allergen-related hazard
- Determine if the identified hazard poses a potential food safety risk, for instance, the food may contain harmful levels of pathogenic bacteria
- Determine what action needs to be taken to manage the food safety risk

The food safety team

- Gathers all necessary information, facts, and data to enable a conscious decision to proceed with a mock recall; a mock recall must have a clearly defined goal and

MOCK RECALL

objective because these exercises can validate specific processes and confirm suspected weaknesses

- Defines the communication with employees, the sales force, customers or consumers, and other stakeholders
- Determines subsequent steps involving the removed products
- Considers all other elements that might affect the FBO

6.2 The Decision to Conduct a Mock Recall

The decision to conduct a mock recall is taken by the food safety manager. A product mock recall should be conducted twice a year. It may occur more frequently if requested by the primary stakeholders.

The decision-making process is carried out according to crisis management procedures and takes into account especially the following:

- The situation and actions to be undertaken in markets where the same material is commercialized (intermarket supply)
- Foreign markets must be examined in making decisions or approving decisions; specific guidelines may apply

Where a food safety issue has been identified for the mock recall, the food safety manager should also consider the possibility of the same problem occurring in the following:

- Different package sizes of the same line
- Different flavors or varieties of the same product
- Food products with a different batch number or date marking
- Different food products processed on the same line or in the same plant
- The same or similar food products packaged under a generic label

If the food safety issue is present in other foods, batches, sizes, or brands, all these foods will need to be considered for inclusion in the recall. The food safety team may also decide to limit the scope of a mock recall.

The food safety manager should also consider whether there are other products on the market or in the food supply chain that may have been affected by the same hazard as the food subjected to the mock recall. This is referred to as a traceback. For example, if the problem is found to be linked to one or more raw materials supplied to the FBO, then the FBO needs to notify the supplier of the raw materials to enable the supplier to notify other customers of the raw materials. This may then result in the launch of additional mock recalls for more food products by other food businesses. Alternatively, the food safety manager may choose to limit the scope of the mock recall.

6.3 Mock Recall Communication

Communication is critical to the success of a mock recall as well as for the image of brands. Communication is based on the following:

- The position statement prepared by the food safety team/recall team
- The sensible and workable mock recall plan

MOCK RECALL

- Test procedures and plans with mock recalls
- The identification of risks and problem areas
- Statutory and regulatory requirements related to mock recall communication, if relevant
- Questions and answers to be used by consumer services

Communication that is adequate to reach customers may be considered an adequate test. Communication must be simple and factual:

- Why is the FBO conducting a mock recall?
- What product is involved in the mock recall?
- What should the FBO do to eliminate the defect and put the product back on the market?

The same principles must be applied for communication with other stakeholders (employees, customers, authorities, and so on).

6.4 Actions in the FBO Factory

The factory provides the traceability data necessary for defining the material and quantities to be removed from the entire supply chain/distribution. All affected batches must be restricted in the FBO computer system.

The accuracy of the traceability system must be considered, and a safety margin on either side of the concerned batch must be added if necessary.

6.5 Actions in FBO Distribution/Logistics

Upon receiving instructions to block a particular product quantity, the warehouse staff must immediately remove it from assembled loads in the warehouse. The blocked stock must be physically marked and segregated.

If advised by the food safety/recall team, distribution will coordinate urgent material pickups from identified warehouses and stores if necessary.

The material received back must be registered in the FBO computer system with the status indicated as blocked as with all returned material.

On request, warehouse personnel can check and sort the suspected stock or hold the affected product until the product is authorized to be released. The food safety manager provides instructions on how to examine the product and gather adequate resources (training, specialists, and so on).

A detailed report must be prepared on the fate of the mock recalled batches. Other goods must be included if relevant (for example, nonrecalled goods, other FBO products, or even the products of competitors).

MOCK RECALL

6.6 Actions in Trade

Materials in warehouses must be blocked and physically marked. Traceability must be performed, and, if required, a pickup schedule must be agreed with the FBO distribution.

Materials in shops (supermarket shelves or back-room storage) must be fully traced and, where required, removed from shelves, blocked, physically marked, and placed in back-room storage to await pickup, destruction, or authorized release (as agreed between the FBO and retailer). Sales or merchandising staff may be called to assist as needed.

The retailer will communicate the actual quantities to be picked up to facilitate transport. The material must be returned as soon as possible to the FBO or to dedicated warehouses.

Disposal at customer sites is possible if there is mutual agreement about what is to be disposed. The method of disposal must be defined and properly documented.

6.7 Return Transport

The return transport of affected material requires special attention and appropriate organization. This must be accomplished without delay.

6.8 Handling of the Returned Product

The returned product must be controlled, registered, marked, and segregated from normal stocks. At a minimum, the product should be obtained for laboratory analysis.

Precise inventories must be kept. Regulatory authorities may have additional requirements on records and information.

The returned product must be handled as a nonconforming product; the rules for responsible destruction or disposal must be followed.

In line with the FBO accounting procedure, all costs related to mock recalls must be charged to production-related overhead, not to bad products.

6.9 Postreview Action Reviews

A postreview action review must be conducted when the mock recall is over and potential improvements implemented.

At a minimum, an analysis of the quantities of the materials involved must be carried out (product produced, sold, returned, destroyed, authorized for release, or not accounted for or consumed).

The simple goal of the mock recall is ideally 100 percent of the product (raw material, ingredients, intermediate, or finished product) is accounted for within two hours or less.

6.10 Mock Recall Frequency

Mock recalls must be practiced. At least two mock recall exercises per year are recommended. These should encompass raw materials, ingredients, intermediate products, product contact materials, or finished products. A postreview action of a real case cannot replace a mock recall. An actual recall is not the occasion to test the FBO recall/traceability system.

7 Records

Document	Location	Duration of record	Responsibility
Mock recall log	Food Safety Office	Indefinitely	Food safety manager
Communication records	Food Safety Office	Indefinitely	Food safety manager
Root cause analysis	Food Safety Office	Indefinitely	Food safety manager
Mock recall report	Food Safety Office	Indefinitely	Food safety manager
Postreview minutes	Food Safety Office	Indefinitely	Food safety manager

CORRECTION AND CORRECTIVE ACTION	
An FBO Procedure	
Document No.	Standard operating procedure SOP-009
Created	April 20, 2018
Updated	January 13, 2019
Controller	Document Controller
Owner	Food Safety Manager
Confidentiality Statement Information in this document must be kept confidential as per the document's classification below and the rules of disclosure.	
All FBO documents are classified in the following way. PUBLIC documents are intended for anyone. COMMERCIAL IN CONFIDENCE documents are to be kept confidential among restricted individuals within the FBO and partner organizations. COMPANY CONFIDENTIAL documents are to be kept confidential within the FBO and used for normal business activities by the general office population. HIGHLY CONFIDENTIAL documents are to be kept confidential among restricted individuals within the FBO.	
© Copyright FBO. All rights reserved. No part of this publication may be reproduced, stored in a retrieval system, or transmitted in any form or by any means, electronic, mechanical, photocopying, recording, or otherwise, without the written permission of the FBO.	
Classification	Company Confidential

Revision History

Date	Version	Author	Comments (including review history)
April 20, 2018	Draft 01	Joe Bloggs	Initial draft for review and discussion
April 24, 2018	V1.0	Mary Cahill	Original issue and update after technical review
January 13, 2019	V1.1	Mary Cahill	Updated document title, related documents; introduced corrective action risk assessment, validation/verification, and risk assessment of implementation prior to closure

Contents

CHAPTER 4

CORRECTION AND CORRECTIVE ACTION

1 Summary

Purpose	The purpose of this procedure is to describe the methodology utilized within the organization to manage the correction and corrective action process.
Scope	This procedure applies to the generation of correction and corrective actions and the associated root cause analysis and the effective closure of correction and corrective actions.
Functional responsibility	The functional responsibility for this procedure lies with the food safety manager/management representative.

2 Related Documents

Policies	Food Safety Policy, POL-001
Processes	Compliance, PRO-004
Procedures	Complaint Management, SOP-015 Strategic Planning, SOP-019 Management Review, SOP-021
Work instructions	Not applicable
Forms	Corrective and preventive action form
Other	Not applicable

3 Definitions

Term or acronym	Description
Correction	Action taken to eliminate a detected nonconformity
Corrective action	Action taken to eliminate the cause of a nonconformity and prevent recurrence
DMS	document management system
FBO	food business operator
FSMS	food safety management system
Root cause analysis	A method of problem solving that involves attempting to identify the root cause of faults or problems

4 Introduction

The identification of issues affecting the food safety management system (FSMS) and the implementation of correction and corrective actions are core requirements for continual improvement within a management system. For such corrective actions to be effective, a rigorous root cause analysis process must be followed to ensure that the actual cause of the issue is identified and eliminated and recurrence prevented.

This procedure outlines the process implemented within the organization to ensure that this is achieved.

CORRECTION AND CORRECTIVE ACTION

5 Procedure Flowchart

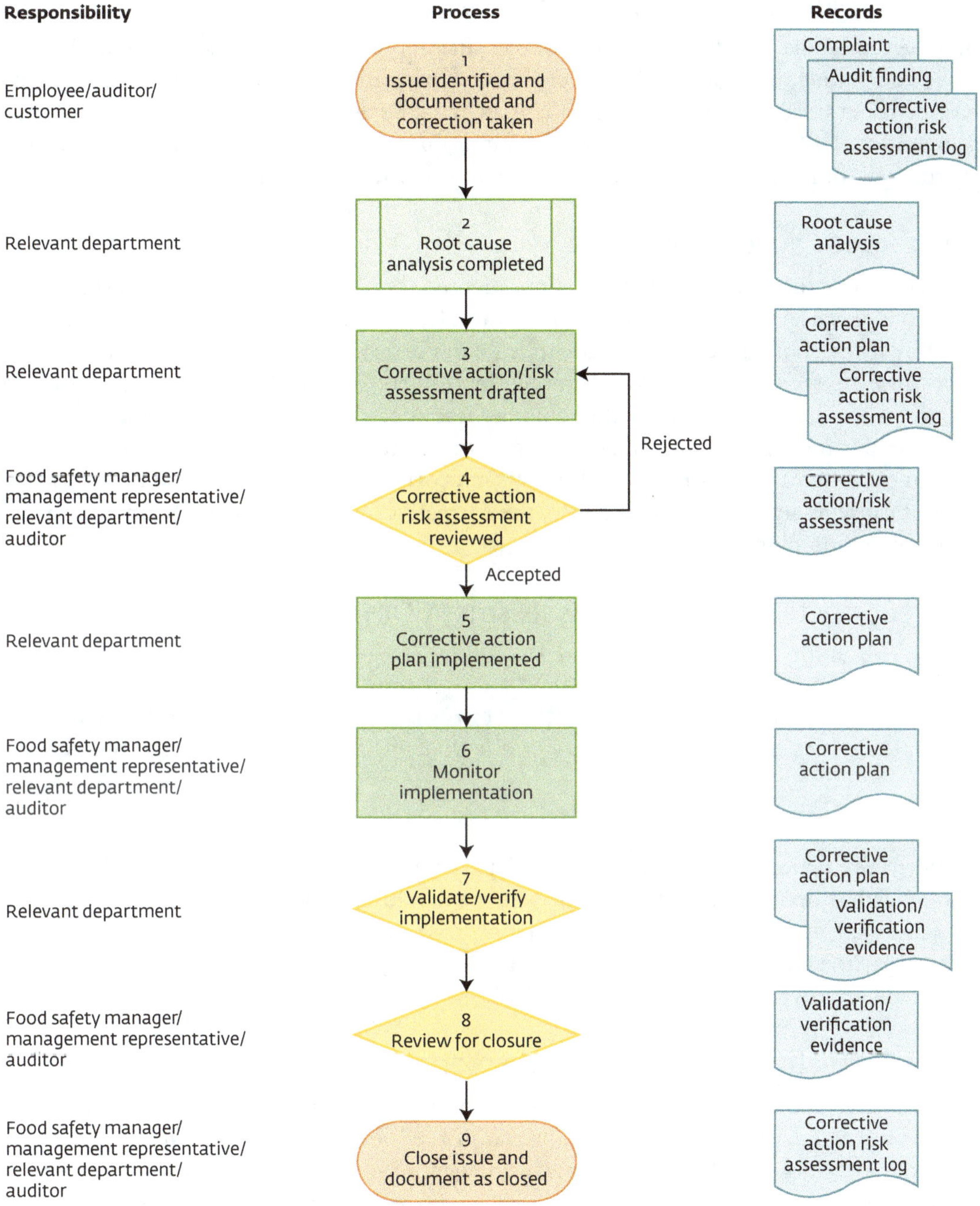

CORRECTION AND CORRECTIVE ACTION

6 Procedure Notes

Step 1 Problem Definition/Record Creation

An issue can be identified from several sources, including auditing (both internal and external), customer complaints, or legal/regulatory problems. Once an issue is identified, immediate correction must be taken to resolve the issue, and the issue must be documented within the document management system (DMS) software. The appropriate resources should then be put in place to manage the investigation of the issue in line with the flowchart on the previous page.

Step 2 Root Cause Analysis

It is mandatory that all issues raised be investigated thoroughly through the utilization of a recognized root cause analysis methodology, for example, the 5 whys, the 8Ds, Go See Think Do, and so on. Only if the root cause has been identified can correction or corrective action be implemented. Root cause analysis may be undertaken only by trained personnel. Under no circumstances should human error or a restatement of the issue be described as the root cause. If this occurs, the root cause analysis must be rejected and carried out again. Root cause analysis must be completed by the department in which the issue arose. If necessary, corrective actions may also be determined during the root cause analysis. If this is the case, the corrective actions must be documented as part of the corrective action plan.

Step 3 Corrective Action Risk Assessment Drafted

A corrective action plan should be created as follows:

Issue description	Root cause	Corrective action(s)	Corrective action(s)	Assigned implementer	Expected completion date

The corrective action plan is created by the department in which the issue arose. Once the corrective action plan has been developed, a risk assessment should be conducted to ensure no unintended consequences that may be reasonably foreseen may arise during implementation. It is the responsibility of the department to generate the plan, conduct a risk assessment, and submit the risk assessment for review and approval.

Step 4 Corrective Action Risk Assessment Reviewed

The corrective action risk assessment must be submitted to the food safety manager/management representative/auditor for review and approval. If the food safety manager/management representative/auditor decides that the assessment is insufficient or unacceptable, they will return it for rework. The assessment may be rejected on the grounds of a poorly completed root cause analysis, unrealistic time frames, the lack of assignment of responsibilities, the lack of identification in the assessment of the risks and risk mitigations that, most reasonably, should have been foreseen, or other grounds as deemed appropriate by the review team.

CHAPTER 4

CORRECTION AND CORRECTIVE ACTION

If the corrective action risk assessment is deemed approved, the review team will notify the department to proceed with the corrective action plan.

Step 5 Implement the Corrective Action Plan

The relevant department will implement the corrective action plan as documented.

Step 6 Monitor Implementation

Implementation will be monitored in accordance with the documented plan on a regular basis to ensure that timely corrective action is taken and that any issue arising is dealt with.

Step 7 Verify Implementation

If the implementing department is satisfied that the corrective action has been completed, a test to determine the effectiveness of the corrective action must be undertaken and relevant evidence recorded. If the results show that the expected outcome has not been achieved, that is, the elimination of the root cause, the department must redo the root cause analysis. Only when this evidence objectively shows that the root cause of the issue has been eliminated, may the department request that the issue be closed out.

Step 8 Review for Closure

The food safety manager/management representative/auditor and other interested parties as necessary must review the objective evidence related to the effectiveness of the corrective action.

After suitable time has elapsed following the implementation of the corrective action, the effectiveness of the corrective action must be determined. The corrective action is aimed at eliminating the cause of the nonconformity and also preventing recurrence. Hence, the validation and verification of effectiveness must test the possibility of the recurrence of the nonconformity and must be completed before the corrective action can be closed. If the test is satisfied, the corrective action can be closed. The test that is performed or the data reviewed as part of this process must be documented.

The assistance of external sources may be utilized in carrying out the review of effectiveness. If the test shows that the corrective action has not been effective, then the root cause analysis must be revisited to ensure that the correct cause has been identified, and the process must be repeated.

Only when the review team is satisfied that the root cause has been eliminated may the team allow the issue to be closed. If any doubt persists, the review team may request extra verification activities to be undertaken and results submitted again, or they may request a complete resubmission of the corrective action plan.

Step 9 Close the Corrective Action Risk Assessment

If the review team is satisfied that the root cause has been eliminated, it will authorize the closure of the issue on the corrective and preventive action system.

Classification	Company Confidential	Correction and Corrective Action Procedure	
Doc ID: SOP-009 Created: April 20, 2018	Printed: Updated: January 13, 2019	Controller: Document Controller Owner: Food Safety Manager	Page 5 of 6

CORRECTION AND CORRECTIVE ACTION

7 Records

Document	Location	Duration of record	Responsibility
Complaint	Food Safety Office	Indefinitely	Food safety manager/management representative
Audit finding	Food Safety Office	Six years	Food safety manager/management representative
Corrective action plan	Food Safety Office	Six years	Food safety manager/management representative
Validation/verification evidence	Food Safety Office	Six years	Food safety manager/management representative

TRACEABILITY	
An FBO Procedure	
Document No.	Standard operating procedure SOP-012
Created	April 20, 2018
Updated	January 13, 2019
Controller	Document Controller
Owner	Food Safety Manager
Confidentiality Statement Information in this document must be kept confidential as per the document's classification below and the rules of disclosure.	
All FBO documents are classified in the following way. PUBLIC documents are intended for anyone. COMMERCIAL IN CONFIDENCE documents are to be kept confidential among restricted individuals within the FBO and partner organizations. COMPANY CONFIDENTIAL documents are to be kept confidential within the FBO and used for normal business activities by the general office population. HIGHLY CONFIDENTIAL documents are to be kept confidential among restricted individuals within the FBO.	
© Copyright FBO. All rights reserved. No part of this publication may be reproduced, stored in a retrieval system, or transmitted in any form or by any means, electronic, mechanical, photocopying, recording, or otherwise, without the written permission of the FBO.	
Classification	Company Confidential

Revision History

Date	Version	Author	Comments (including review history)
April 20, 2018	Draft 01	Joe Bloggs	Initial document for review and discussion
April 24, 2018	V1.0	Joe Bloggs	Approved and released by process owner
January 13, 2019	V1.1	Joe Bloggs	Updated related documents section and formatting

Contents

TRACEABILITY

1 Summary

Purpose	The purpose of this procedure is to describe the ability of the process to trace each ingredient back to the source and to track dairy products after they leave the dairy plant.
Scope	This instruction covers all products manufactured or distributed by the food business operator (FBO). Local regulations and laws take precedence over this guideline.
Functional responsibility	The functional responsibility for this procedure lies with the food safety manager and traceability/recall prerequisite program (PRP) team, who are responsible for the effective implementation and maintenance of this procedure.

2 Related Documents

Policies	Food Safety Policy, POL-001 Customer/Consumer Complaints Policy, POL-002
Processes	Departmental process descriptions
Procedures	Control of Nonconforming Product, SOP-003 Correction and Corrective Action, SOP-009 Communication, SOP-020 Product Recall and Withdrawal, SOP-023 Crisis Management, SOP-029
Work instructions	Not applicable
Forms	Recall/withdrawal log Communication log Root cause analysis/corrective action
Other	Not applicable

3 Definitions

Term or acronym	Description
Complaint	Expression of dissatisfaction communicated to an organization related to the organization's products or services or the complaints-handling process if a response or resolution is explicitly or implicitly expected
Correction	Action to eliminate a detected nonconformity
Corrective action	Action to eliminate the cause of a nonconformity and to prevent recurrence
Critical tracking event	An event that identifies those core business processes in which traceability data capture is vital to a successful traceability process
Dilution	The ability to separate products that may have a large amount of a contaminant from products that may have only possible traces

TRACEABILITY

Term or acronym	Description
Exclusion	The ability to exclude the products that do not contain any contaminant
FBO	food business operator
Inclusion	The ability to include any products that may contain any trace of a possible contaminant
Key data element	The data captured during a critical tracking event to support a successful traceability process
Nonconformity	Nonfulfillment of a requirement
Product	Output that is a result of activities that are not necessarily performed at the interface between the provider and the customer; for the food business operator (FBO), this may be an ingredient, raw material, intermediate product, or finished product supplied to a customer or consumer
Recall	The process by which a product is removed from the external supply chain/distribution and consumers are publicly advised to take specific actions with the product (for instance, do not consume the product or return the product to the shop or manufacturer); this includes U.S. Food and Drug Administration (FDA) class I and class II recalls
Regulatory requirement	Obligatory requirement specified by an authority mandated by a legislative body
Risk	The effect of uncertainty on an expected result
Root cause	A cause that, once removed from the problem fault sequence, prevents the final undesirable event from recurring
Root cause analysis	A method of problem solving that involves the attempt to identify the root cause of a fault or problem
Statutory requirement	Obligatory requirement specified by a legislative body
Traceability	The ability to track a food through all stages of production, processing, and distribution, including importation and retail; traceability should mean that movements can be traced one step back and one step forward at any point in the supply chain
Tracing	The capability to identify the origin and characteristics of a product based on criteria determined at each point of the supply chain
Tracking	The capability to locate a product based on specific criteria no matter where the product is located along the supply chain

4 Introduction

Traceability systems are designed to trace and track products and their components along the supply chain. Although traceability must be an end-to-end process, it is accomplished along a supply chain consisting of independent firms with separate stand-alone information systems. To ensure continuity in the flow of traceability information, each partner in the food supply chain must pass on information about the identified lot or product group to the next partner in the food supply chain.

Thus, to accomplish end-to-end traceability, supply chain partners must undertake three key activities.

Data collection: The system must be able to capture required data. Although this may be accomplished using paper-based methods, more effective technologies, such as bar code scanners, radio frequency identification, handheld computers, and specially engineered input devices, are simplifying data collection and allowing more data to be captured.

Data storage: Once collected, the data must be organized and stored in a database that allows various options for retrieval and search.

Data transmission and sharing: The system is effective only if data can be exchanged between supply chain intermediaries. Thus, traceability systems must have systems integration capabilities to connect hardware and software, thus allowing diverse corporate systems to communicate.

In dairy processing, traceability requires information collection, filing, and sharing on:

- Product ingredients
- Processing
- Packaging
- Labeling
- Storage
- Distribution

5 Procedure Flowchart

Not applicable.

6 Procedure Notes

6.1 Identify and Record Lot IDs or Key Data Elements

The food business operator (FBO) product flow diagram should identify the places in the FBO facility where bulk products, ingredients, or packaging materials are added to make the final product. It should also identify key points in the physical process where the product is transformed or where product lots can be discretely separated, that is, within critical product flows.

Create a method to record lot IDs at each of these places. For example, recording the batch or the lot and the batch number or lot number typically starts at the manufacturing plant. Batches should be maintained and recorded according to the following:

- Quantity
- Manufacturing cycle
- Expiry date
- Weight of the active ingredient
- Excipient(s)

TRACEABILITY

Decide which identifying mark will be used for the lot ID on the various materials.

Train FBO employees to be consistent and accurate in recording lot IDs.

Keep FBO records in a way that makes the lot IDs easy to find. Identify and record flows (critical tracking events).

The following specific areas are common in the dairy foods industry and should be considered in listing key data elements–lot entry points:

- *Raw milk receiving:* When receiving raw milk, the receiving facility should consider each farm on a truck as a lot of product received. The facility should have or have access to the farm name and the address of the farmer. Model the receiving bay as a lot entry point and record the products of each farm received and the silo where the products were received. This can be accomplished in three ways:
  - The receiver records the load information only and turns the dairy farm tickets into the office, where the individual tickets are correlated with the load information and farm milk samples. This system would be used if multiple dairy farms pickups are accumulated in a single delivered load.
  - Only the route information is recorded by the receiver because the load is collected and comingled by a cooperative. In this case, the cooperative would need to have the dairy farm information for each load and raw milk sample that may become involved in the traceability effort if a recall were required.
  - The receiver records the individual dairy farm tickets and raw milk samples that are received with the load information.
- *Milk hauler/driver responsibility:* The records of the milk hauler/driver performing the dairy farm pickups are essential if a recall is to work properly and represent the first step in creating a successful traceability program. Accurate identification of the dairy farm and the product quantities and records on cleaning in place and on milk samples are critical and must reference the manifest or e-manifest or digital recorder, if this technology is used.
- *Using dairy farm IDs:* Dairy farm IDs are often used as identifiers of the dairy farm loads. This can be helpful in tracing the loads because this number is issued by a national department of agriculture and is used in inspections and other records. However, many cooperatives and other dairy businesses assign their own dairy farm IDs as well. Hauler/driver and receiver records must be consistent and accurate.
- *Raw milk pooling:* When milk is picked up from the dairy farm, loaded into silos or tanks, and reshipped to dairy food plants, it is the responsibility of the milk pooling facility to keep the records of the dairy farm loads and farm raw milk samples according to the tankers shipped. This facility is modeled in the same way as any other facility.
- *Rework:* Rework is common, but complicates traceability. Rework should be viewed and modeled in the same way as any other ingredient or product. Rework is best handled in the following manner:
  - List the points where rework could be collected during the process. Identify and label the rework as a final product.

- If the rework is not a final product, create a lot identifying mark for the rework. If it is a bulk rework situation, create a lot identifying mark and mark or tag the tank with this identifier.
- If the rework is a final product, use the appropriate lot identifying mark.
- List the points where the rework is added back into the process; record the lot identifying marks as would be done with any other ingredient (key data elements–lot entry point).
- Take note of the rework narrative in the critical tracking events section.
- Limit the addition of the rework from one day into the rework from another day as much as possible to reduce the comingling of lots.

- *Packaging materials:* Any packaging materials that touch the product should be recorded, including the following:
  - Bags and liners for product packaging
  - Vitamins and small-quantity additives
- *Disposed ingredients or products:* Records should be maintained on ingredients, products, and packaging materials that are disposed. The quantity disposed and the lot identifying marks should be recorded as with any final product.

6.2 Identify and Record Flows or Critical Tracking Events

Identify the main flow paths in the dairy plant that the product passes through from beginning to end.

- Create a method of recording each of these flows.
- Train FBO employees to be consistent and accurate in recording these flows.
- Keep FBO records in a way that facilitates relating the recorded lot IDs with the flows.
- Track FBO flows between the facilities within individual corporations or cooperatives. Keeping good records of FBO interplant transfers or a system that can link the traceability of FBO products between facilities will reduce the time it takes to identify products or exclude an FBO from a recall.

There are a few areas of special consideration in modeling the critical tracking events in a dairy foods facility. This process may be automated by the dairy processing plant within the dairy where the FBO follows instructions on the dairy processing equipment or supported by documented procedures, as follows:

- Products in storage that are not frequently cleaned in place: Oils, sugars, and other bulk ingredients are stored for long periods without being completely emptied or cleaned in place. This is common and safe, but is not in line with a granular model of traceability.
- Reset the trace for this vessel using a calculated first in, first out method. For example, if 65,000 pounds of oil are delivered, the first 65,000 pounds used exhausts that lot. On a reoccurring basis (possibly monthly), true up the calculated inventory to the actual inventory.

TRACEABILITY

- Reset the trace based on a periodic schedule. This is common practice in city water utilities because there is never really an interruption. In the case of government-supplied water, many reset the trace every 24 hours.
- Continuous processes: Some processes run longer than is practical for consideration as one lot of finished product. Spray dryers, powder silos, and other processes may run for several days without stopping for cleaning in place. Yet, the flows through these processes need to be documented either manually or automatically to provide good traceability.
- Reset the critical tracking event whenever a source or destination changes. For instance, on a dryer, create a new flow record when the powder bin selection changes. In the case of an evaporator, change the flow record whenever the silo feeding the evaporator changes. If these two are combined, the quantity of product under one critical tracking event becomes much smaller, thereby reducing the size of the lot that will be considered for a recall.

If the critical tracking event is reset as described, the following traceability can be accomplished:

Inclusion: Depending on the risk of the contaminant, the entire list of final product lot identifying marks can be maintained, recalled, or tested during cleaning in place to the cleaning in place run of the dryer.

Exclusion: Depending on the risk of the contaminant, the final products that are within the narrowest scope of a single silo crossing to a single powder bin can be isolated. This may be the highest risk product.

Dilution: Depending on the risk of the contaminant, a final product that contains items such as a common silo, powder bin, a common rework lot identifying mark can now be isolated to find those product lots with trace amounts of the contaminant.

In fact, this method can be used, especially in an automatically collected traceability solution, to find the source of the contaminant.

- *Adding rework into the process:* A rework addition should be handled in the same way as any other ingredient addition. However, where the creation of rework is possible, the points in the process should be modeled as a critical tracking event, with a final lot identifying mark so that, when the rework is added, it can be traced.

6.3 Place a Standard, Human-Readable Lot ID on the FBO Products

Label the FBO final products with a simple, human-readable lot ID so anyone using the products in manufacturing can also maintain consistent and accurate records.

- Use this lot ID in FBO records as a primary identity or, at least, a searchable field in an FBO electronic or enterprise resource planning system.
- Use this lot ID in every record in both manual and electronic enterprise resource planning.
- Add the lot or lot ID label near the human-readable lot ID so the operators in the facilities of the FBO's customers can easily record the correct identity.

6.4 Product Labeling

A simple, readable lot ID should be accurately recorded as the key element in a successful traceability system. To allow efficient and expedient traceability, the lot identifying mark should:

- Be human-readable for customers that use manual lot tracking records
- Stand out on a package, pallet label, or bill of lading so that customers can clearly determine the lot identifying mark they should use in traceability records.

If the FBO is incorporating a bar code that is used by all customers into records, ensure that both the distributors and the final customers are bar code scanning the lot identifying mark and integrating it into their traceability records as well.

The lot identity should be obvious on every package, container, pallet, and bill of lading that leaves the FBO.

If the product is meant for use by another manufacturer or processor, the text lot or lot ID should be printed boldly and visibly next to the lot identifying mark.

Alternatively, at a small manufacturer, the number should be applied in human-readable form. The text lot or lot ID should appear near the code.

If a customer requests or has accepted more extensive lot identifying marks, this is also appropriate as long as the mark is clear. The lot identifying marks should be used in all correspondence.

The recommended lot identifying mark content should consist of the following:

- The dairy plant number, the date, and a process identifier. The plant numbers are typically 4–6 digits long.
- The date—for example, July 26, 2012—should be in number form, such as 20120726 or 20122607.
- An additional identifier for the product created on a specific day is a line identity.

6.5 Dairy Milk Traceability Records

GENERAL INFORMATION

- Any final product, bulk or packaged, should have a listing of the lot numbers it contains.
- The lot numbers that these records contain should match the lot numbers in the warehouse records.
- If the FBO traceability system is stored on a database, the lot identifying marks should link or associate all the records.

Traceability records should enable the FBO to find a lot identifying mark and any contributing lot identifying marks quickly and accurately. The traceability records need to contain only the information necessary to accomplish this.

For internal records, it is recommended that the basic traceability information be linked with the full record of the process and the quality assurance records.

TRACEABILITY

The following is the contents of the basic record content set.

- *Key data elements–lot entry points:* An up-to-date listing of the key data elements–lot entry points for the facility or process area. This shows that one may track where other lot identifying marks enter the process. It will also correlate with the daily records that are kept, either manually or electronically, of the lot identifying marks that are incorporated into the final products. These records can be either textual or in flowcharts.
- *Critical tracking events:* An up-to-date listing of the physical flows in the process, or critical tracking events. This will correlate to the daily records of the flows in your facility, and will be used to find the path of the lot identifying marks through the process. These records can be either textual or flowcharts.
- *Lot identifying mark:* This record is only a short written description of the structure of the lot identifying mark and what the digits represent.

Among the basic records to be maintained by the FBO, the farm milk records should contain at least the following:

- Farm number
- Carrier/hauler identification
- Driver identification
- A list of the farm identifications in the load
- The time the load was received
- The therapeutic drug (antibiotic) test result
- The name of the receiver/tester
- Milk temperature
- Silo destination of the load

Bulk receipt records should contain at least the following:

- The bill of lading number
- Carrier information
- The lot identifying mark of the supplier
- The time received

Ingredient addition records should contain at least the following:

- The lot identifying mark of the supplier
- Carrier information
- Manufacturer name if the system is manual; if the system is electronic, the manufacturer name can be joined on the database with the lot identifying mark
- Ingredient name if the system is manual; if the system is electronic, the ingredient name can be joined on the database with the lot identifying mark
- The time of the addition
- The operator

TRACEABILITY

Final product records should contain at least the following:

- The lot identifying mark
- The product name
- The time of the product run start
- The time of the product run end

PERIPHERAL AREAS (WAREHOUSE, DISTRIBUTION CENTERS, SHIPPING)

Outside the physical processing environment within the supply chain, traceability becomes discrete, meaning each product that may be contaminated is contained in one package. If an easily identifiable lot identifying mark is included on the bill of lading, shipping records, receiving records, warehouse system, and so on, each suspect product can be quickly held, tested, removed from the food chain, or destroyed once it is traced and identified.

RECORD RETENTION, SECURITY, AND BACKUP

Traceability records are retained for the same duration as other regulatory records, such as cleaning in place and pasteurization records. Until regulatory documents begin listing traceability record retention, assume that retention should be for the same amount of time as the U.S. pasteurized milk ordinance specifies for high-temperature/short-time record retention.

It is important that these records not be lost or edited.

- If the records are manual, they should be stored in locked file cabinets or in rooms that are locked when they are not staffed or after business hours.
- If the records are electronic, they should be backed up once every 24 hours and stored in a database or data archival system in a write once, read many format.

6.6 Testing and Validation of the Traceability System

The testing and validation of the FBO traceability system should cover at least two scenarios through the FBO product recall procedure:

- Using one or more final product key data element–lot identification mark(s) to locate the contributing bulks, dairy farms, ingredients, additives, or packaging materials that the product contains
- Using a suspect or possible adulterated alert of a bulk, dairy farm, ingredient, additive, or packaging material and finding the final products that contain the possible containment.

The results of the traceability system testing and validation should be confirmed through quality assurance/laboratory results. The quality assurance/laboratory results should be maintained in the laboratory information management system.

6.7 Traceability System Testing and Validation Frequency

It is the policy of the FBO that the frequency of testing and validation of the traceability system should be at least twice a year, or following a serious food incident/event, or following a significant change in the FBO or food chain partner traceability system.

CHAPTER 4

TRACEABILITY

6.8 Postreview Actions

Postreview actions must be conducted when the mock recall is over and potential improvements implemented. Any subsequent actions occurring should be monitored and tracked through the FBO corrective and preventive action procedure.

At a minimum, an analysis of the quantities of materials involved must be undertaken, whether the materials have been produced, sold, returned, destroyed, authorized for release, not accounted for, or consumed.

The ideal goal of the mock recall is to account for 100 percent of the product (bulk, dairy farm, ingredient, additive, intermediate product, or finished product) within two hours or less.

7 Records

Document	Location	Duration of record	Responsibility
Dairy plant records (various)	Food Safety Office	Indefinitely	Food safety manager
Mock recall log	Food Safety Office	Indefinitely	Food safety manager
Communication records	Food Safety Office	Indefinitely	Food safety manager
Root cause analysis	Food Safety Office	Indefinitely	Food safety manager
Mock recall report	Food Safety Office	Indefinitely	Food safety manager
Postreview minutes	Food Safety Office	Indefinitely	Food safety manager

IDENTIFICATION AND EVALUATION OF COMPLIANCE	
An FBO Procedure	
Document No.	Standard operating procedure SOP-013
Created	April 20, 2018
Updated	January 13, 2019
Controller	Document Controller
Owner	Food Safety Manager
Confidentiality Statement	
Information in this document must be kept confidential as per the document's classification below and the rules of disclosure.	
All FBO documents are classified in the following way. PUBLIC documents are intended for anyone. COMMERCIAL IN CONFIDENCE documents are to be kept confidential among restricted individuals within the FBO and partner organizations. COMPANY CONFIDENTIAL documents are to be kept confidential within the FBO and used for normal business activities by the general office population. HIGHLY CONFIDENTIAL documents are to be kept confidential among restricted individuals within the FBO.	
© Copyright FBO. All rights reserved. No part of this publication may be reproduced, stored in a retrieval system, or transmitted in any form or by any means, electronic, mechanical, photocopying, recording, or otherwise, without the written permission of the FBO.	
Classification	Company Confidential

Revision History

Date	Version	Author	Comments (including review history)
April 20, 2018	Draft 01	Joe Bloggs	Initial document for review and discussion
April 24, 2018	V1.0	Joe Bloggs	Approved for release by process owner
January 13, 2019	V1.1	Joe Bloggs	Updated related documents and updated to reflect changes in International Organization for Standardization (ISO) 22000:2018

Contents

IDENTIFICATION AND EVALUATION OF COMPLIANCE

1 Summary

Purpose	The purpose of this procedure is to outline the management process within the food business operator (FBO) of the identification and evaluation of compliance with statutory, regulatory, and other requirements (hereafter referred to simply as compliance).
Scope	The procedure is initiated with the identification of a new or changed compliance requirement. It proceeds to recording, the collection of information, assessment of the relevance of impacts, and the establishment of the degree of compliance. Gaps, if any, are also identified and resolved. The compliance register is updated and improved, and the ongoing monitoring and evaluation of compliance are carried out.
Functional responsibility	The functional responsibility for this procedure lies with the food safety manager, who is responsible for the effective implementation and maintenance of the procedure. Departmental managers are responsible for ensuring that the records under their control are managed in accordance with this documented procedure.

2 Related Documents

Policies	Food Safety Policy, POL-001
Processes	Food Safety Process Description, PRO-002
Procedures	Control of Documents, SOP-001 Control of Nonconforming Product, SOP-003 Internal Audit, SOP-006 Correction and Corrective Action, SOP-009 Management Review, SOP-021 Product Recall and Withdrawal, SOP-023 Food Safety Legal Register, REG-001
Work instructions	Not applicable
Forms	Master document register
Other	Document management system (DMS)

CHAPTER 4

IDENTIFICATION AND EVALUATION OF COMPLIANCE

3 Definitions

Term or acronym	Description
Compliance	Statutory and regulatory compliance, including compliance with other legal obligations and requirements
Compliance register	Food safety legal register
Enforcement agency	Any person or organization delegated with vested or statutory authority, capacity, or power to perform a designated function. The enforcement agency may also be any agency that enforces the law, for example, the Food Standards Agency of the United Kingdom or the U.S. Food and Drug Administration.
FBO	food business operator
Interested party	Any external person or group—for example, an external unit of the food business operator (FBO), consumers, or regulatory agencies—with an interest in the performance or success of the organization

4 Introduction

A corporate vision for food safety compliance is a defined and documented strategy for mapping out the business's objectives in meeting the business's compliance obligations now and in the future. It is focused on future-proofing the business's need to meet a dynamic compliance framework, maintain a high level of consumer protection, and support business development objectives.

This procedure outlines the steps in the identification and evaluation of a food business operator's (FBO) legal obligations, primarily statutory and regulatory and especially toward the customer.

IDENTIFICATION AND EVALUATION OF COMPLIANCE

5 Procedure Flowchart

Responsibility	Process	Records
Interested party	1 Identify changes	Subscription
Food safety manager	2 Record information	Compliance register/meeting minutes
Food safety manager	3 Collect additional information	Compliance register
Food safety manager	4 Assess relevance/ impact	Compliance register
Food safety manager/ internal auditors	5 Evaluate compliance	Compliance report
Food safety manager/ management team	6 Define and document a gap resolution plan	Change management form
Food safety manager/ management team	7 Implement resolution	Change management form
Food safety manager/ management team	8 Update/improve, if required	Document management system (DMS)
Food safety manager	9 Monitor and measure	Compliance register

6 Procedure Notes

Step 1

Any new compliance requirements or changes in compliance requirements are identified through a combination of the FBO, the enforcement agency, industry representatives, and a legal register subscription communication and update service.

Step 2

The food safety manager shall record the information, including updating the compliance register or food safety legal register as required.

Step 3

The food safety manager shall collect additional information on the new or changed compliance requirement, where necessary, to achieve a better understanding and evaluation. The relevant legal register shall be updated and maintained as required.

Step 4

Once the necessary information and data have been collected, the relevance and impact of the new or changed compliance requirement shall be identified. The relevant legal register shall be updated and maintained if required. The food safety manager shall communicate the information to the relevant internal parties through a combination of e-mail, report, or meeting. The management review meeting shall review all new or changed compliance requirements according to the management review procedure.

Step 5

Based on the information collected, the food safety manager shall determine the best strategy for evaluating the degree of compliance, for example, document review, monitoring and measurement data, audit, or a combination of one or more, and so on, referencing the relevant legal register and updating this if required.

Step 6

If the periodic evaluation results show there is a gap, a gap resolution plan shall be defined and documented. This may include a corrective and preventive action plan if required. Reference the Control of Nonconforming Product Procedure and the Correction and Corrective Action Procedure.

Step 7

The gap analysis plan shall be implemented in a timely manner to ensure full compliance.

Steps 8 and 9

The relevant compliance register, including the food safety management system (FSMS) document facility, shall be reviewed and updated as required.

7 Records

Document	Location	Duration of record	Responsibility
Food safety legal register	Food Safety Office	Indefinitely	Food safety manager
Internal audit file	Food Safety Office	Three years	Food safety manager
Management review package	Food Safety Office	Three years	Food safety manager

TRAINING AND DEVELOPMENT	
An FBO Procedure	
Document No.	Standard operating procedure SOP-014
Created	April 20, 2018
Updated	April 24, 2019
Controller	Document Controller
Owner	Human Resources Manager
Confidentiality Statement	
Information in this document must be kept confidential as per the document's classification below and the rules of disclosure.	
All FBO documents are classified in the following way. PUBLIC documents are intended for anyone. COMMERCIAL IN CONFIDENCE documents are to be kept confidential among restricted individuals within the FBO and partner organizations. COMPANY CONFIDENTIAL documents are to be kept confidential within the FBO and used for normal business activities by the general office population. HIGHLY CONFIDENTIAL documents are to be kept confidential among restricted individuals within the FBO.	
© Copyright FBO. All rights reserved. No part of this publication may be reproduced, stored in a retrieval system, or transmitted in any form or by any means, electronic, mechanical, photocopying, recording, or otherwise, without the written permission of the FBO.	
Classification	Company Confidential

Revision History

Date	Version	Author	Comments (including review history)
April 20, 2018	Draft 01	Joe Bloggs	Initial document for review and discussion
April 24, 2019	V1.0	Mary Cahill	Format changes and reviewing and approval by process owner

Contents

CHAPTER 4

TRAINING AND DEVELOPMENT

1 Summary

Purpose	The purpose of this procedure is to describe the methodology used by the company to enable individuals, business units, and the company overall to fulfill performance requirements through the provision of training and development.
Scope	This procedure applies to the training and development of all employees, from initial entry to the identification of training and development needs following a performance evaluation or mandatory corporate training and ending with the evaluation and conformation of performance.
Functional responsibility	The functional responsibility for this procedure lies with the Human Resources Department, specifically the human resources manager, who is responsible for the effective implementation and maintenance of this procedure.

2 Related Documents

Policies	Food Safety Policy, POL-001
Processes	Human Resources Process Description, PRO-003
Procedures	Recruitment and Selection Procedure, SOP-025 Performance Appraisal Procedure, SOP-026 Disciplinary Procedure, SOP-027 Purchasing Procedure (for provision of external training), SOP-028
Work instructions	Not applicable
Forms	Job descriptions Training Attendance Form Training Request Form Logging data from the learning management system
Other	Training-the-trainer training course

3 Definitions

Term or acronym	Description
FBO	food business operator
Job description	A formal account of an employee's responsibilities

TRAINING AND DEVELOPMENT

4 Introduction

The overall objective of training and development is to develop a trained workforce that can deliver superior customer service using the latest technology and expert domain knowledge. To achieve this goal, the food business operator (FBO) provides several types of training, including new joiner onboarding, domain training, food safety compliance training, and on-the-job training.

New joiner onboarding (induction) training helps new employees integrate quickly and effectively into their new work environment.

Domain training refers to the industry-specific knowledge training that is required of individuals for the individuals to be successful in the role that they hold.

Food safety compliance training is mandatory and plays an important part in the process of educating employees on industry laws and regulations and company food safety policies and procedures. Every new employee must go through this training immediately after joining, and every employee must complete food safety compliance training on a yearly basis.

TRAINING AND DEVELOPMENT

5 Procedure Flowchart

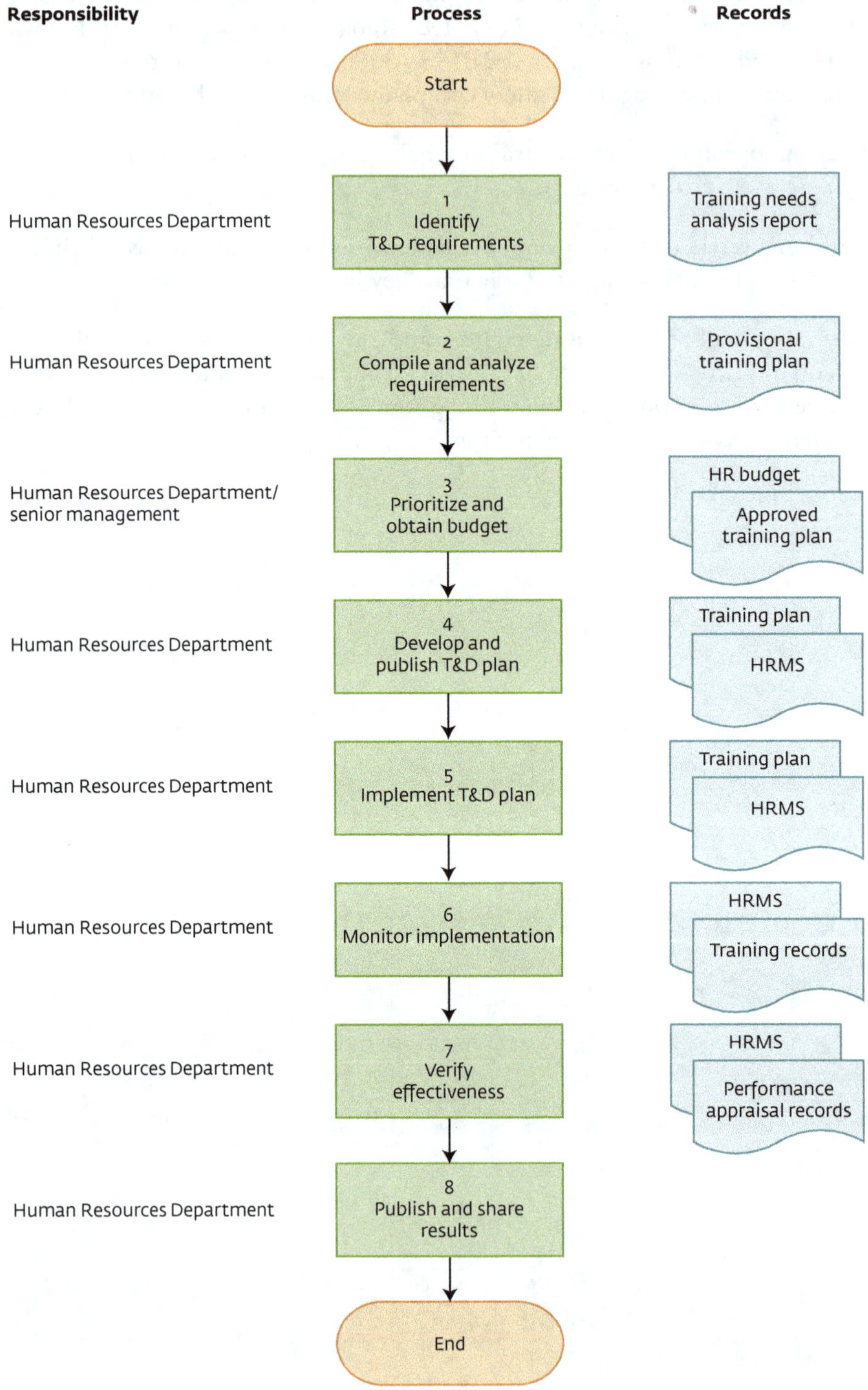

Note: HR = Human Resources; HRMS = human resources management system; T&D = training and development.

Classification	Company Confidential	Training and Development Procedure	
Doc ID: SOP-014 Created: April 20, 2018	Printed: Updated: April 24, 2019	Controller: Document Controller Owner: Human Resources Manager	Page 4 of 9

TRAINING AND DEVELOPMENT

6 Procedure Notes

Step 1 Training and Development Requirements

The Human Resources Department will identify the training and development needs across the company. This will be achieved through a review of corporate mandatory training requirements, training requirements identified through the recruitment and selection process, and the outcomes of performance appraisals. Each department will be consulted during this process.

Step 2 Training and Development Analysis

Based on the needs identified in step 1, the Human Resources Department will compile and analyze the requirements. This will result in a provisional training plan.

Step 3 Training and Development Prioritization/Budget

The provisional training plan, including prioritization requirements, will be submitted to top management for approval. Once approved, the necessary resources will be provided as part of the human resources budget.

Step 4 Training and Development Plan

Once budget approval has been received, the Human Resources Department will develop and publish the approved training and development plan through the human resources management system. The plan will outline both the mandatory and optional training and development that will be provided during the coming period.

Steps 5 and 6 Training and Development Plan Implementation and Monitoring

The Human Resources Department, in association with other departments as appropriate, will implement the training and development plan. It will continuously monitor the implementation of the training plan, using the human resources management system and the learning management system, to ensure that it is flawlessly executed. The human resources management system and the learning management system will indicate the training that has been completed by each employee. Training attendance sheets and training evaluation records will also be maintained. In cases in which it becomes evident that the training and development plan is not being followed, the Human Resources Department will take the necessary actions, including reviews conducted with senior management, to bring the plan back on track or take other measures to ensure training is completed properly.

Step 7 Training and Development Effectiveness Verification

The Human Resources Department will determine the effectiveness and impact of the training and development courses on the performance of individuals, business units, and the company. If analysis shows that training and development are not having the desired effect, a review of the training and development plan and its implementation will be conducted, and the necessary actions identified, taken, and recorded. The outputs of the performance appraisal process will be a direct input to the determination of the overall effectiveness of training and development and drive the creation of the next training and development plan.

TRAINING AND DEVELOPMENT

Step 8 Training and Development Published Results

The Human Resources Department will publish and share the results achieved through the implementation of the training and development plan with all interested parties to ensure that decisions related to ongoing training and development are based on precise factual information.

7 Management of Training and Development

7.1 Selection, Approval, and Evaluation of Trainers

INTERNAL TRAINERS

All employees selected to act as company trainers are required to meet the following minimum criteria:

- They must have been working in the area covered by the training for a significant period, two to three years minimum.
- They must be subject matter experts in the required subjects or areas.
- They must have successfully completed a training-the-trainer training course.
- They must have successfully presented the training course to their peers and the Human Resources Department.

EXTERNAL TRAINERS

If it is necessary to employ the services of external trainers to provide training, they shall be selected in accordance with a defined process. The Human Resources Department shall ensure that these trainers are competent to complete the training task. All external trainers shall meet the following criteria:

- They must be subject matter experts in the required subjects or areas.
- They must have successfully completed a training-the-trainer training course.
- They must hold the necessary educational qualifications related to the training course.
- They must have several years of work experience related to the training course and, ideally, still be working in a related area.
- They must provide written references and approvals with respect to the provision of training.
- If required, they must hold the necessary certifications from recognized certification bodies or work on behalf of a certified or accredited training organization.

Documented evidence that trainers meet the above criteria must be maintained on file by the Human Resources Department for all external training organizations and the related trainers.

In the event that an approved external trainer is unable to attend scheduled training and a substitute external trainer is recommended by the supplier, the substitute external trainer must also meet the above requirements.

A panel of approved trainers and training organizations will be maintained by the Human Resources Department.

TRAINING AND DEVELOPMENT

7.2 Training Course Evaluation

All training course material and trainers will be subject to evaluation. This is required to ensure that the level of course materials and course delivery does not deteriorate and remains relevant and that the trainees are receiving a high standard of training.

Evaluations will be implemented as follows:

- Training course evaluation forms, completed at the end of the course by the trainees, will outline their ratings of course delivery, course materials, trainers, and other relevant criteria.
- Defined internal trainer presentation reviews: this will involve witnessed and documented evaluations of the presentation of internal trainers completed by the Human Resources Department.

The Human Resources Department will review the output of these evaluation processes and ensure that, where standards are not being met, the appropriate actions are taken to ensure that there is no negative impact on the trainees or the company.

7.3 Trainee Evaluation

Depending on the type of training delivered, trainees will be evaluated to ensure that they have both received and understood the information being delivered and can implement the training in their day-to-day roles. This evaluation and assessment process can take several forms, including, but not necessarily limited to, the following:

- Written examinations on the subject matter
- Documented continuous assessments throughout course delivery
- Trainer assessments of trainees through role playing or similar exercises
- On-the-job mentoring and review
- Performance appraisals

Trainee evaluations must be documented and maintained on file.

7.4 Training Materials

If training materials, such as PowerPoint presentations, training manuals, examinations, or quizzes, are developed in-house, these must be assessed for quality and technical content prior to use and following any updates.

The Human Resources Department will review the materials from a quality perspective to ensure the following:

- The materials are in a form, manner, and language that are likely to be understood.
- They are grammatically correct.
- They are clear, concise, and visually acceptable.
- They meet company requirements on templates or notes, for example, for PowerPoint presentations.

Classification	Company Confidential	Training and Development Procedure	
Doc ID: SOP-014 Created: April 20, 2018	Printed: Updated: April 24, 2019	Controller: Document Controller Owner: Human Resources Manager	Page 7 of 9

- They do not contain any unauthorized language or content.
- Revisions are controlled.

The tutors or subject matter experts will review the materials from a technical perspective to ensure the following:

- The training course content is technically correct, accurate, and up-to-date.
- The information and examples presented are compliant with all necessary rules and regulations; if possible, training materials provided by external providers will be reviewed prior to course delivery.

The Human Resources Department will be responsible for the maintenance of internal course materials. However, it is the responsibility of the subject matter experts to ensure that the courses and course materials are updated as necessary in line with any changes in food safety compliance requirements, regulatory requirements, or other significant changes affecting course content.

All internal training materials will be held by the Human Resources Department and issued to the trainers as required.

7.5 Training Course Attendance

Once a training course has been scheduled, it is the responsibility of management to release staff to attend the training course. It is required that all trainees attend the full duration of relevant courses. If, for any reason, trainees must leave a course, they must retake the entire course. The Human Resources Department may amend this requirement on a case-by-case basis.

7.6 Poor Performance or the Unsuccessful Completion of Training

A training matrix will be maintained by the Human Resources Department to identify both the mandatory and optional training courses available. The matrix may be used by other departments and the management team to identify potential training solutions available where an employee is found not to be performing to expected levels.

In the event an employee does not successfully complete a mandatory training course, they may be offered the option of retaking the course or course assessment. If an employee has not successfully completed a mandatory training course after numerous attempts or the performance of the employee in the job role does not improve, then both the Human Resources Department and the departmental manager will meet to determine the best course of action to be taken with regard to the employee. A decision will be made and communicated to the employee. This decision will be documented and monitored by the Human Resources Department.

TRAINING AND DEVELOPMENT

8 Records

Document	Location	Duration of record	Responsibility
Induction pack forms	Human Resources Department	Indefinitely	Human resources manager
Training needs analysis	Human Resources Department	One year, then archive	Human resources manager
Training plan	Human Resources Department	One year, then archive	Human resources manager
Training attendance sheet	Human Resources Department	One year, then archive	Human resources manager
Training record	Human Resources Department	One year, then archive	Human resources manager
Education records	Human Resources Department	Indefinitely	Human resources manager
Learning management system login records	Learning management system	Indefinitely	Human resources manager
Learning management system evaluation results	Learning management system	Indefinitely	Human Resources Department

<table>
<tr><th colspan="2">COMPLAINT MANAGEMENT</th></tr>
<tr><td colspan="2">An FBO Procedure</td></tr>
<tr><td>Document No.</td><td>Standard operating procedure SOP-015</td></tr>
<tr><td>Created</td><td>April 20, 2018</td></tr>
<tr><td>Updated</td><td>January 13, 2019</td></tr>
<tr><td>Controller</td><td>Document Controller</td></tr>
<tr><td>Owner</td><td>Food Safety Manager</td></tr>
<tr><td colspan="2">Confidentiality Statement
Information in this document must be kept confidential as per the document's classification below and the rules of disclosure.</td></tr>
<tr><td colspan="2">All FBO documents are classified in the following way. PUBLIC documents are intended for anyone. COMMERCIAL IN CONFIDENCE documents are to be kept confidential among restricted individuals within the FBO and partner organizations. COMPANY CONFIDENTIAL documents are to be kept confidential within the FBO and used for normal business activities by the general office population. HIGHLY CONFIDENTIAL documents are to be kept confidential among restricted individuals within the FBO.</td></tr>
<tr><td colspan="2">© Copyright FBO. All rights reserved. No part of this publication may be reproduced, stored in a retrieval system, or transmitted in any form or by any means, electronic, mechanical, photocopying, recording, or otherwise, without the written permission of the FBO.</td></tr>
<tr><td>Classification</td><td>Company Confidential</td></tr>
</table>

Revision History

Date	Version	Author	Comments (including review history)
April 20, 2018	Draft 01	Joe Bloggs	Initial draft for review and discussion
April 24, 2018	V1.0	Joe Bloggs	Technical review and update of Correction and Corrective Action Procedure notes
January 13, 2019	V1.1	Joe Bloggs	Updated document reference section

Contents

COMPLAINT MANAGEMENT

1 Summary

Purpose	The purpose of this procedure is to describe the methodology used by the organization to manage complaints and maintain customer (retail and commercial) and consumer satisfaction and trust.
Scope	This procedure applies to the receipt, review, investigation, and resolution of complaints.
Functional responsibility	The functional responsibility for this procedure lies with the food safety manager.

2 Related Documents

Policies	Food Safety Policy, POL-001
Processes	Food Safety, PRO-001
Procedures	Control of Nonconforming Product, SOP-003 Correction and Corrective Action, SOP-009 Management Review, SOP-021
Work instructions	Not applicable
Forms	Complaint form
Other	Not applicable

3 Definitions

Term or acronym	Description
Correction	Immediate action to correct an actual or potential problem
Corrective action	Action to eliminate the root cause of a problem and prevent recurrence
FBO	food business operator
Root cause	A cause that, once removed from the problem fault sequence, prevents the final undesirable event from recurring
Root cause analysis	A method of problem solving involving the attempt to identify the root cause of a fault or problem

4 Introduction

The food business operator (FBO) has implemented a set of processes for gathering customer/ consumer complaints, reviewing these complaints, conducting investigations, determining the root cause of the complaints, and taking action to resolve the complaints with a view to the prevention of a recurrence.

CHAPTER 4

COMPLAINT MANAGEMENT

A customer/consumer complaint can be defined as any expression of dissatisfaction communicated by the FBO's customer or consumer regarding any product or services provided by the company. This policy covers all written complaints, serious or unresolved telephone complaints, and complaints raised in a face-to-face meeting or by a third party acting for that customer, such as an intermediary, a legal representative, or a food safety regulatory body.

5 Procedure Flowchart

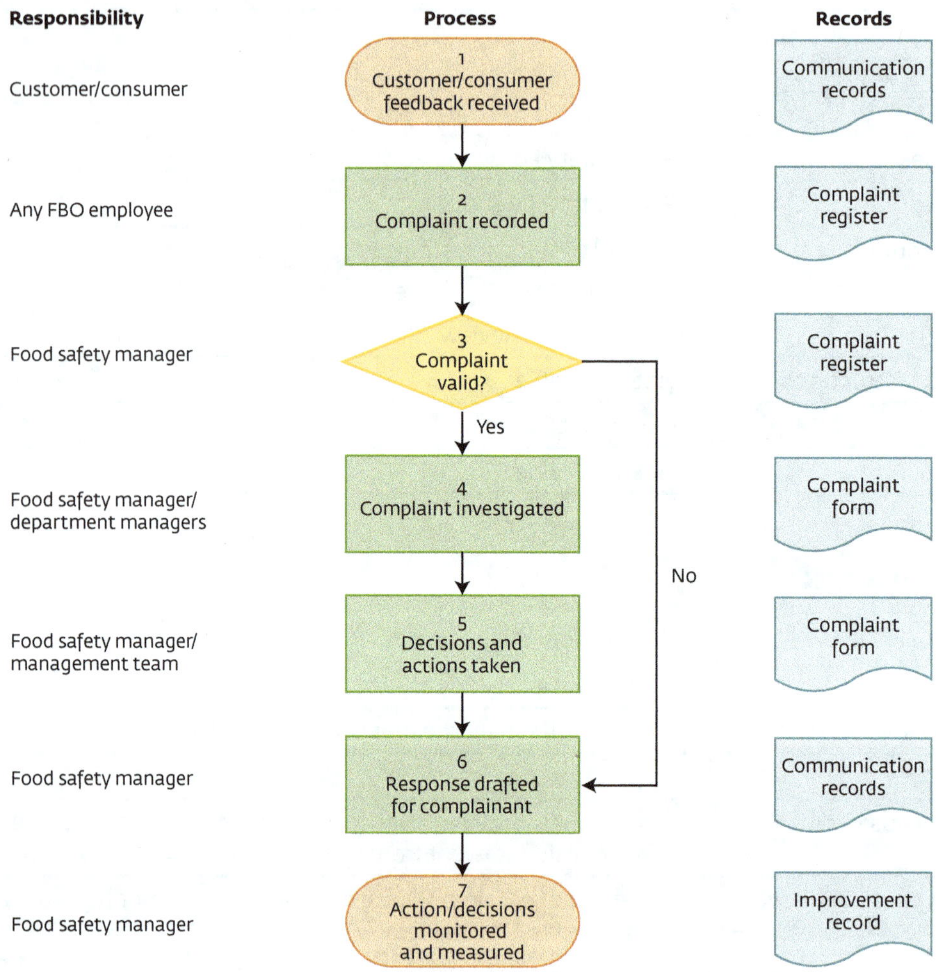

6 Procedure Notes

Step 1 Receipt of Complaint

A complaint can be made by a customer/consumer face-to-face, over the phone, by e-mail, or by other method. Where relevant food safety regulations also apply, these must be completed in conjunction with this procedure.

COMPLAINT MANAGEMENT

If you receive a client complaint,

1. Listen to the client as the complaint is being communicated verbally if face-to-face.
2. Establish what the complaint is and record the complaint on a complaint form.
3. Clarify with the client that you have understood the complaint correctly.
4. As a matter of good practice, apologize for the occurrence of the issue that they have experienced.
5. Outline to the client that the company has a complaints policy, that the complaint will be investigated, and that a formal response will be issued to them. Explain that this process may take time.
6. Establish the client's contact details.

Retain copies of any documentation provided by the client and keep these with the complaint form.

Step 2 Recording of Complaint

Formally complete the complaint form, including the following:

1. Date
2. Reference number
3. Customer name
4. Customer contact number
5. Certificate of product registration number
6. Customer complaint—description
7. Action taken
8. Final status

Attach all documentation relating to the complaint. Forward the complaint details to the food safety manager.

The food safety manager formally completes the complaint register. The client should be contacted by phone/mail to advise the client that the complaint will be considered within a maximum of five working days. The complaint is forwarded to the food safety manager. A deputy may also carry out this work on behalf of the food safety manager.

Step 3 Review of Complaint—Validity

The complaint resolution officer carries out an initial assessment on whether the complaint is valid or not. If it is, the complaint is moved on to step 4. If it is not, a formal response outlining the reasons for this outcome is communicated according to step 6. The complaint is forwarded to the relevant department manager for investigation.

Step 4 Investigation of Complaint

The department manager carries out a detailed investigation using the staff resources available, the branch manager, the member of staff who took the initial complaint, and other staff members as required. The department manager uses the corrective action procedure to investigate the root cause and determine initial containment actions and corrective actions.

COMPLAINT MANAGEMENT

Step 5 Action and Decision

Appropriate actions and decisions are taken following the complaint investigation and documented as a correction and corrective actions, with reference to the corrective action procedure. The corrective actions are verified for effectiveness as per the corrective action procedure. It may take time to verify the effectiveness of any corrective actions.

Step 6 Closure of Complaint

The department manager drafts a response for the complainant. This is agreed with the food safety manager if required by the circumstances and then released to the customer/consumer. The food safety manager retains a copy of the formal response with the complaint form. Complaints are filed by reference number and date. This should occur within 20 working days of reception of the complaint. If required, the complaint response is communicated to the relevant food safety regulatory body.

Step 7 Monitoring and Measuring

The food safety manager maintains the complaint files and complaint register for review. The food safety manager carries out an analysis of complaints (specifically, recurrences), a trend analysis, and an analysis of the effectiveness of the complaints system. The food safety manager prepares trending data for the management review process to demonstrate that the complaints are being effectively managed to the satisfaction of the FBO and the client. The complaints and summaries of trending data are inputs for the management review procedure. All complaints are to be completed and audited according to the internal audit procedure. Finally, the food safety manager continues to monitor and measure the effectiveness of actions and decisions to ensure effectiveness and to verify if the same problem and cause occur subsequently.

7 Records

Document	Location	Duration of record	Responsibility
Complaint form	Food Safety Office	Seven years	Food safety manager
Complaint register	Food Safety Office	Seven years	Food safety manager
Complaint investigation notes and formal responses	Food Safety Office	Seven years	Food safety manager
Trend analysis	Food Safety Office	Indefinitely	Food safety manager

PEST CONTROL	
An FBO Procedure	
Document No.	Standard operating procedure SOP-016
Created	November 9, 2017
Updated	January 13, 2019
Controller	Food Safety Manager
Owner	Hygiene Manager
Confidentiality Statement Information in this document must be kept confidential as per the document's classification below and the rules of disclosure.	
All FBO documents are classified in the following way. PUBLIC documents are intended for anyone. COMMERCIAL IN CONFIDENCE documents are to be kept confidential among restricted individuals within the FBO and partner organizations. COMPANY CONFIDENTIAL documents are to be kept confidential within the FBO and used for normal business activities by the general office population. HIGHLY CONFIDENTIAL documents are to be kept confidential among restricted individuals within the FBO.	
© Copyright FBO. All rights reserved. No part of this publication may be reproduced, stored in a retrieval system, or transmitted in any form or by any means, electronic, mechanical, photocopying, recording, or otherwise, without the written permission of the FBO.	
Classification	Company Confidential

Revision History

Date	Version	Author	Comments (including review history)
November 9, 2017	Draft 01	Joe Bloggs	Initial document for review and discussion
November 11, 2017	Draft 02	Joe Bloggs	Updated in line with review comments
November 12, 2017	Draft 03	Joe Bloggs	Updated in line with review comments
March 24, 2018	V1.0	Mary Cahill	Format changes
December 27, 2018	V1.1	Mary Cahill	Update following technical review
January 13, 2019	V1.2	Mary Cahill	Format changes; added form document control numbers and updated document reference section

Contents

PEST CONTROL

1 Summary

Purpose	The purpose of this procedure is to eliminate or minimize conditions that may allow pests into the food plant.
Scope	This procedure applies throughout the dairy.
Functional responsibility	The functional responsibility for this procedure lies with the hygiene manager, who is responsible for the effective implementation and maintenance of the procedure.

2 Related Documents

Policies	Food Safety Policy, POL-001
Processes	Food Safety Process Description, PRO-002
Procedures	Hygiene Procedure, SOP-005 Maintenance Procedure, SOP-088
Work instructions	Not applicable
Forms	Pest Sighting Log 020
Other	Dairy Plant Map SOP-016-1 J&J Pest Control: 608-222.4400 Pest Control Book SOP-016-2

3 Definitions

Term or acronym	Description
Pest sighting log	A log containing details of the pests seen, with entries on where and when the sightings took place

4 Introduction

The Pest Control Procedure documents and identifies target pests and hazards and addresses plans, methods, schedules, and control procedures.

5 Procedure Flowchart

Not applicable.

CHAPTER 4

PEST CONTROL

6 Procedure Notes

6.1 Materials

1. Pest Control Book, document ID SOP-016-2, located on the shelving unit in the loading dock
2. J&J Pest Control: 608 222-4400, list of pest control chemicals

6.2 Hazards

1. Chemical treatments are located throughout the dairy; see the pest control dairy plant map. Plant personnel are not responsible for handling the treatments, but should be aware of their presence.

6.3 Notes

1. A pest control representative from J&J Pest Control Company comes to the plant on a weekly basis to inspect for pests. The representative should also be noting any conditions that are conducive to pest infestation.
   a. The inspection includes the interior and exterior of the building, especially areas that are prone to pest infestation.
   b. The representative must also inspect any areas noted in the Pest Sighting Log 020 and Pest Control Book SOP-016-2.
2. A report is filled out on each visit and is kept in the Pest Control Book SOP-016-2. The pest control representative must notify the food safety manager of inspection results and obtain the food safety manager's signature.
3. It is the pest control representative's responsibility to follow through on any treatment that may be necessary. The plant should utilize the suggestions of the representative on preventing pest infestation.
4. If a plant employee notices any pests, they should undertake the following:
   a. Record in the Pest Sighting Log 020 what pest was seen, where it was seen, and the date of the sighting.
   b. Notify the food safety manager.

6.4 Recordkeeping

All visits and treatments related to pest control must be documented in the SOP-016-2 Pest Control Book.

7 Records

Document	Location	Duration of record	Responsibility
Pest Sighting Log	Loading dock	Indefinitely	Hygiene manager

MANAGEMENT REVIEW	
An FBO Procedure	
Document No.	Standard operating procedure SOP-021
Created	April 20, 2018
Updated	April 24, 2019
Controller	Document Controller
Owner	Food Safety Manager
Confidentiality Statement	
Information in this document must be kept confidential as per the document's classification below and the rules of disclosure.	
All FBO documents are classified in the following way. PUBLIC documents are intended for anyone. COMMERCIAL IN CONFIDENCE documents are to be kept confidential among restricted individuals within the FBO and partner organizations. COMPANY CONFIDENTIAL documents are to be kept confidential within the FBO and used for normal business activities by the general office population. HIGHLY CONFIDENTIAL documents are to be kept confidential among restricted individuals within the FBO.	
© Copyright FBO. All rights reserved. No part of this publication may be reproduced, stored in a retrieval system, or transmitted in any form or by any means, electronic, mechanical, photocopying, recording, or otherwise, without the written permission of the FBO.	
Classification	Company Confidential

Revision History

Date	Version	Author	Comments (including review history)
April 20, 2018	Draft 01	Joe Bloggs	Initial document for review and discussion
April 24, 2019	V1.0	Joe Bloggs	Approved for release by process owner

Contents

CHAPTER 4

MANAGEMENT REVIEW

1 Summary

Purpose	The purpose of this procedure is to describe the following: • The methodology employed by senior management to ensure that the food safety management system (FSMS) remains suitable, adequate, and effective and is continuously improved.
Scope	This procedure applies to the following: • Planning, data gathering, the identification of trends, presentations to the senior management team, and the follow-up of any identified action items, including the updating of the FSMS.
Functional responsibility	The functional responsibility for this procedure lies with the food safety team leader or food safety manager, who is responsible for the effective implementation and maintenance of the procedure.

2 Related Documents

Policies	Food Safety Policy, POL-001
Procedures	Internal Auditing, SOP-006 Correction and Corrective Action, SOP-009 Strategic Planning, SOP-019 Risk Management, SOP-030
Work instructions	Not applicable
Forms	Management review meeting minutes, document template Management review meeting, presentation template
Other	Data reviewed as part of the management review meeting

3 Definitions

Term or acronym	Description
FBO	food business operator
FSMS	food safety management system
Management team	The individual or group who directs and controls an organization at the highest level

4 Introduction

In line with good business practice and the requirements of International Organization for Standardization (ISO) 22000:2018, clause 9.3, the senior management of the company will review the food safety management system (FSMS) at least once a year on a fixed date to ensure that the company remains suitable, adequate, and effective. This review will be a structured process and identify outputs and actions related to continual improvement opportunities, the need for changes to the FSMS, and resource needs.

This review will be held at least once a year and must be attended by the general manager, the heads of departments, and the food safety team leader. A quorum of at least the general manager, all heads of departments, and the food safety manager is required for the meeting to proceed. Minutes must be taken, including any action items identified during the meeting and held on file.

5 Procedure Flowchart

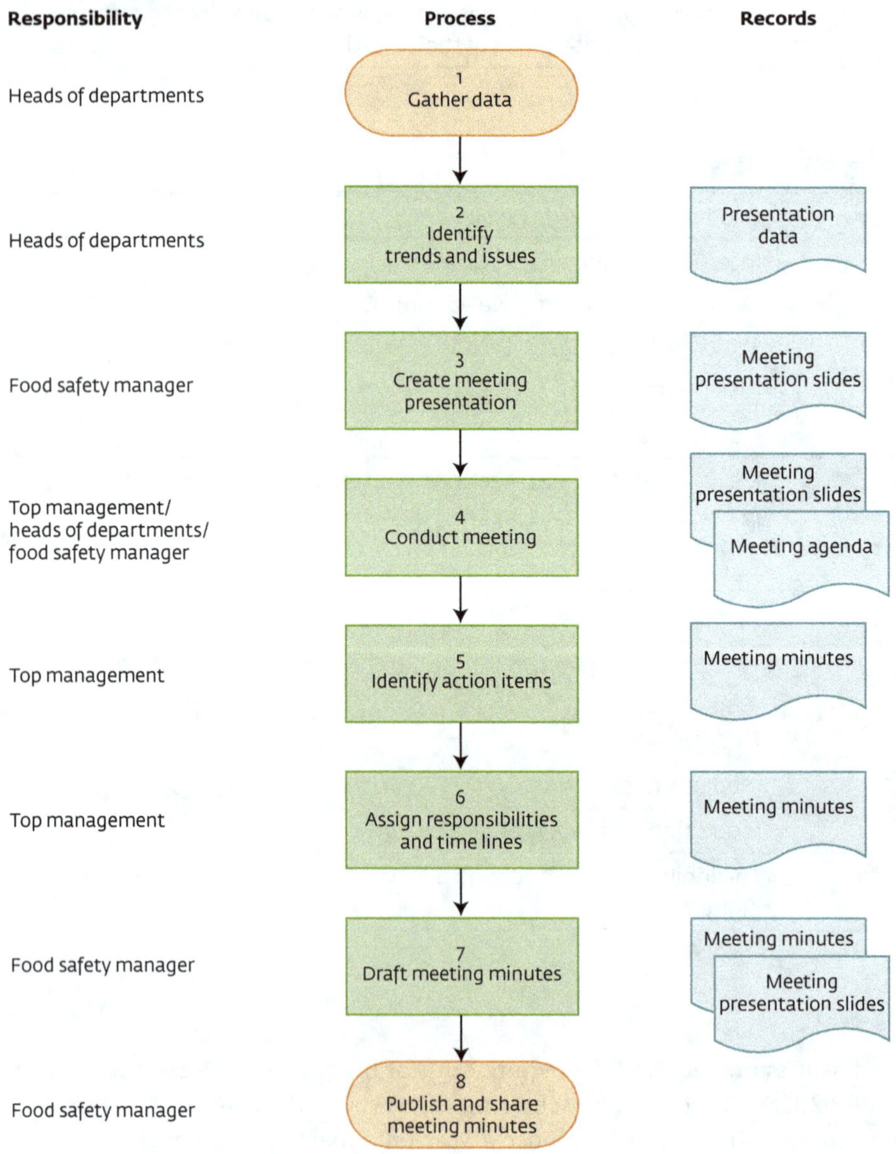

MANAGEMENT REVIEW

6 Procedure Notes

Steps 1 and 2

In advance of the scheduled management review meeting, the heads of departments will bring together data on the performance of the processes and activities of their departments. These data will then be reviewed by them to identify trends, either positive or negative. These trends will then be presented to the management team during the review and sent to the food safety manager in preparation for the management review.

Step 3

Based on the data received from the heads of departments, the food safety manager will create the overall management review presentation slides if necessary or required.

Steps 4, 5, and 6

The general manager of the food business operator (FBO) will chair the meeting, supported by the food safety manager. The two will assign a person to take the minutes of the meeting. The food safety manager may invite other process owners to present specific agenda items at the meeting. Each attendee will be allowed to ask any questions and so on in relation to the data to allow for a full and open discussion. If decisions are taken or action items are identified, these must be agreed by the management team and recorded in the minutes in accordance with section 7.4 of this procedure. If an action is agreed, the specific action, person responsible, and time frame should be recorded.

Steps 7 and 8

The minutes will be taken during the meeting, and the minutes will be verified as follows:

- The food safety manager must review and approve the minutes prior to issuing them to the general manager.
- The general manager or the deputy of the general manager must sign and date the minutes of the meeting to signify approval of the minutes and a commitment to ensure completion and implementation of any decisions or actions identified.

Once approved, the minutes can be circulated to the organization. A copy of the minutes must be held on file for the purposes of recordkeeping.

The minutes of the meeting should be published within five days of completion of the meeting.

7 Management Review Meeting

7.1 Attendees

The following persons are required to attend the meeting:

- General managers
- Heads of departments
- Food safety manager
- Any other persons as required

MANAGEMENT REVIEW

If a deputy attends and stands in as the representative of another person, the deputy is assumed to have the full authority of the other person in relation to making decisions and accepting responsibility to carry out any decisions or actions agreed at the meeting. Deputies should be relied on in this way only as an exception.

7.2 Agenda

The agenda for the management review meeting must include at least the following points:

- The status of actions agreed during previous management reviews
- Changes in external and internal issues that are significant and that are relevant to the FSMS, including, but not limited to, legal, technological, competitive, market, cultural, social, or economic environments; cybersecurity and food fraud; food defense and intentional contamination; and the knowledge and performance of the organization, whether international, national, regional, or local
- Information on food safety performance, including trends and indicators on the following:
  - The results of system updating activities
  - Monitoring and measurement results, including, but not limited to, trends in environmental monitoring, hygiene inspection rounds, glass inspection rounds, product quality (including microbiological data), product shelf life monitoring data, pest control, customer or consumer complaints, supplier complaints, blocked stock, critical control point (CCP) violations, operational prerequisite program (OPRP) violations, and mock recall performance
  - Nonconformities and corrective actions
  - Analysis of the results of verification activities related to prerequisite programs (PRPs) and the hazard analysis critical control point (HACCP) plan
  - The results of both internal and external audits
  - Inspections, for example, regulatory or customer inspections
  - Customer satisfaction survey results
  - Issues concerning external providers and other relevant interested parties
  - The extent to which the objectives of the FSMS have been met
- The adequacy of the resources required for maintaining an effective FSMS
- New or revised statutory and regulatory requirements
- Emergency situations, accidents, and food product withdrawals
- Review of the effectiveness of the food safety team: is the food safety team informed of all relevant changes and assigned sufficient time to provide the necessary input
- Analyses of risks and opportunities and the effectiveness of actions taken to address the risks and opportunities
- New, potential opportunities for continual improvement
- Documentation management system (DMS) performance
- Food safety policy, including potential update and signing by the general manager
- Updating objectives, that is, objectives that are SMART (specific, measurable, attainable (or achievable), realistic, and time-bound)
- Any other business

Classification	Company Confidential	Management Review Procedure	
Doc ID: SOP-021 Created: April 20, 2018	Printed: Updated: April 24, 2019	Controller: Document Controller Owner: Food Safety Manager	Page 5 of 6

MANAGEMENT REVIEW

7.3 Review Output

The output of the management review meeting shall include decisions and actions related to continual improvement opportunities and to the following:

- Assurance of food safety
- Improvement of the effectiveness of the FSMS, including communication if necessary
- Resource needs
- Any need for changes in the FSMS, including revisions in food safety policy and objectives

The overall output of the meeting is a decision on whether the FSMS remains suitable, adequate, and effective. Documented information on the meeting and the decisions and actions of the meeting should be kept as evidence of the results.

7.4 Management Review Minutes

Minutes must be produced following every meeting and be created using the approved template. The minutes must be detailed and accurate and give a clear description of the topics covered. If any decisions or actions are identified as a result of the meeting, these must accomplish the following:

- Clearly describe the decisions taken, including potential implications
- Clearly describe the required actions to be taken
- Identify the individual or group responsible for the completion of the action
- Identify the time frame assigned for the completion of the action

Management review records will be maintained for six years.

7.5 Approval of the Management Review Minutes

The minutes are approved as outlined in steps 7 and 8 in the procedure notes.

7.6 Communication of the Output of the Management Review

An abridged version of the minutes will be communicated to the company through the heads of departments.

8 Records

Document	Location	Duration of record	Responsibility
Management review presentation slides if used	Food Safety Office	Six years	Food safety manager
Management review meeting minutes	Food Safety Office	Six years	Food safety manager

CALIBRATION	
An FBO Procedure	
Document No.	Standard operating procedure SOP-022
Created	April 20, 2018
Updated	January 13, 2019
Controller	Document Controller
Owner	Food Safety Manager
Confidentiality Statement Information in this document must be kept confidential as per the document's classification below and the rules of disclosure.	
All FBO documents are classified in the following way. PUBLIC documents are intended for anyone. COMMERCIAL IN CONFIDENCE documents are to be kept confidential among restricted individuals within the FBO and partner organizations. COMPANY CONFIDENTIAL documents are to be kept confidential within the FBO and used for normal business activities by the general office population. HIGHLY CONFIDENTIAL documents are to be kept confidential among restricted individuals within the FBO.	
© Copyright FBO. All rights reserved. No part of this publication may be reproduced, stored in a retrieval system, or transmitted in any form or by any means, electronic, mechanical, photocopying, recording, or otherwise, without the written permission of the FBO.	
Classification	Company Confidential

Revision History

Date	Version	Author	Comments (including review history)
April 20, 2018	Draft 01	Joe Bloggs	Initial document for review and discussion
April 24, 2018	V1.0	Joe Bloggs	Approved for release by process owner
January 13, 2019	V1.1	Joe Bloggs	Updated related documented section

Contents

CHAPTER 4

CALIBRATION

1 Summary

Purpose	The purpose of this procedure is to describe the calibration program requirements for measurement and test equipment.
Scope	This procedure applies to measurement and test equipment, that is, devices used to test, measure, evaluate, inspect, or otherwise examine materials, supplies, equipment, and systems or to determine compliance with specifications. This includes process control devices with the potential to impact food safety. This procedure provides instructions for the management of calibrations performed on measurement and test instruments by service organizations, original equipment manufacturers, contractors, or laboratories (herein referred to as contractor) and to ensure traceability to national or international standards. This procedure states the documentation requirements for equipment calibrated by in-house personnel. All measurement and test equipment is to be enrolled in the Calibration Program or the Preventive Maintenance Program. Enrollment includes measurement and test equipment designated "Reference Only" and "No Calibration Required."
Functional responsibility	The functional responsibility for this procedure lies with the food safety manager, who is responsible for the effective implementation and maintenance of this procedure. Departmental managers are responsible for ensuring that records under their control are managed in accordance with this documented procedure.

2 Related Documents

Policies	Food Safety Policy, POL-001
Processes	Food Safety, PRO-001
Procedures	Control of Documents, SOP-001 Control of Nonconforming Product, SOP-003 Correction and Corrective Action, SOP-009
Work instructions	Not applicable
Forms	Test equipment installation qualification Measurement instrument status change form
Other	TEM Manuals ISO/IEC 17025:2017 Internal calibration report

CALIBRATION

3 Definitions

Term or acronym	Description
Accuracy	The relative agreement of a measured value with an accepted standard
Calibration	Verification of a measurement instrument's performance against a traceable standard
Calibration interval	The duration of time between calibrations
FSMS	food safety management system
Loop calibration	The calibration of measurement instruments as installed in a total system; represents the calibration of the instruments as used
Measurement equipment	Any instrument that monitors or controls a critical parameter of a manufacturing process or controlled environment or that is used to measure a product or component specification
National standard	A reference tool utilized by an internationally recognized standards laboratory representing the country that operates the laboratory
Precision	Also known as repeatability; the variation in readings obtained by repeating the exact same measurement(s)
Range	The breadth or span of an instrument's capability of measurement
Reproducibility	A measure of the ability of a measuring instrument to produce the same readings if the instrument is used by a different operator
Resolution	The power of discrimination of an instrument; for analogue instruments, this is limited to one-half of a minor scale graduation
Standard	A defined reference tool that is traceable to a national standard
Test accuracy ratio	The amount of uncertainty in a measurement with respect to an absolute standard
Traceability	The documented reference of calibration results to a recognized standard

4 Introduction

Calibration defines the accuracy and quality of measurements recorded using a piece of equipment. Over time, there is a tendency for results and accuracy to drift, especially during the use of particular technologies or the measurement of particular parameters, such as temperature and humidity. If one is to be confident in the results being measured, equipment needs to be calibrated, serviced, and maintained throughout its lifetime to produce reliable, accurate, and repeatable measurements.

The goal of calibration is to minimize any measurement uncertainty by ensuring the accuracy of the test equipment. Calibration quantifies and controls errors or uncertainties within measurement processes to an acceptable level.

Calibration is thus vitally important wherever measurements are important. It enables users and businesses to have confidence in the results that they monitor, record, and subsequently control.

CALIBRATION

5 Procedure Flowchart

Not applicable.

6 Procedure Notes

6.1 Enrollment of Equipment in the Calibration Program

The requester will notify the Calibration Department of new equipment by completing and returning the test equipment installation qualification form to the Calibration Department.

The requester shall deliver the following to the Calibration Department:

- Measuring equipment (if portable).
- The test equipment installation qualification form.
- The test equipment installation qualification form shall detail the measurement instrument's suitability for its intended use prior to enrollment. The determination of suitability must consider accuracy, the test accuracy ratio, precision, range, resolution, and conditions of use, including environmental conditions. A test accuracy ratio of at least 4:1 is required; the rationale for any exceptions must be documented and approved.
- A test accuracy ratio of at least 10:1 shall be required for the standards used for in-house calibration; the rationale for exceptions must be documented and approved.
- A copy of the equipment specifications (if available from the manual/catalogue) will be included; otherwise, the calibration requirements will be listed in the special instructions section.
- Operation or service manual(s) for equipment that is to be or that can potentially be calibrated in-house (if available).
- Calibration certificate(s).
- The certificates for new measurement and test equipment require, at a minimum, a statement of traceability to national, international, or consensus standards and conformity with published specifications.

Active measurement and test equipment that is not calibrated over the entire measurement range or capabilities shall be identified with a "Limited" label or equivalent. The limitations on use shall be affixed on or near the measurement and test equipment. Limitations shall be listed in the "Special Instructions" section of the test equipment installation qualification form.

6.2 Calibration Intervals

The interval assignment should be established as recommended below in descending order of preference:

- The calibration history and intended use of the equipment under evaluation
- Similar measurement and test equipment enrolled in the calibration system
- Documented engineering rationale based on usage

CHAPTER 4

CALIBRATION

- Manufacturer's recommendation
- In the event none of the above information is available, the initial interval shall not be greater than six months.

Interval changes may be requested by the owner department by completing the interval change form. The rationale must be documented on the form. Approvals should consider the risk of using out-of-tolerance measurement and test equipment in the production or inspection process(es).

Interval increases greater than half (1/2) the current calibration cycle require justification based on the recommendations above.

The initial introduction of new measurement instruments that have not been used since the initial calibration performed by the original equipment manufacturer may be extended another full cycle if this is permitted by the original equipment manufacturer as documented on the original equipment manufacturer's calibration certificate within the calibration system software.

Calibration intervals shall be evaluated and documented on an annual basis by the calibration coordinator within the calibration system software.

6.3 Change of Equipment Status

The equipment owner will request changes in the equipment/calibration status using the measurement instrument status change form. Equipment/calibration status categories include, but are not limited to, the following:

- *Active:* measurement and test equipment that is calibrated over the entire measurement range or capabilities; this equipment shall be labeled with a "Calibrated" label
- *Inactive:* measurement or test equipment that is currently not in use and, consequently, should not be an active part of the calibration program; this equipment shall be labeled with a "Do Not Use–Out of Service" label and made inoperable if possible
- *Discontinued:* measurement and test equipment that has been discontinued or destroyed
- *Reference only:* measurement and test equipment that has a measurement capability, but is currently not used for any measurement or test activities to determine conformity with any equipment, product, process, design verification/validation, or environmental specifications; this equipment shall be labeled with a "Not Calibrated–For Reference Only" label
- *No calibration required:* measurement and test equipment that, by nature or application, does not require periodic calibration; equipment in this category includes intrinsic standards and equipment used in specific applications in which output values are verified by other calibrated measurement and test equipment; this equipment shall be labeled with a "No Calibration Required" label
- *Lost:* equipment that cannot be located by the owner department

DISCONTINUED/DISPOSED EQUIPMENT

The department owning the equipment to be disposed of/discontinued will complete the measurement instrument status change form.

CALIBRATION

The calibration ID label will be removed from the equipment by the owner department and affixed to the measurement instrument status change form.

The equipment shall be appropriately identified for disposal/destruction by the owner department.

The owner department will obtain approval for disposal of the asset.

The completed measurement instrument status change form will be forwarded to the Calibration Department.

Equipment transfers—measurement and test equipment the primary use/ownership of which is being permanently transferred between departments or divisions. The original owner department is responsible for completing the measurement instrument status change form for any measurement and test equipment that is being transferred to another department or division and for obtaining the signature of the new owner department.

6.4 Calibration Database and Reporting

Quality assurance/engineering/document control shall maintain a system for tracking and controlling measurement and test equipment that will prevent the use of expired or unfit measurement and test equipment.

The calibration database shall outline the calibration method in the comments section of the equipment history record, for example, subcontract on-site calibration (performed on-site by an approved supplier).

Subcontractor: typically measuring and test equipment sent out to an approved supplier.

The calibration database shall distinguish internal company standards from measurement instruments.

The Calibration Department will issue a calibration status report to the product department supervisors, department calibration representatives, food safety manager, and production manager once every month.

The monthly calibration status report shall consist of the following:

- Equipment due for calibration in the next 30 days
- Equipment overdue for calibration
- Equipment on hand, that is, on-site and properly calibrated
- Remedial/corrective action form status

The calibration manager has two main sets of records that store all pertinent information: the equipment master and equipment history. Records are entered into these two corresponding screens by the calibration coordinator(s). The equipment master stores general information, such as ID description and scheduling information (called events) for each piece of equipment. The equipment history maintains historical information for specific equipment. Events can be calibrations, repairs, operations, and so on. Each time an event is performed, the result of the event, including any measurement information, is entered as a history record by the calibration coordinator(s).

All data are entered by the calibration coordinator(s) and administrator; other users, such as the calibration representatives, have "User" or "Read Only" status within the calibration system software. All information, whether deleted or entered, is mapped through an audit trail in the database.

CALIBRATION

6.5 Remedial/Corrective Action Process

The Calibration Department will issue a remedial/corrective action form to the equipment's owner department supervisor when measurement and test equipment is returned from calibration with an identified out-of-tolerance condition before calibration. A description of the specific out-of-tolerance parameters will be included or attached to the form.

Any equipment with out-of-tolerance occurrences before calibration will be issued a "Do Not Use–Out of Service" label or will be quarantined in the calibration area, pending completion of the remedial/corrective action form by the owner department.

All remedial/corrective action forms will address the impact of the out-of-tolerance condition on the product(s)/process(es). A concise and detailed explanation of this decision shall be documented. The following should be addressed in the remedial response:

- How important the affected feature is to the end user
- How the out-of-tolerance condition relates to the product specification(s)
- Any potential product impact
- If the product impact has been identified, the product failure mode effect analysis or a risk analysis report shall be used to define the potential patient or user safety impact
- This may include ancillary documents, such as handwritten notes, calculations, graphs, tables, sketches, or photographs

Remedial actions should also address measurement and test equipment disposition, as follows:

- The fitness of the equipment for continued use
- The calibration interval of the equipment if a change to the interval is being made as a result of the evaluation
- Other changes to prevent recurrence, including the appropriateness of the equipment for the measurement/test function and operator handling of the equipment

Any remedial/corrective actions open for more than four weeks will be reported by the calibration coordinator to the food safety manager and department supervisor.

6.6 Labeling, Identification, and Storage

A calibration label must be attached to or posted within visual range of the measurement and test equipment.

Calibrated measurement and test equipment shall be marked with a label displaying the following:

- The date of the most recent calibration
- The date when the next calibration is scheduled
- The aforementioned dates shall be of the format type requirements of SOP-024, for example, Jan/5/2015 or 5/Jan/15 to avoid confusion between calibrations performed in Europe and the United States
- The initials of the personnel or subcontractor who performed the calibration or the name of the subcontractor

If the item is too small for this type of marking, a color code or smaller identifying mark shall be employed and cross-referenced on the test equipment installation qualification form for that specific item.

Calibration seals shall be affixed to measurement and test equipment if a possibility of alteration of calibrated settings might occur. A tamper-proof seal is affixed to the setting adjustment area or access screw; this acts as a safeguard against any internal or external adjustments that could invalidate the calibration settings. Acceptable methods of sealing are as follows:

- Tamper-proof labels
- Inspection lacquer
- Low-strength Tread Loc

To avoid damage, measurement instruments and standards, where applicable, such as vernier calipers, shall be stored in suitable packaging when not in use.

Spare/backup measurement and test equipment (portable) shall be stored in locked cabinets.

Cabinets identified as "Calibrated Test Equipment" contain standards and measurement equipment that are currently suitable for use. Cabinets identified as "Test Equipment Not Calibrated" contain items due for calibration and inactive measurement and test equipment.

Only the calibration coordinator(s) and administrator shall have access to these storage cabinets.

6.7 Battery Replacement

The calibration coordinator shall perform battery replacement on any measurement and test equipment that requires battery replacement. The calibration coordinator shall use appropriate electrostatic devices and practices and subsequently replace any tamper-proof seals/labels as required.

6.8 Calibration Procedures

Calibration procedures must be application specific and must prescribe step-by-step instructions for the calibration of measurement and test equipment or categories thereof. These shall be prepared internally, by another agency, the manufacturer, or a composite of any of these. Internal calibration procedure part numbers and current revisions shall be referenced on the related calibration record/form.

Calibration procedures must state the acceptable limits of accuracy and precision, the standards required, and sufficient information to enable qualified personnel to perform the calibration.

Equipment used for calibration(s) shall have a test accuracy ratio of at least 10:1, that is, calibration equipment uncertainty will be 10 times greater than the uncertainty of the measurement and test equipment being calibrated. The rationale for exceptions must be documented and approved. The rationale may include an increase in the calibration frequency to compensate for this lack of compliance.

Calibration procedures and internal calibration records must state "Calibration performed by trained personnel only."

6.9 Calibration—Internal

Requirements for calibrations performed by company personnel are as follows:

- The calibration standards used to perform internal calibrations shall be traceable to a national or international standard(s).
- Calibrations are to be performed per application-specific written procedures at the most current revision level and describing the step-by-step method of calibrating specific instruments or categories of instruments.
- For company-manufactured equipment, calibrations will be performed at the revision level applicable to the equipment.
- Calibrations conducted by company personnel require a cross-check to be performed prior to commencing to ensure that proper documentation/procedure(s) is/are used.
- Company personnel performing calibrations must be trained in the proper procedure and revision level based on evidence in the individual's training record.
- Calibration procedures shall clearly state the ranges of acceptable tolerances or limits.
- Recorded calibration data shall be recorded to the significant digit expressed in the limits.
- The environmental conditions for test and measurement equipment calibration, such as lighting, vibration, and so on, other than temperature and humidity, unless these are defined by the manufacturer's specification or user manual, shall comply with the manufacturer's published specification.
- Environmental conditions shall be monitored by calibration personnel to ensure that requirements are met during the performance of in-house calibration.
- Upon completion of the calibration of an item, the personnel performing the calibration will indicate environmental compliance by checking off the appropriate section on the company calibration report.
- If temperature or humidity exceeds the specified limits for a particular calibration type, work for that type will be suspended and a supervisor will be notified to assess the impact.

Documentation on calibration within the calibration system software or equivalent shall include a completed internal calibration report, showing the following:

- Equipment ID number
- Description of equipment
- Part number or manufacturer of the equipment
- Revision of the equipment (if applicable)
- Calibration/test procedure/drawing numbers used
- Revision of the procedure used
- Indication of the cross-check performed (if applicable)
- Identification of the person performing the calibration/test
- Calibration standard(s) or equipment used
- Due date(s) of the standard(s) used
- Date the calibration was completed
- Next calibration due date
- Indication of the condition of the equipment (pre- and postcalibration)

CALIBRATION

Personnel training to perform calibrations must also include the trainer's signature in the section labeled "Approved by" and indicating verification of the following:

- Training on and the use of correct procedures
- Cross-check (if applicable)
- Acceptability of data

Pre- and postcalibration data, including acceptable tolerances/limits may be recorded on the internal calibration report or on a data sheet specific to the equipment's calibration procedure; the completed data sheet will be attached to the internal calibration report.

The Calibration Department will perform a cursory review of the completed internal calibration report form and applicable data sheets to carry out the following:

- A review for completeness
- A review for out-of-tolerance conditions

If the results indicate that the precalibration condition was out-of-tolerance, issue a remedial/corrective action form. If the equipment is not fully calibrated to the manufacturer's or procedural specifications, the equipment may be used in a limited status. In these circumstances:

- Equipment will be identified using the limited calibration label
- Limitations on use will be clearly identified on or near the equipment.

If the calibration is found to be acceptable, the Calibration Department shall do the following:

- Sign or stamp and date the calibration report as evidence of completion in the section labeled "Reviewed by"
- Apply, or issue, an updated calibration label
- If the equipment status is being changed, the equipment owner must complete a measurement instrument status change form
- The Calibration Department shall file the internal calibration report and relevant data sheets in the equipment's history folder

6.10 Calibration—External

Calibration method: Calibration performed by contractors shall be conducted by approved suppliers (registered on the approved supplier list).

The methods and criteria used to perform the calibration of measurement and test equipment shall comply with the manufacturer specifications and shall be traceable, through certification, to a national or international standard, for example, the U.S. National Institute of Standards and Technology, the United Kingdom Accreditation Service, or equivalent.

The method of calibration for linear measurement instruments, such as external micrometers, vernier calipers, and dial gauges, may rely on the methodology outlined in the British Standards in Engineering Metrology, for example, BS 870, BS 887, BS 907, and so on.

Special instructions for calibration shall be detailed in the test equipment installation qualification form, where applicable.

CALIBRATION

Documentation requirements: all documentation provided by the contractor shall include, at a minimum, the following:

- Measurement instrument identifier
- The date of calibration
- Tolerances or the specified accuracy
- Precalibration data
- Postcalibration data (if adjusted)
- The identity of the standards used
- The calibration due date of the standards
- Ancillary measurement documentation (graphs, tables, photos, and so on), if applicable
- Statement of acceptability (pass/fail)
- Signature or stamp of person performing the calibration, or the contractor's name and address

Repairs: for equipment identified as requiring repairs by the contractor, the Calibration Department will do the following:

- Request that the contractor provide a quote for the cost of the repair and provide an estimated time for completion of the repair
- Notify the owner department of the need for equipment repair and request approval for the repairs

Approval of repairs: The owner department will provide a signed and dated purchase request for the cost of the repair. The Calibration Department will tell the contractor to proceed with repairs and provide an account number or purchase order number.

Lack of approval for the repairs: Tell the contractor to return the equipment unrepaired if offsite

Receiving equipment: Upon receipt of the equipment from the contractor, the Calibration Department will do the following:

- Physically examine the measurement and test equipment for any damage.
- Review all calibration documentation for required information by checking off blocks on the calibration return checklist as conformity with the requirements is verified.
- Place measurement and test equipment with documentation missing or insufficient information in the calibration storage cabinet ("Test Equipment Not Calibrated") or labeled "Do Not Use–Out of Service."
- The approval of measurement and test equipment with documentation missing or insufficient information shall be required of Calibration Department personnel prior to the release of the equipment for use. Calibration Department personnel will print their names, sign, or stamp and date the discrepancy approval section of the calibration documentation return checklist upon acceptance or approval of documentation, as well as the document rationale in the remarks section for any deficient certificates.

CALIBRATION

- If Calibration Department personnel approval is denied, contact the subcontractor, requesting the deficient information. Repeat the process from the start.
- Compare specific values (data) with acceptance criteria (tolerances/accuracy specifications) or review the statement of acceptability for out-of-tolerance conditions.
- Owner departments of equipment with a precalibration out-of-tolerance condition shall be issued a remedial/corrective action form.

If the calibration certificate indicates that the equipment is not calibrated over the entire range of measurement or the postcalibration condition was out of tolerance, the equipment may be treated as follows:

- Discontinued
- Placed in "Not in Use" or "Inactive" status
- Used as "Reference Only"
- Used in a limited status. In these circumstances, equipment will be identified using a limited calibration or special calibration label; limitations of use will be clearly identified on or near the equipment
- Verify that the dates on the calibration label and calibration certificate concur and compare the due date with the calibration interval.
- Check for calibration seals, where appropriate.

6.11 Finalizing

- Print name, sign, or stamp and date the form and return checklist as evidence of review and availability for use; the form will be placed with equipment calibration certificate records in the designated cabinet.
- Update the calibration database to include all newly received information, such as next calibration due date and status, and so on.
- File the certificate of calibration and relevant documents as part of the equipment's calibration history records.
- Place the equipment in the calibrated equipment storage cabinet if it is not required for immediate use.
- Notify the owner department if applicable.

External calibration company supplier survey/audits: Accreditation by a recognized body, for example, the International Laboratory Accreditation Cooperation, may be accepted in lieu of an audit; if an audit is not deemed necessary, a copy of the current certificate of accreditation will be maintained in the supplier audit file.

7 Records

Document	Location	Duration of record	Responsibility
Equipment master and history list	Calibration Department	Indefinitely	Calibration coordinator
Calibration program/schedule	Calibration Department	Three years	Calibration coordinator
Equipment calibration report	Calibration Department	Three years	Calibration coordinator
Equipment calibration certificate	Calibration Department	Indefinitely	Calibration coordinator

PRODUCT RECALL AND WITHDRAWAL	
An FBO Procedure	
Document No.	Standard operating procedure SOP-023
Created	April 20, 2018
Updated	January 13, 2019
Controller	Document Controller
Owner	Food Safety Manager
Confidentiality Statement Information in this document must be kept confidential as per the document's classification below and the rules of disclosure.	
All FBO documents are classified in the following way. PUBLIC documents are intended for anyone. COMMERCIAL IN CONFIDENCE documents are to be kept confidential among restricted individuals within the FBO and partner organizations. COMPANY CONFIDENTIAL documents are to be kept confidential within the FBO and used for normal business activities by the general office population. HIGHLY CONFIDENTIAL documents are to be kept confidential among restricted individuals within the FBO.	
Classification	Company Confidential

Revision History

Date	Version	Author	Comments (including review history)
April 20, 2018	Draft 01	Joe Bloggs	Initial document for review and discussion
April 24, 2018	V1.0	Joe Bloggs	Approved and released by process owner
January 13, 2019	V1.1	Joe Bloggs	Updated terms and definitions in accordance with International Organization for Standardization (ISO) 22000:2018, for example

Contents

PRODUCT RECALL AND WITHDRAWAL

1 Summary

Purpose	The purpose of this procedure is to describe the process for effectively removing a product from the external supply chain/distribution.
Scope	This instruction covers all products manufactured or distributed by the food business operator (FBO). Local regulations and laws take precedence over this guideline.
Functional responsibility	The functional responsibility for this procedure lies with the food safety manager, who is responsible for the effective implementation and maintenance of this procedure.

2 Related Documents

Policies	Food Safety Policy, POL-001 Customer/Consumer Complaints Policy, POL-002
Processes	Departmental process descriptions
Procedures	Control of Nonconforming Product, SOP-003 Mock Recall, SOP-008 Correction and Corrective Action, SOP-009 Communication, SOP-020 Crisis Management, SOP-029
Work instructions	Not applicable
Forms	Recall/withdrawal log Communication log Root cause analysis/corrective action
Other	Not applicable

3 Definitions

Term or acronym	Description
Complaint	An expression of dissatisfaction communicated to an organization in relation to the organization's products or services or the complaints-handling process, during which a response or resolution is explicitly or implicitly expected
Correction	Action to eliminate a detected nonconformity
Corrective action	Action to eliminate the cause of a nonconformity and to prevent recurrence
FBO	food business operator
FSMS	food safety management system
Nonconformity	Nonfulfillment of a requirement

PRODUCT RECALL AND WITHDRAWAL

Term or acronym	Description
Product	Output that is a result of activities none of which is necessarily performed at the interface between the provider and the customer; for the food business operator (FBO), this may be an ingredient, raw material, intermediate product, or finished product supplied to a customer or consumer
Recall	The process by which a product is removed from the external supply chain/ distribution and consumers are publicly advised to take specific actions with the product, for example, to not consume the product, or to return the product to the shop or manufacturer; this includes the U.S. Food and Drug Administration (FDA) class I and class II recalls
Regulatory requirement	Obligatory requirement specified by an authority mandated by a legislative body
Risk	The effect of uncertainty on an expected result
Root cause	A cause that, once removed from the problem fault sequence, prevents the final undesirable event from recurring
Root cause analysis	A method of problem solving that involves an attempt to identify the root cause of faults or problems
Statutory requirement	Obligatory requirement specified by a legislative body
Update	Immediate or planned activity to ensure application of the most recent information
Withdrawal	The process by which a product is removed from the external supply chain/ distribution, but which does not require any action by the consumer

4 Introduction

Even within the most well-managed food businesses, an issue involving the safety and suitability of a food may occur. This may be the result, for example, of a packaging defect, a product formulation error, a manufacturing or storage problem, or a problem with the food ingredients. It is important that food business operators (FBOs) be aware that food safety issues can become associated with their products and therefore recognize that there is a need to plan ahead.

European Union (EU) food law requires that all FBOs be able to trace the food they receive back to the immediate supplier of the food. Then, following food handling, preparation, or processing, FBOs must be able to track the distribution of food forward from their own business to their immediate customer.

FBOs are also required to withdraw unsafe food from the market after it has left their immediate control. If it has reached consumers, FBOs must inform consumers of the reason for the removal of the food from the market and, if necessary, recall the food. FBOs should therefore develop documented food traceability and food recall/withdrawal systems and integrate these into their food safety management systems (FSMSs).

5 Procedure Flowchart

Not applicable.

PRODUCT RECALL AND WITHDRAWAL

6 Procedure Notes

6.1 Data Collection and Management

The food safety team

- Gathers all necessary information, facts, and data to enable a conscious decision to confirm the validity of the claim and proceed to a withdrawal or recall
- Informs regulatory authorities according to crisis management rules and local regulations
- Defines the communication with employees, the sales force, customers or consumers, and other stakeholders
- Determines subsequent steps involving the removed products
- Considers all other elements that might affect the FBO

6.2 Decision to Recall or Withdraw

The decision to recall or withdraw is taken by the food safety manager. The decision-making process is carried out according to crisis management procedures and takes into account especially the following:

- The situation and actions to be undertaken in markets where the same material is commercialized (intermarket supply)
- Foreign markets must be examined in making decisions or approving decisions; specific guidelines may apply

6.3 Communicating the Decision to Recall or Withdraw

Communication is critical to the success of a recall as well as to the image of brands. Communication is based on the following:

- The position statement prepared by the food safety team and the FBO public relations/legal advisor recalling a product
- Questions and answers to be used by consumer services

The media used for communication must be adequate to reach the potential consumers of the product to be recalled.

Communication must be simple and factual:

- Why are we recalling the product?
- What are we recalling?
- What do we do as an FBO to eliminate the defect and put the product back on the market?
- What is our refund policy?

The same principles must be applied for communication with other stakeholders (employees, customers, authorities, and so on).

Classification	Company Confidential	Product Recall and Withdrawal Procedure	
Doc ID: SOP-023 Created: April 20, 2018	Printed: Updated: January 13, 2019	Controller: Document Controller Owner: Food Safety Manager	Page 4 of 6

6.4 Actions in the FBO Factory

The factory provides the traceability data necessary to define the material and the quantities to be removed from the entire supply chain/distribution. All affected batches must be restricted in the FBO computer system.

The accuracy of the traceability system must be considered, and a safety margin on either side of the concerned batch must be added if necessary.

The incident must be investigated, the root cause analyzed, and corrective actions taken.

6.5 Actions in FBO Distribution/Logistics

Upon receiving instructions to block a particular product quantity, the warehouse staff must immediately remove the product from assembled loads in the warehouse. The blocked stock must be physically marked and segregated.

If advised by the food safety team, distribution will coordinate urgent material pickups from identified warehouses and stores if necessary.

The material received back must be registered in the FBO computer system with the status indicated as blocked as with all returned material.

On request, warehouse personnel can check and sort the suspected stock. The food safety manager provides instructions on how to examine the product and gather adequate resources (training, specialists, and so on).

A detailed report must be prepared on the fate of the recalled batches. Other goods must be included if relevant (for example, nonrecalled goods, other FBO products, or even the products of competitors).

6.6 Actions in Trade

The food safety team establishes clear instructions for shops and retailers on how to proceed with the affected material.

Materials in warehouses must be blocked and physically marked and a pickup schedule agreed with FBO distribution.

Materials in shops (supermarket shelves or back-room storage) must be removed from shelves, blocked, physically marked, and placed in back-room storage to await pickup or destruction (as agreed between the FBO and the retailer). Sales or merchandising staff may be called to assist as needed.

The retailer will communicate the actual quantities to be picked up to facilitate transport. The material must be returned as soon as possible to the FBO or to dedicated warehouses.

Disposal at customer sites is possible if there is mutual agreement about what is to be disposed. The method of disposal must be defined and properly documented.

PRODUCT RECALL AND WITHDRAWAL

6.7 Return Transport

The return transport of affected material requires special attention and appropriate organization. This must be accomplished without delay.

6.8 Handling of the Returned Product

The returned product must be controlled, registered, marked, and segregated from normal stocks.

Precise inventories must be kept. Regulatory authorities may have additional requirements on records and information.

The returned product must be handled as a nonconforming product; the rules for responsible destruction or disposal must be followed.

In line with the FBO accounting procedure, all costs related to recalls and withdrawals must be charged to production-related overhead, not to bad products.

6.9 Postreview Action Review

A postreview action review must be conducted when the incident is over and potential improvements implemented.

At a minimum, an analysis of the quantities of the materials involved must be carried out (product produced, sold, returned, destroyed, and not accounted for or consumed).

6.10 Postreview Action Exercises

Recalls and withdrawals must be practiced. An annual mock recall exercise is mandatory (see the Mock Recall Procedure). A postreview action on a real case cannot be considered a substitute for a mock recall. An actual recall is not the occasion to test the FBO recall/traceability system.

7 Records

Document	Location	Duration of record	Responsibility
Recall/withdrawal log	Food Safety Office	Indefinitely	Food safety manager
Communication records	Food Safety Office	Indefinitely	Food safety manager
Root cause analysis	Food Safety Office	Indefinitely	Food safety manager
Recall/withdrawal report	Food Safety Office	Indefinitely	Food safety manager
Postreview minutes	Food Safety Office	Indefinitely	Food safety manager

FOOD DEFENSE	
An FBO Procedure	
Document No.	Standard operating procedure SOP-044
Created	April 20, 2018
Updated	January 13, 2019
Controller	Document Controller
Owner	Food Safety Manager
Confidentiality Statement Information in this document must be kept confidential as per the document's classification below and the rules of disclosure.	
All FBO documents are classified in the following way. PUBLIC documents are intended for anyone. COMMERCIAL IN CONFIDENCE documents are to be kept confidential among restricted individuals within the FBO and partner organizations. COMPANY CONFIDENTIAL documents are to be kept confidential within the FBO and used for normal business activities by the general office population. HIGHLY CONFIDENTIAL documents are to be kept confidential among restricted individuals within the FBO.	
© Copyright FBO. All rights reserved. No part of this publication may be reproduced, stored in a retrieval system, or transmitted in any form or by any means, electronic, mechanical, photocopying, recording, or otherwise, without the written permission of the FBO.	
Classification	Company Confidential

Revision History

Date	Version	Author	Comments (including review history)
April 20, 2018	V1.0	Joe Bloggs	Original draft
April 24, 2018	V1.1	Joe Bloggs	Approved for release by process owner
January 13, 2019	V1.2	Joe Bloggs	Introduction of TACCP/VACCP assessments and templates

Contents

CHAPTER 4

FOOD DEFENSE

1 Summary

Purpose	The purpose of this procedure is to document the measures taken by the food business operator (FBO) to protect food and the food production process from intentional harm.
Scope	This procedure is applicable to products, processes, storage and production environments, and suppliers across the food chain of the FBO. It addresses the risks to the people, products, assets, and brand of the FBO.
Functional responsibility	The functional responsibility for this procedure lies with the food safety manager, who is responsible for the effective implementation and maintenance of this procedure.

2 Related Documents

Policies	Food Safety Policy, POL-001
Processes	Food Safety, PRO-001
Procedures	Control of Documents, SOP-001 Traceability, SOP-012
Work instructions	Not applicable
Forms	Master document register
Other	Document management system (DMS) Food defense plan review form

3 Definitions

Term or acronym	Description
Electronic security	Procedures implemented to protect electronic systems from sources of threat, such as malware and hackers, intent on misusing the systems, corrupting them, or putting them out of use
FBO	food business operator
Food defense	Security of food and drink and their supply chains from all forms of malicious attack, including ideologically motivated attack, leading to contamination or supply failure
Food supply	Any and all elements of the food supply chain, network, or web, including drink and support services and allied services
Personnel security	Procedures used to confirm an individual's identity, qualifications, experience, and right to work; the procedures are also used to monitor the conduct of employees or contractors
Product security	Techniques used to make food products resistant to contamination or misuse, including tamper-evident closures and lot marking

FOOD DEFENSE

Term or acronym	Description
Protective security	All the measures related to physical, electronic, and personnel security that any organization takes to minimize the threat of malicious attack
TACCP	Threat assessment critical control point; the systematic management of risks through the assessment of threats, the identification of vulnerabilities, and the implementation of controls on raw materials, packaging, finished products, processes, premises, distribution networks, and business systems by a knowledgeable and trusted team with the authority to implement changes to procedures
VACCP	Vulnerability assessment critical control point; a management process to defend a food supply chain from any form of dishonest conduct that impacts detrimentally on the quality or authenticity of food and drink

4 Introduction

Multinational organizations are driving the need for suppliers globally to institute food defense programs, thereby minimizing the risk of intentional contamination and tampering. If suppliers work with or want to work with a multinational company, they will probably be required to develop a food defense plan. A food defense plan builds on existing food safety plans, hazard analysis critical control point (HACCP) plans, and crisis management plans and incorporates audits of the security of premises, shipping and receiving, and personnel to help ensure safe and secure food supplies.

Risks can originate from various sources—internal (employees, temporary workers, cleaning staff, and so on) or external (visitors, delivery personnel, suppliers, terrorist groups, activists, and so on). Malicious acts may originate outside, but the scope of the identification of risks must be understood in a broader sense. Internal risks should not be overlooked: 70 percent to 80 percent derive from staff—disgruntled employees, for example. The risks can take many forms: fraud, vandalism, sabotage, terrorist acts, theft, blackmail, and so on. The probability is high, and such occurrences have a relatively strong impact on business.

The scope of food defense can be represented as shown in the figure below.

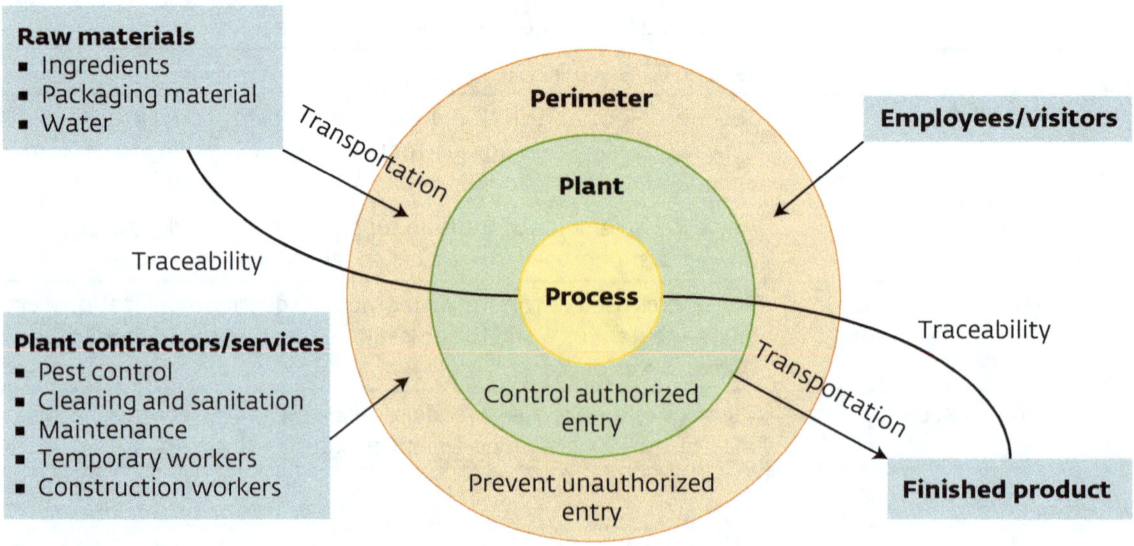

FOOD DEFENSE

5 Procedure Flowchart

Not applicable.

6 Procedure Notes

The food safety manager/management representative should establish a cross-functional team to conduct a threat assessment critical control point (TACCP)–vulnerability assessment critical control point (VACCP) plan for the organization. Where appropriate, external experts may be used to facilitate the process.

All participants should receive relevant training on TACCP/VACCP based on publicly available specification PAS 99:2012 of the British Standards Institution or training providers approved by the U.S. Food and Drug Administration (FDA).

For the TACCP, the TACCP team should undertake the following: (1) document a TACCP standard operating procedure (SOP); (2) identify points in the food supply chain at which threats are possible against staff, operations, and products; (3) conduct assessments of the critical points to identify risks using a TACCP template that is similar to the TACCP template provided in the food safety training kit; (4) analyze the risks, establish appropriate threat controls, and continue to monitor the control points; (5) create action plans in anticipation of the possible breach of controls; and (6) continue to improve the TACCP process by reviewing documentation and anticipating new threats, including through a robust internal and external review of industry sector threats.

For the VACCP, the VACCP team should undertake the following: (1) assess the possibility of food fraud in the supply chain using a VACCP template that is similar to the VACCP template provided in the food safety training kit; (2) determine the point in the supply chain at which fraud can become an economic incentive; (3) check the measures being used to control food fraud; (4) create action plans in anticipation of a possible breach of controls; (5) continue to improve the VACCP process by reviewing documentation and anticipating new possibilities of food fraud, including a robust internal and external review of industry sector food fraud vulnerabilities.

Both the TACCP and the VACCP should be reviewed formally at least once a year.

The sample food defense plan below is organized into four sections, as follows: (1) outside security measures, (2) inside security measures, (3) personnel security measures, and (4) incident response security measures. The plan should include both the TACCP and VACCP assessments and related documentation.

6.1 Outside Security Measures

(Examples: door locks, lighting, the monitoring of loading and unloading)

GOAL: To prevent unauthorized access to the facility by people with unapproved materials

FOOD DEFENSE

The FBO has implemented at least one of the following sets of measures to establish outside security.

PHYSICAL SECURITY

- Manufacturing plant boundaries are clear and secured to prevent unauthorized entry (for example, fences installed, "No Trespassing" signs posted).
- Entrances are secured (for example, locks and alarms are installed and operating).
- The plant perimeter is periodically monitored for suspicious activity.
- Outside lighting is present to deter unauthorized activities.
- Other access points such as windows and vents are secured.
- Outside storage on the premises is protected from unauthorized access.
- Other____

SHIPPING/RECEIVING SECURITY

- Incoming shipments are examined for potential tampering.
- Incoming and outgoing vehicles are examined for suspicious activity.
- Loading and unloading are scheduled and monitored.
- Loading dock access is controlled (for example, monitored or locked).
- Incoming shipments are secured with locks or seals.
- Outgoing shipments are locked or sealed.
- Other____

MAIL HANDLING SECURITY

- Mail is handled away from food, including ingredients and packaged food products.
- Employees who handle mail are aware of the proper handling of suspicious mail and U.S. Postal Service guidelines.
- Other____

6.2 Inside Security Measures

(Examples: signs, observation, restricted access)

GOAL: To protect products from intentional contamination throughout the production process

The FBO has in place at least one of the following sets of measures for inside security.

GENERAL INSIDE SECURITY

- Suspicious packages are reported to appropriate personnel.
- Restricted areas of the establishment are clearly identified.
- Previously unattached materials are checked before use.
- Unexpected changes in inventory (product or equipment) are reported to appropriate personnel.
- Emergency lighting is in place.
- An emergency alert system is identifiable, tested, and reviewed with emergency contacts (for example, police or fire personnel).
- Other____

FOOD DEFENSE

PROCESSING AREA SECURITY

- Access to ingredients and packaged products is restricted.
- Access to process control equipment, such as ovens and mixers, is restricted.
- Ingredients are examined for possible tampering.
- Records ensure traceability one step backward, one step forward, or both.
- Other____

STORAGE SECURITY

- Access to storage areas is restricted.
- Stock rotation (first in, first out) is practiced.
- Labels and packaging materials are controlled to prevent theft and misuse.
- Periodic examinations for tampering of materials in storage are preformed.
- Other____

INGREDIENTS/WATER/ICE SECURITY

- Access is restricted to storage tanks for potable water and to the water reuse system.
- Access to lines that transfer water or ingredients is examined and restricted.
- Access to plant ice-making equipment is controlled.
- Restricted ingredients (for examples, nitrates) are controlled.
- Supplier food safety/security information is requested.
- Other____

CHEMICAL/HAZARDOUS MATERIAL CONTROL SECURITY

- Chemicals/hazardous materials, including pesticides, cleaning or laboratory materials, and sanitizers, are in a restricted area or secured by a lock.
- An up-to-date inventory of hazardous materials and chemicals is maintained, and discrepancies are investigated.
- Potential hazardous waste (biological or chemical) is controlled and disposed of properly.
- Other____

INFORMATION SECURITY

- Access to sensitive information, such as site plans and processing details, is controlled.
- Access to computer systems is protected through firewalls and passwords.
- Other____

6.3 Personnel Security Measures

(Examples: check references, use visitor log or sign-in, or check IDs)

GOAL: To ensure that only authorized personnel are in the facility at any time

FOOD DEFENSE

The FBO has in place at least one of the following sets of measures for personnel security.

EMPLOYEE SECURITY

- A method to recognize or identify employees in the facility is implemented.
- Background or reference checks are conducted on new hires.
- Employees are restricted in what they can bring in or take from the facility (for example, cameras).
- Other________________________________

NONEMPLOYEE SECURITY (EXAMPLE: VISITORS, CONTRACTORS, GUESTS, CUSTOMERS, TRUCK DRIVERS)

- A log is maintained of nonemployees and persons working for or on behalf of the FBO who enter the establishment.
- A method to recognize or identify nonemployees and persons working for or on behalf of the FBO in the establishment is in place.
- Nonemployees and persons working for or on behalf of the FBO are chaperoned on-site.
- Nonemployees and persons working for or on behalf of the FBO are restricted to appropriate areas.
- Nonemployees and persons working for or on behalf of the FBO are restricted on what they can bring in or take from the facility.
- Other________________________________

SECURITY TRAINING

- Awareness training on security measures is provided to new employees and persons working for or on behalf of the FBO.
- Periodic refresher awareness training on security measures is offered to employees and persons working for or on behalf of the FBO.
- Employees or persons working for or on behalf of the FBO are trained to report suspicious activities or unusual behavior they have noticed.
- Other________________________________

6.4 Incident Response Security Measures

(Examples: reference the emergency plan, security plan, or other)

GOAL: To respond quickly to a product contamination threat or event using planned measures

The FBO has in place at least one of the following sets of measures for incident response security.

INVESTIGATING A SECURITY CONCERN

- Procedures have been established to ensure that adulterated or potentially harmful products are held.
- Customer/consumer comments are investigated.
- Reporting on unusual activities is encouraged.
- Information is available to employees on how to respond to phone or other threats.

FOOD DEFENSE

- Employees have the ability to stop activities to minimize a potential food defense incident.
- Reported security breaches (for example, alarms, suspicion of tampering) are investigated.
- Other____

EMERGENCY CONTACT SECURITY

- Plant personnel contact information is kept up-to-date.
- Emergency contact information is kept up-to-date.
- Other____

OTHER PLAN SECURITY

- A product recall plan is maintained and periodically reviewed.
- Key personnel are trained in product recall/withdrawal procedures.
- Other____

6.5 Outside Security Tools

Below is a list of tools or additional security measures. These are provided to assist in tailoring the plan to meet the specific needs of the FBO.

PHYSICAL SECURITY TOOLS

- Ensure proper lighting to monitor the establishment outdoors at night and in the early morning.
- Install self-locking doors or alarms at emergency exits.
- Ensure the following are secured with locks, seals, or sensors when unattended (after hours/weekends) to prevent unauthorized entry:
  - Outside doors and gates
  - Windows
  - Roof openings
  - Vent openings
  - Trailer (truck) bodies
  - Tanker truck hatches
  - Railcars
  - Bulk storage tanks/silos
  - Loading ports
  - Hose/pump stations
- Regularly conduct and document security inspections of storage facilities, including temporary storage vehicles.
- Restrict outdoor access to water wells and other water sources.

SHIPPING/RECEIVING SECURITY

- Closely monitor the loading and unloading of vehicles transporting raw materials, finished products, or other materials used in food processing.
- Inspect tanker trucks and railcars to detect the presence of any material, solid or liquid, in tanks prior to loading liquid products. Load only if appropriate. Report/record the results.
- Control access to loading docks to avoid unverified or unauthorized deliveries.
- Require advance notification from suppliers for all deliveries.

FOOD DEFENSE

- Immediately investigate suspicious changes in shipping documents.
- Check all deliveries outside establishment premises pending verification.
- If off-hour delivery is accepted, require prior notice of the delivery and an authorized person to be present to verify and receive the delivery.
- Check less-than-truckload or partial-load shipments for content and condition.
- Require incoming shipments of raw product, ingredients, and finished products to be sealed with tamper-evident or numbered, documented seals and verify the seals prior to entry. Reject if the seal is broken or missing.
- Select transportation companies and suppliers on the basis of consideration of the security measures that they use.
- Examine returned goods at a separate location for evidence of tampering before salvage or use in rework.
- Maintain records of the disposal of returned goods.
- Require drivers or delivery personnel to provide identification, preferably with a photo ID. Record names.
- Minimize the time a truck is unlocked during loading or delivery.

6.6 Inside Security Tools

GENERAL INSIDE SECURITY

- Install and monitor security cameras.
- Increase visibility within the establishment (for example, improve lighting, openness, increase supervision, add cameras).
- Regularly take inventory of the keys to secured/sensitive areas of the establishment.
- Restrict access to controls (by locked door/gate or limiting access to designated employees) for the following systems:
  - Heating, ventilation, and air conditioning
  - Propane, natural gas, water, and electricity
  - Disinfection systems
  - Clean-in-place systems or other centralized chemical systems

PROCESSING AREA SECURITY

- Maintain records to allow efficient traceability backward or forward of materials and finished products.
- Reduce the time an area is left unmonitored.
- Reduce access to product containers or processing equipment.
- Do not allow unnecessary personal items within the production area.

STORAGE SECURITY

- Maintain an access log for product and ingredient storage areas.
- Regularly check the inventory of finished products for unexplained additions and withdrawals from existing stock.
- Restrict access to external storage facilities to designated employees only.

FOOD DEFENSE

INGREDIENTS/WATER/ICE SECURITY

- Before they are used, examine packages of ingredients for evidence of tampering.
- Restrict access to product, ingredient, and packaging storage areas to designated employees only (for example, by lock or gate).
- Ensure that water is from a municipally or local authority controlled source.
- Inspect water lines for possible tampering (perform visual inspection of the integrity of infrastructure, proper connections).
- Make arrangements with local health officials to ensure immediate notification of the establishment if the potability of the public water supply is compromised.

CHEMICAL/HAZARDOUS MATERIAL CONTROL

- Restrict access to the in-plant laboratory.
- Have procedures in place to control the receipt of samples.
- Have a procedure in place to receive, securely store, and dispose of reagents.

INFORMATION SECURITY

- Track customer and consumer complaints/comments for trends.
- Keep details of food defense procedures confidential, as necessary.
- Have up-to-date establishment layout/blueprint/drawings for local law enforcement, including the fire department.

6.7 Personnel Security Tools

- Authorize appropriate employees and persons working for or on behalf of the FBO to stop a process on which there are significant concerns.
- Control the access of employees, nonemployees, and persons working for or on behalf of the FBO to the FBO establishment during working and nonworking hours (use coded doors, receptionist on duty, swipe cards).
- Restrict temporary employees, nonemployees, and persons working for or on behalf of the FBO to areas relevant to their work.
- Implement systems to associate personnel with their specific functions, assignments, or departments (for example, corresponding colored uniforms or hair covering).
- Prohibit employees from removing company-provided uniforms or protective gear from the premises.
- Maintain an up-to-date shift roster for each shift.

6.8 Incident Response Tool

- Establish evacuation procedures and include them in the food defense plan.
- Establish procedures for responding to threats as well as actual product contamination events.
- Preestablish communication with local, state, and federal incident response personnel to foster more efficient response.

CHAPTER 4

FOOD DEFENSE

FBO

Food Defense Plan Review Form

Complete this form to document the annual review of the food defense plan.

Not all measures are required or need to be reviewed each time this form is completed.

Date of annual review	Person who conducted annual review (name and title)	Was the food defense plan tested?[a] (yes/no)

a. Testing can be carried out using simple measures, such as checking locked doors or making unannounced perimeter checks.

7 Records

Document	Location	Duration of record	Responsibility
Food defense plan review record	Food Safety Office	Indefinitely	Food safety manager

FOOD FRAUD AND VULNERABILITY	
An FBO Procedure	
Document No.	Standard operating procedure SOP-046
Created	April 20, 2018
Updated	January 13, 2019
Controller	Document Controller
Owner	Food Safety Manager
Confidentiality Statement Information in this document must be kept confidential as per the document's classification below and the rules of disclosure.	
All FBO documents are classified in the following way. PUBLIC documents are intended for anyone. COMMERCIAL IN CONFIDENCE documents are to be kept confidential among restricted individuals within the FBO and partner organizations. COMPANY CONFIDENTIAL documents are to be kept confidential within the FBO and used for normal business activities by the general office population. HIGHLY CONFIDENTIAL documents are to be kept confidential among restricted individuals within the FBO.	
© Copyright FBO. All rights reserved. No part of this publication may be reproduced, stored in a retrieval system, or transmitted in any form or by any means, electronic, mechanical, photocopying, recording, or otherwise, without the written permission of the FBO.	
Classification	Company Confidential

Revision History

Date	Version	Author	Comments (including review history)
April 20, 2018	V1.0	Joe Bloggs	Original draft
January 13, 2019	V1.1	Joe Bloggs	Introduction of vulnerability assessment critical control point (VACCP) assessments and templates

Contents

CHAPTER 4

CHAPTER 4

FOOD FRAUD AND VULNERABILITY

1 Summary

Purpose	The purpose of this procedure is to document the measures taken by the food business operator (FBO) to protect food and the food production processes from intentional harm.
Scope	This procedure is applicable to products, processes, storage and production environments, and suppliers across the food chain of the FBO. It also addresses the risks to the people, products, assets, and the brand of the FBO.
Functional responsibility	The functional responsibility for this procedure lies with the food safety manager, who is responsible for the effective implementation and maintenance of the procedure.

2 Related Documents

Policies	Food Safety Policy, POL-001
Process	Food Safety, PRO-001
Procedures	Control of Documents, SOP-001 Traceability, SOP-012 Food Defense, SOP-044
Work instructions	Not applicable
Forms	Master document register
Other	Document management system (DMS) Food fraud and vulnerability assessment tool, vulnerability assessment critical control point (VACCP)

3 Definitions

Term or acronym	Description
Economically motivated adulteration	The U.S. Food and Drug Administration's (FDA's) working definition of economically motivated adulteration is the fraudulent, intentional substitution or addition of a substance in a product for the purpose of increasing the apparent value of the product or reducing the cost of the production of the product, that is, for economic gain
Electronic security	Procedures implemented to protect electronic systems from sources of threat, such as malware and hackers, intent on misusing the systems, corrupting them, or putting them out of use
FBO	food business operator
Food defense	Security of food and drink and their supply chains from all forms of malicious attack, including ideologically motivated attack, leading to contamination or supply failure

FOOD FRAUD AND VULNERABILITY

Term or acronym	Description
Food fraud	Any act whereby food is deliberately placed on the market for financial gain, with the intention of deceiving the consumer; although there are many kinds of food fraud, the two main types are as follows: (1) the sale of food that is unfit and potentially harmful, such as the recycling of animal by-products back into the food chain; packing and selling beef or poultry of unknown origin; knowingly selling goods that are past the use by date; and (2) the deliberate misdescription of food, such as products substituted with cheaper alternatives, for example, farmed salmon sold as wild salmon or basmati rice adulterated with cheaper varieties, or making false statements about the source of ingredients, that is, the geographical, plant, or animal origin. Food fraud may also involve the sale of meat from animals that have been stolen or illegally slaughtered, as well as wild game animals, such as deer, that have been poached.
Food supply	Any and all elements of the food supply chain, network, or web, including drink and support services and allied services
Personnel security	Procedures used to confirm an individual's identity, qualifications, experience, and right to work; the procedures are also used to monitor the conduct of employees or contractors
Product security	Techniques used to make food products resistant to contamination or misuse, including tamper-evident closures and lot marking
Protective security	All the measures related to physical, electronic, and personnel security that any organization takes to minimize the threat of malicious attack
VACCP	Vulnerability assessment critical control point: a management process to defend a food supply chain from any form of dishonest conduct that impacts detrimentally on the quality or authenticity of food and drink

4 Introduction

Food fraud is a crime and an emerging risk, given the complexity of global food supply chains. It is estimated to cost the food and drink industry up to €50 billion a year. It also has the potential to become a major food safety issue. An extreme example of this is fake alcohol. Fraudulent booze can contain substitutes for ethanol, including chemicals used in cleaning fluids and automobile windshield cleaner, as well as methanol and isopropanol, which are used in antifreeze and some fuels. Drinking alcohol containing these chemicals can lead to serious health problems. In recent years, high-profile international food fraud has harmed and even killed many innocent people.

The challenges associated with food fraud are (1) defining the nature of food fraud, (2) treating food fraud as a criminal activity, (3) forging closer cooperation and partnerships to combat food fraud, and (4) predicting the likelihood of food fraud more accurately.

This procedure outlines how food business operators (FBOs) may take positive steps to prevent food fraud in their supply chains with a focus on the three generic types of food fraud, namely, (1) product substitution, (2) product addition, and (3) false statements about the geographical origin of products.

FOOD FRAUD AND VULNERABILITY

5 Procedure Flowchart

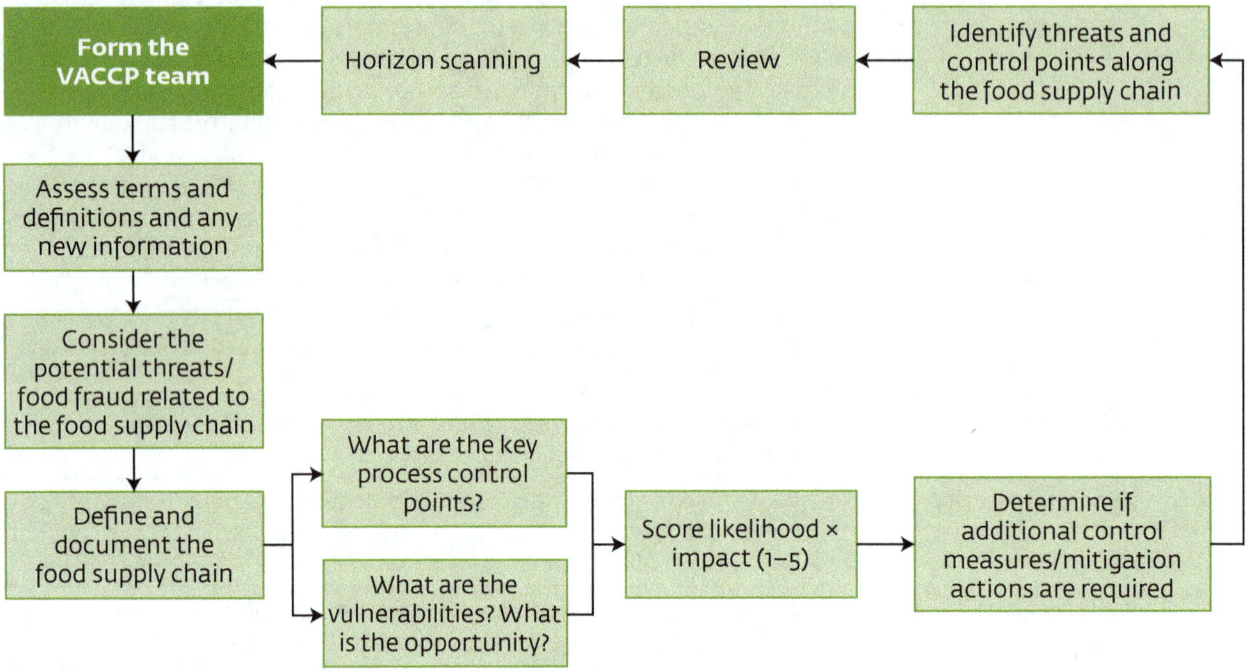

6 Procedure Notes

6.1 The VACCP Team

The food safety manager/management representative should establish a cross-functional team to conduct a vulnerability assessment critical control point (VACCP) study on the organization. External experts may be used to facilitate and support the VACCP study and the process.

All participants should receive relevant training in VACCP based on publicly available specification PAS 99:2012 of the British Standards Institution or training providers approved by the U.S. Food and Drug Administration (FDA) or equivalent. The training program should cover at least the principles outlined in box 4.1.

Box 4.1 Training Topics

Ensure all VACCP team members are trained in the following principles:

- Scope of the assessment
- Terms and definitions
- Aim of the VACCP
- Types of threats to consider
- Understanding the attacker, that is, the motivation, the opportunity, and the realization
- How to assess the threats and undertake a risk assessment procedure
- Critical controls in relation to the VACCP
- Response to an incident

FOOD FRAUD AND VULNERABILITY

The VACCP team should consider new terms and definitions, plus any new relevant information.

6.2 Consider the Food Supply Chain and the Chain of Custody

The VACCP team should conduct an initial review of the food supply by mapping out and verifying the food supply chain. This involves examining the unbroken path a product takes from the first stage in the food supply chain to the end customer, including raw commodity materials, conversion, transformation, distribution, and logistics.

6.3 Consider Food Threats and Vulnerabilities

Assess the possibility of food fraud in the supply chain using a VACCP template that is similar to the VACCP template provided in the food safety training kit.

Step 1: develop the key control points. After the food supply chain has been analyzed, key control points should be developed. A key control point is an area in the supply chain containing more than one product. The key process control indicates where contamination or the mixing of materials can occur or where there may be an economically motivated activity. In an ideal world, all products are fully tested and certified prior to processing. However, the reality of any supply chain operation suggests that these key process control points are points of vulnerability. Once key process controls have been developed, tracking and quality assurance can become drivers of a uniform marking system. To support determination of food provenance or geographical origin, periodic laboratory testing should be implemented.

Step 2: product identification and the uniform marking system. The simplest way to ensure there is no mixing of certified and uncertified materials is to create a marking and identification system that is fail-safe for even the simplest FBO. The marking system must be clear in every part of the process and cover raw materials, work in progress, finished goods, distribution, and logistics to account for the entire supply chain. A classification strategy for certified and uncertified materials would require the materials to be segregated at separate locations. It is essential that policies and regulations address third parties that are responsible for the logistics and warehousing of the components.

Step 3: recordkeeping and document programs. A chain of custody program requires detailed records and record systems to track all the activities involving a product down to the lot, batch, minute, and second of the food supply chain. The documentation process will keep track of this supply chain activity from the first producer through to the end consumer. Heavily regulated industries, such as the pharmaceutical and aerospace industries, can lead the way because they have established this process.

6.4 Score the Likelihood and Consequences

Score the likelihood of the threat occurring and the consequences if the threat is realized using a VACCP template similar to the VACCP template provided in the food safety training kit. The rating is 1 to 5. The consequence category is selected from table 4.1 on the basis of the most likely impact if the threat is realized, and the likelihood category is selected from table 4.2 on the basis of the corresponding likelihood that the threat will be realized.

FOOD FRAUD AND VULNERABILITY

Table 4.1 Categories of Threat Likelihood

Rating	Criteria
Almost certain, rating 5	99 percent probability, or impact is occurring now, or it is expected to occur within days to weeks
Likely, rating 4	>50 percent probability, or balance of probability it will occur, or it is expected to occur within weeks to months
Possible, rating 3	>20 percent probability, or may occur, but this is against short-term probabilities, or it is expected to occur within months to years
Unlikely, rating 2	>1 percent probability, or may occur, but not anticipated, or it is expected to occur in years to decades
Rare, rating 1	<1 percent probability occurrence requires exceptional circumstances exceptionally unlikely, even in the long-term future it may only occur as a 100-year event

The matrix in table 4.2 is used to determine the relative magnitude of the residual risk, ranging from trivial (E) to very high (A).

Table 4.2 Threat Risk Matrix

Impact of threat						
	5	C	B	A	A	A
	4	D	C	B	B	A
	3	E	D	C	C	B
	2	E	D	D	C	B
	1	E	E	D	C	C
		1	2	3	4	5
		Likelihood of threat happening/detection				

Very high risk	Threat A
High risk	Threat B
Moderate risk	Threat C
Low risk	Threat D
Trivial risk	Threat E

6.5 Describe the Current Control Measures

The VACCP team will define and maintain the current control measures, that is, the critical controls in place against food fraud.

Classification	Company Confidential	Food Fraud and Vulnerability Procedure	
Doc ID: SOP-046 Created: April 20, 2018	Printed: Updated: January 13, 2019	Controller: Document Controller Owner: Food Safety Manager	Page 6 of 7

FOOD FRAUD AND VULNERABILITY

6.6 Create Strategies and Actions to Mitigate in the Event of a Breach

The VACCP team will create action plans and strategies in anticipation of a possible breach of current control measures.

6.7 Verify Effectiveness and Strategies to Mitigate Vulnerabilities in the Event of a Breach

The VACCP assessment and controls should be reviewed formally at least once a year. The goal is to continue to improve the VACCP process by reviewing documentation and anticipating new possibilities of food fraud. This should include horizon planning, a robust internal and external review of food fraud vulnerabilities in the food industry.

7 Records

Document	Location	Duration of record	Responsibility
Food fraud and VACCP assessment record	Food Safety Office	Indefinitely	Food safety manager

CHAPTER 5

Food Safety Training

Introduction

Food business operators (FBOs) and others working in the food industry are legally required to undertake food safety training and undergo supervision according to their level of activity. Thus, for example, food managers and food servers receive different sorts of training. Individuals responsible for the development and maintenance of an FBO food safety management system (FSMS) must be trained in the application of hazard analysis critical control point (HACCP) system principles to the FSMS.

No required frequency is set out in the standard legislation on training. It is up to FBOs to decide if staff need new or refresher training.

This chapter does not seek to recommend a specific path in training or a specific type of training provider an FBO may select. It does offer guidelines and tips to help FBOs derive maximum benefit from food safety training. The aim is to supply sufficient information about useful considerations in organizing training so informed decisions can be made based on individual needs.

When embarking on training, FBOs often consult with learning professionals within their own training departments or externally. A learning professional is an entity that provides learning services to clients. Learning professionals may be engaged as trainers, coaches, instructional designers who design and develop training courses and programs, or performance consultants.

The practical effectiveness of an HACCP system or FSMS depends on the skills of the people who have developed the system and who operate the system and the prerequisite programs (PRPs) that support the system. If the system is to be successful, then there must be an overriding internal belief in the benefits for an FBO of a properly implemented HACCP or FSMS approach. Factors such as the variable quality of the education and training available and of the impact on the development of an FBO can bear directly on the ability of HACCP or food safety teams to conduct and maintain appropriate hazard analysis. This potential for problems may be exacerbated by the generally weak understanding of the relationship between PRPs and HACCP systems or FSMSs, especially during implementation and in maintenance.

When is training necessary?

An FBO would not purchase new equipment without first checking whether the equipment has the correct specifications to meet the needs of the FBO or whether the benefits of the equipment will outweigh the associated costs. Training may be viewed as an investment to ensure that equipment specifications match needs and that the returns on the equipment are likely to be satisfactory. Too many organizations regard training only as a mandatory expense and therefore hunt for ways to minimize the cost, rather than recognizing training as an essential part of an overall business strategy.

Alignment of needs

Every FBO experiences gaps between desired and current organizational performance. These gaps may be said to represent the needs of the organization. To be effective, training interventions must strategically align with categories of FBO needs, including the following.

Business needs are the highest order needs from which all other FBO needs derive. Not attending to these needs may threaten the existence of an FBO. Typical business needs include raising revenues, reducing costs, enhancing productivity, boosting efficiency, ensuring compliance with legislation and international food safety standards, providing better customer service, and attracting and retaining qualified personnel. Business needs may also be referred to as business or organizational goals, business objectives, or operational needs.

Performance needs are the on-the-job accomplishments and behavior of individuals in an FBO who perform specific functions contributing to the achievement of organizational goals. They represent what these individuals must do to achieve the goals of the FBO. They are often described in terms of performance parameters in quality, timeliness, dependability, flexibility, and cost. Performance needs may also be referred to as performance objectives, behavioral requirements, or accomplishments and behavior.

Learning needs are capability shortcomings individuals must offset if they are to perform sufficiently to fill gaps in knowledge, skills, or attitudes that are holding back the effectiveness of an FBO.

The learner needs of individual staff members are the particular knowledge characteristics of staff members that may influence the ability of the staff members to perform, including, for example, learning preferences, age, spoken languages, and literacy.

These categories of needs are expressed differently across FBOs, and they are often associated with key performance indicators with observable, measurable values that gauge the effectiveness of an organization in achieving key business objectives. The indicators are measures of the gap between performance and goals. High-level performance indicators focus on the overall performance of an FBO, while low-level indicators focus on internal FBO processes. The metrics applied may include (1) financial metrics, such as profit, costs, and sales by region; (2) customer metrics, such as customer satisfaction and customer retention; (3) process metrics, such as the number of product defects or the frequency of incidents of noncompliance; and (4) people metrics, such as employee turnover or employee satisfaction.

Key performance indicators inform an FBO's objectives (the specific results of an activity, such as training, to be achieved with available resources and within a time frame) and desired outcomes (the hoped-for benefits of the achievement of a business goal) (box 5.1; figure 5.1).

In the same way that misaligned equipment functions poorly, misaligned training will not achieve the desired results. FBOs are more likely to achieve greater returns on training investments if they adopt a strategic view of training, clarify the specific groups of their needs, and ensure that training is targeted on these needs.

Box 5.1 Examples of Needs Fulfillment through Training

To fulfill regulatory requirements, an FBO may wish to ensure that its food safety policies and procedures are regularly updated to promote the utmost efficiency. This is an example of a business need.

To solve the business need, the FBO may require that supervisors store documentation related to food safety policies and procedures correctly so that the documentation may be readily retrieved. This is an example of a performance need aligned with a business need.

To meet the performance need, the FBO may decide to train the FBO's supervisors in correctly uploading the latest information on food safety policies and procedures and adjusting version control numbers and dates of issue. This is an example of a learning need aligned with a performance need.

To offset the learning need, the FBO may be obliged to take into account specific characteristics of the supervisors who will undertake the training required to achieve the learning. For example, the supervisors may not be fluent in the language of the documentation, and some may not be computer savvy. This is an example of learner needs that must be taken into consideration in aligning training with learning needs.

A suitable training intervention in alignment with the above needs might involve a short course on the digital capture of documents and the collection of the necessary metadata on document versions and dates of issue. The training might include examination of well-chosen examples and dos and don'ts. The training could be supplemented by electronic performance support, for instance, post-training coaching

continued

Box 5.1 Examples of Needs Fulfillment through Training *(continued)*

or the creation of an interactive computer-based form that provides guidance to supervisors as they scroll over blank forms that must be completed.

Related to the training intervention would be nontraining interventions, that is, considerations that are not part of a training program per se, but can affect greatly training outcomes, such as securing convenient access to computers by supervisors to update and store documents.

Training that is in alignment with the needs of an FBO is likely to be effective and achieve the desired outcomes. As a counterexample, imagine how ineffective training would be in the aforementioned scenario if the following were true:

- The learning need has been defined too generally, for instance, the need for broad knowledge about seven key factors in a document management system: records creation; retention and disposition; storage and maintenance; access and retrieval; appropriate use of technology; promotion and support (archival records); and management program integration (Norris 2002).
- Learner needs have not been specifically defined, and the targeted learner group encompasses all FBO staff, from transport drivers to food handlers and managing directors.
- The proposed training intervention consists of a rehashed slideshow presentation about the pros and cons of three kinds of document management systems—paper, electronic, and hybrid—followed by a paper-and-pencil true-false quiz.
- The proposed nontraining intervention is a memorandum issued by management on the importance of storing documents properly.

These learning needs and solutions are not aligned well with the performance and business needs; nor are the proposed training and nontraining interventions directly applicable to the needs.

Figure 5.1 Effective Training: Aligning Business, Performance, Learning, and Learner Needs

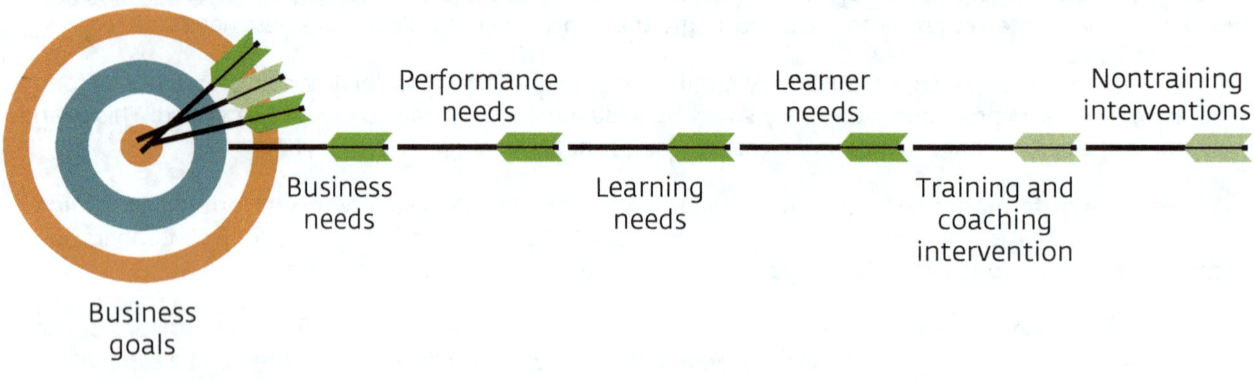

When can training help?

The inadequate performance of FBO staff is often the first symptom that some sort of training intervention is required. While training can indeed help resolve many performance issues, it may not be appropriate in all instances. The best balance depends on the source of the performance problem. Does the performance problem arise because of one of the following?

The performers do not know what to do or how to do it. The staff members lack the knowledge or skills to perform particular tasks. This is an issue on which training may provide a solution. If the staff members once knew how to perform the tasks, they may merely require a reminder (such as a job aid) or reinforcement (such as more practice); otherwise, training may be necessary. A job aid may be sufficient in cases in which the task is occasional and the performers do not need to internalize the procedure; rather, they may simply follow the appropriate steps whenever necessary. If the job aid is self-explanatory, no training is required.

The performers know what to do, but do not want to do it. The staff members have a poor attitude or lack motivation to perform appropriately. The staff members need to understand why proper performance matters. Training, such as realistic simulations or role playing, may help, or the staff members may need an incentive.

The performers want to do what they need to do and know how to do it, but a lack of resources or some other roadblock is preventing successful accomplishment. Training will not resolve this problem. Roadblocks to performance may stem from a host of issues in the FBO environment, for example: (1) inadequate physical resources, such as a lack of office space, tools, time, or budget; (2) inadequate structure or process, such as unclear processes or policies, poor workflow, or structural or process deficiencies; (3) lack of adequate information, lack of timely or accurate feedback; and (4) wellness issues, such as poor physical health, emotional instability, or a toxic company culture (figure 5.2).

In a perfect world, performance issues would divide evenly among these probable causes. In a complex world, training may not be the sole answer to a performance issue, and training should be offered alongside

Figure 5.2 Is Training the Answer?

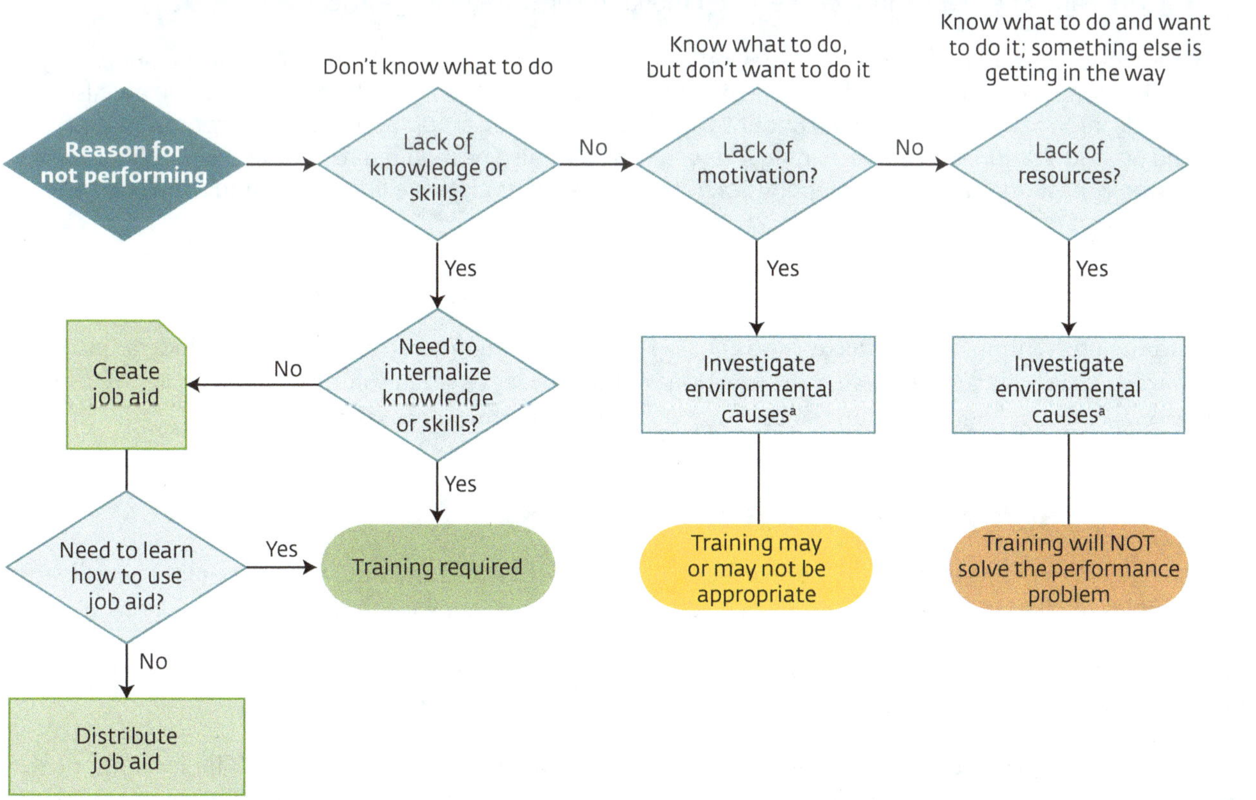

a. Examples of environmental causes: lack of physical resources; unclear policies and procedures; lack of information; wellness issues.

Box 5.2 Aligning Training with Needs: A Sample Case

Ferdinand is the food safety manager of a medium-size meat processing company. The company prepares and packages local favorite specialty grilling products. Over the past several months, two cases of *Listeria monocytogenes* contamination have been detected in the company's meat preparation and packaging areas. The local regulatory authority has investigated and made significant recommendations regarding the company's *Listeria* control strategy, including practices related to product processing, intensive cleaning, sanitation, and additional environmental and end product testing. Ferdinand is conscious of his company's stated business goal to become the trusted brand supplier for all grocery stores in the region. The contamination incidents have negatively affected the company's key business outcomes, particularly key performance indicators related to safety, quality assurance, and marketing. Ferdinand recognizes that he needs to train food handling staff more effectively and hires Marie, a learning professional, to help him sort through the problem.

In this case, any proposed training must align with the following needs:

- The business need is to reestablish trust in the company brand, as indicated by specific food safety, quality assurance, and marketing measures.
- The performance need is to ensure that staff apply thorough sanitation procedures during product processing to avoid *Listeria* contamination. The hazard control plan must also be updated, and control measures and the entire system must be revalidated.
- The learning need may include food handler understanding of microbiological contamination, personal hygiene, and the proper use of cleaning equipment and materials. It may also encompass some motivational aspect, such as understanding the consequences of inadequate sanitation and the serious infections or even death that may result among friends and family from eating foods with *Listeria monocytogenes*.
- The learner needs may include the accommodation of local language in learning.

The proposed training intervention may reflect a variety of strategies, such as formal step-by-step instruction, accompanied by a job aid that can be posted in the workplace. Marie will likely also include information in the training about the serious consequences of microbiological contamination and poor sanitation. Some nontraining interventions may also be required to resolve the problem, for example, ensuring the provision of physical resources, such as cleaning equipment and supplies, and convenient food handler access to these resources.

nontraining interventions to be effective (box 5.2). The nontraining interventions might include adjustments to resources, structures and processes, information and feedback, or attention to wellness issues. A learning professional can supply guidance on the best way to resolve a performance issue.

What to look for in a training provider

It is advisable to survey several training providers to find one that suits the FBO requirements. The following are examples of characteristics that might be wanted among learning professionals.

Does the training provider have appropriate food safety qualifications? The learning professional must have a background in food safety and relevant experience in the food industry.

Does the training provider have appropriate training and educational qualifications? The learning professional must have knowledge about how people learn about and acquire training skills. This may include the following:

- Credentials from accredited international professional training organizations, such as the Association for Talent Development (https://www.td.org/), the Institute for Performance and Learning (https://performanceandlearning.ca), the International Society for Performance Improvement (https://ispi.org/), and the Learning and Performance Institute (https://www.thelpi.org/), or related skill certifications, such as in training and facilitation, instructional design, adult learning, or coaching.
- Formal educational degrees or diplomas in adult education; educational technology; instructional design, training, and development; or performance improvement.

Does the training provider use a systematic approach? Most learning professionals follow a well-defined problem-solving approach in undertaking a training project, such as the training life-cycle approach. The approach often embodies a process for developing, delivering, and managing a training product or service.

Does the training provider adopt a collaborative approach? Is management participation encouraged? Management involvement is crucial to the success of the training, particularly during pre- and post-training phases. The learning professional and management should work together to ensure that all food safety issues at the business are addressed by the training course or program. A seasoned learning professional would partner with the FBO to clarify performance needs and expectations as well as the parameters of the training project. This is also an opportunity to expand thinking and mutually consider possible innovative solutions.

Is the training provider interested in identifying specific needs and appropriate solutions? Does the learning professional take time to explore the context of the FBO's needs? Is the learning professional willing to spend time listening to FBO management and explaining the best method for meeting requirements? Is the learning professional able to pinpoint problems and likely causes, suggest appropriate solutions, and provide guidelines in measuring success?

Is the learning professional willing to demonstrate training materials or previous work? Does the learning professional supply cost-effective, learning-efficient, learner-focused solutions? The answers to these questions will offer an indication of the quality of the training one may expect and of whether the training is generic or may be tailored to needs. Management should examine the relevance of the training content to the needs of the FBO. Is the training content cost-effective? Is it replicable or customizable for potential future training requirements? Is the training process structured appropriately and efficiently? Does it keep learners engaged, accommodate individual learner preferences, and provide sufficient practice and feedback so learners may acquire the desired performance skills and knowledge?

Does the learning professional provide follow-up support? Does the learning professional address the application of training in the workplace? Post-training follow-up is essential in putting the training theory into practice in the workplace and can help clarify issues that may arise among participants after the training has taken place. Management should consider possible post-training issues such as the fit of the proposed follow-up with the work environment, the opportunities available for retrieving and practicing the skills and knowledge learned during the training, the provision of periodic reminders and practice on key training elements, and the mechanisms included to monitor post-training performance.

What do previous clients say about the learning professional? Can the training provider demonstrate the outcomes of previous training engagements? Talking to previous clients will help management determine the quality of the learning professional's interventions. Former clients may be able to answer questions such as the following: Did the training solution help resolve staff performance issues? Were the outcomes attributable to the training alone, or did other factors contribute or interfere? How did the learning professional evaluate the effectiveness of the training? What modifications would improve the effectiveness of the training? Were the previous clients satisfied with the benefits of the training relative to the cost?

Other desirable qualities worth considering in hiring a training provider include (1) excellent communication skills, including written and oral presentation skills; (2) strong interpersonal and teamwork skills; (3) the ability to work effectively across organizational boundaries; and (4) a sense of humor.

For more guidance on what to look for in the selection of learning professionals, see GLC (Grow Learn Connect), International Finance Corporation, Washington, DC, https://www.growlearnconnect.org/.

Where can FBO management find learning professionals or training courses?

Other than global directories available through international accreditation bodies, there may be no centralized list of learning professionals or training courses in the local area of an FBO. One should therefore check among local training providers or search online for courses or learning professionals nearby.

The International Finance Corporation (IFC) has developed several relevant FSMS training courses, as follows:

- *IFC Food Safety Foundation Course*. This is an entry-level course aimed at the FBO processing, catering, and retail sectors. The course is recommended as a prerequisite for participation in the IFC Food Safety Handbook Training Course. The course covers the basics of food safety management and focuses on the prerequisites for establishing an HACCP-based FSMS. The course can also be tailored to an industry sector.
- *IFC Food Safety Handbook Training Course*. This is an intermediate-level course aimed at providing an FBO with appropriate knowledge and skills. It involves components on access to best practice HACCP tools and techniques and useful links to enable the FBO to establish and develop an FSMS based on the HACCP requirements in most schemes recognized by the Global Food Safety Initiative (GFSI).
- *IFC Food Safety Internal Audit Course*. This is a two-day advanced-level course aimed at providing FBOs with the knowledge and practical skills for conducting food safety internal audits. The course covers planning, preparing for an internal audit, performing an internal audit, communicating the audit results, and the conclusion of the auditing process, including corrective action and root cause analysis of the findings. The course also involves a mock recall exercise to allow the participants to put their knowledge into practice. Courses have been conducted in, for instance, Myanmar and Vietnam.

For more information on the first two courses, go to "Food Safety Training Courses," International Finance Corporation, Washington, DC, https://www.ifc.org/wps/wcm/connect/industry_ext_content/ifc_external_corporate_site/agribusiness/resources/food+safety+training+courses.

Developing a training program

Most learning professionals will implement a well-defined problem-solving approach in realizing a training program. One approach is the training life cycle, which typically consists of six steps, as follows:

- Investigating performance needs and identifying suitable performance improvement solutions

- Designing curricula to undertake a comprehensive training strategy, especially in larger programs
- Designing and developing learning experiences, that is, training courses, especially in smaller programs or within a larger training scheme
- Facilitating learning, that is, conducting the training
- Supporting the transfer of learning, that is, ensuring that the knowledge and skills acquired through training are applied in the workplace
- Evaluating learning, that is, making sure the training successfully promotes learning

Some professionals add a seventh step: revising the training. This step signals that any training represents a continuous cycle of improvement, whereby the results of evaluation feed into enhancements in future training interventions. Some learning professionals also incorporate a rapid prototyping of training solutions into the above process as a way of helping clients visualize more concretely the final training scheme.

Below are descriptions of each step, including the purpose, the process, the deliverables, the approximate time required, and ways management can help the training program succeed.

Assessing performance needs

PURPOSE

A learning professional will typically partner with an FBO to accomplish the following: (1) investigate performance and training needs; (2) ensure that appropriate performance problems and causes are being addressed; (3) determine whether the problems can be resolved through training and, if so, whether additional nontraining interventions need to be considered, such as interventions directed at motivational issues, the adequacy of physical resources, adjustments to organizational structure or processes, clear information and feedback on performance, or the cultural climate of the FBO; and (4) derive the best learning solution to align with the FBO's business and performance needs to maximize impact.

WHAT TO EXPECT

In assessing an FBO's performance needs, a learning professional may take only a few hours to clarify what participants should be able to do after completing a proposed training course, or the professional may take several weeks to analyze complex performance issues that involve many work processes, functions, and roles. This may entail assessing performance needs across FBO staff, processes, and the entire organization.

FBO management should be prepared to respond to wide-ranging questions from the learning professional, who will seek to obtain a full picture of the FBO's performance issues and may ask questions related to the following: (1) trends affecting the region, the industry, and the FBO; (2) the FBO's business goals, particularly the strengths and opportunities on which management wishes to build and the weaknesses and threats that may be hindering progress; (3) performance issues at an individual, process, or organizational level and the perceived gap between current and desired performance; (4) learning needs, that is, the staff's knowledge, skills, and attitudes that, if improved, could help close the performance gap; and (5) specific learner needs that would necessarily shape any proposed training intervention, such as literacy, familiarity with computers (if considering e-learning), and scheduling matters (for instance, shifts and caregiving responsibilities outside work).

The responses and other information may be collected in several ways, including the following:

- One-on-one interviews or focus groups with key management staff, representatives of the target learners, other stakeholders, perhaps even customers, depending on the nature of the performance issues identified

- Research into human resource records and past training and training results
- Site visits and observation of target learners at work

KEY DELIVERABLES

There are two key deliverables involved in this process. First is the survey questions and responses during interviews and focus group discussions and research during the performance needs assessment or training needs assessment. Second is a report on the performance needs assessment. The report provides a summary of findings and recommendations on ways to close the performance gap. It may contain a training program proposal, which should not only describe the proposed training intervention (for example, classroom learning, e-learning, mentoring, coaching, on-the-job training, and job aids), but also comment on nontraining interventions that may need to be implemented to maximize the impact of training. For example, if food handlers receive excellent training on the steps to ensure good food hygiene, but have difficulty accessing the necessary equipment or protective ware to carry out these steps, the results will be unsatisfactory. Report recommendations should also include a discussion of the way learning will be transferred to the workplace to ensure sustained, high-quality performance and the way the results of learning will be measured and evaluated. The performance needs assessment report will help FBO management make informed decisions and fix priorities among the proposed training interventions, including whether to opt for a standardized course or program and customize it to fit the organization or for a custom course or program (box 5.3).

APPROXIMATE TIME

It is worthwhile to spend a fair amount of time on this step because all other steps hinge upon it. To complete this step, the learning professional should be allowed at least three weeks or even more, depending on the size and complexity of the training program.

Box 5.3 A Sample Assessment Process on Performance Needs

Marie conducts a performance needs assessment for Ferdinand, and, through focus group interviews with supervisors and employees, she uncovers some underlying causes for the performance issues Ferdinand is facing. While each employee receives thorough orientation training in the foundations of food safety when they are hired, including hygiene and sanitation practices, the training only occurs on the first of each month. New hires who miss the orientation cycle sometimes need to wait a full month to receive the training. Marie also discovers that cleaning equipment and supplies are located far from the meat processing area and are sometimes not adequately stocked. Moreover, because sanitation procedures are usually performed at the end of shifts, they are often carried out hurriedly by employees anxious to get home. Finally, she remarks that a quarter of the food handlers exhibit low literacy.

Marie proposes the following training solution:

- The use of a short training video in which a respected work colleague demonstrates the sanitation procedure step by step. The video will be provided in the local language and subtitled. The video will be the property of the FBO. It can be played on demand on multiple devices and used by employees who wish to obtain a refresher on the procedure and by supervisors or assigned buddies (mentors) who wish to show it in conjunction with the coaching of new hires or employees who are underperforming.
- A pictorial job aid accompanies the video. It highlights the important aspects of each step and will be posted at key locations throughout the meat processing area.

continued

Box 5.3 *(continued)*

Marie also proposes some additional nontraining interventions, as follows:

- The relocation of cleaning equipment and supplies to a place that is more convenient for employees
- Adjustments in document management in the company's FSMS, particularly in the restocking procedure for cleaning equipment and supplies to avoid stockouts
- The provision of sufficient time for sanitation procedures to be completed prior to the end of a shift
- Random inspections by managers to ensure adherence to procedures
- Laboratory validation of cleaning activities
- Reward incentives for staff teams that achieve 100 percent compliance with standards and rules over the year

HOW MANAGEMENT CAN HELP

This step can be expedited if management readily provides the learning professional with access to information, such as the following:

- The background of the FBO and the decision to implement training
- The estimated training budget and any training infrastructure
- Previous training that has been conducted
- Technical documentation, including step-by-step processes and procedures
- Demographic information on the target learners
- The technical environment in which the target learners work and where the training will take place
- The arrangements for requested interviews and focus groups

Designing curricula (training strategy) and designing and developing learning experiences (training courses)

PURPOSE

During these steps, the learning professional builds a plan and develops learning solutions to meet the needs identified during the first step.

For a complex training program, this step involves the preparation of a training strategy report, which is essentially a blueprint describing the interrelated courses and materials that will be used to build the competence of staff over an extended period. Designing the curricula involves determining the content to be included and defining the strategy of the curricula, for the training process, and to support the learners.

For smaller training programs or for an individual course within a complex training program, this step requires detailed planning of the training course and the learning experience, including content, structure, instructional delivery, and evaluation methods.

Box 5.4 A Sample Course Design on Sanitization of Meat Processing Equipment

For her training program for Ferdinand's food handlers, Marie focuses on the design of a learning experience (or training course) on the sanitization of meat processing equipment. Such a course could also conceivably be part of a much larger curriculum (or training strategy) designed to support food safety management across many functions at Ferdinand's company. The curriculum might include many other courses, such as an introduction to food safety, personal hygiene, pest and waste control, food storage, introductory and advanced HAACP, FSMS policies and procedures, document and record control, and internal auditing, some or all of which may be mandatory depending on the job of the employee. The curriculum would also specify the level of the content required, ranging from introductory to intensive, the best sequence in which to take the courses, and the staff for whom each course is ideally intended.

Once the plans are agreed, the learning professional proceeds to development and (1) creates training materials for each course for both the trainer and the participants; (2) may help select trainers and organize training the trainers, if necessary, to orient them to the courses they will teach; and (3) prepares logistics related to course scheduling, participant registration, and course delivery (box 5.4).

WHAT TO EXPECT

During planning, whether the training program is large or small, the learning professional will want to gain clarity from management on the following:

- The profile of the target learners: the knowledge and abilities they already possess, their demographic background and educational attainment, and their motivations, aspirations, and concerns; what makes them tick?
- The establishment of clear learning objectives that are SMART—specific, measurable, attainable (or achievable), realistic, and time-bound—and tie directly to the FBO's performance and business goals
- The selection and sequencing of training content
- The best training delivery format, that is, classroom, live online, or self-study e-learning
- The identification of suitable practice activities to reinforce skills and concepts
- The identification of appropriate post-training support for learners

At the development stage, as training content is prepared, the learning professional may require additional input from content experts and reviews by content consultants and other stakeholders FBO management may designate to ensure content accuracy. During this stage, management may also receive prototypes for review, such as sample facilitator and participant guides in the case of classroom-based learning or sample e-learning modules.

KEY DELIVERABLES

Planning stage

First is course plans for simpler programs, including profiles of intended target learners, a description of the learning objectives and how these will be measured, a content outline, and a decision about the delivery format. Second is a training strategy report for more complex programs, including profiles of the intended target learners and a curriculum roadmap, that is, the interrelationship of courses, their sequencing, and the delivery format. At this stage, management may also expect prototypes, templates, and guidelines.

Development stage

First is training materials such as facilitator guides for classroom or live online learning, participant workbooks and handouts, presentation slides, reference materials, job aids, posters, and e-learning or paper-based materials for self-study. Second is the organization of a web-hosting platform for e-learning. Third is the creation of a training registration database for participants. Fourth is the establishment of training schedules. Fifth is training the trainers, if necessary. Sixth is the organization of announcements on upcoming training.

APPROXIMATE TIME

Management should allow 2–12 weeks for the planning stage, depending on the number of courses, the availability and quality of source documents, and the availability of content experts able to guide the instructional designer in preparing the plan.

The amount of time required for the development of training materials will vary considerably depending on the duration of the course and the training delivery format. Table 5.1 supplies general guidelines derived from a survey conducted in 2017 by the Association for Talent Development that can help management estimate how long it will take to develop a course. The table assumes the development of a custom course, rather than the adaptation of a preprepared course.

Table 5.1 Time Required to Produce One Hour of Instruction

Delivery format	Number of hours
Instructor-led classroom training	40
Live online web-based training	30
E-learning with limited learner interaction	70
E-learning with complex learner interaction	130
E-learning with real-time simulations	140

Source: Defelice 2018.

HOW MANAGEMENT CAN HELP

Management can expedite these stages by (1) providing the learning professional with access to content experts and technical documentation; (2) reviewing and providing feedback on plans in a timely manner; (3) preparing FBO staff by informing them of the reason for and the importance of the upcoming training; research has shown that managers who establish among staff the context and the expectations associated with training prior to the delivery of training have the most influence on the successful impact of training, even more influence than the trainers or the target learners; and (4) securing the resources necessary to deliver the training, such as web-based platforms or training rooms.

Facilitating learning

PURPOSE

This stage focuses on conducting the training effectively and efficiently so that participants can achieve the expected learning outcomes in as little time as possible. During this stage, the trainer seeks to create a positive learning environment for participants during the course and foster learning by encouraging the active engagement of learners. The trainer will use activities to monitor and evaluate the progress of the participants toward the learning objectives.

WHAT TO EXPECT

During this stage, management may expect the following. First, a pilot course is conducted with representatives of the target learners, content experts, and representatives of the FBO management team. The test allows management and the learning professional to determine if the course is likely to achieve what has been proposed and if any modifications are necessary before rolling the course out among staff more widely. Second, the training course or program is launched. Third, learning supports are implemented, such as e-mail reminders, job aids, coaching by supervisors or the trainer, on-site follow-up with participants, and recognition and incentives for good performers.

KEY DELIVERABLES

The key deliverables are (1) the course and (2) evaluations of participant satisfaction with the course and the achievement of the course learning objectives, possibly through a test, a demonstration event, or a new project.

APPROXIMATE TIME

The time required for this stage is contingent on the duration of the course.

HOW MANAGEMENT CAN HELP

Management can support this phase during training by encouraging participants to take the training seriously and giving participants sufficient time to acquire new skills and knowledge and obtain feedback on their performance.

Supporting the transfer of learning

PURPOSE

Learning does not stop after a training course ends. Supporting the transfer of learning helps participants apply in the workplace what they have learned during the course. During this stage, the learning professional wants to ensure that participants are able to use what they have learned during their courses and seeks ways to reinforce the learning achieved during the course. Although the support for the transfer of learning is considered a separate stage, it is a focus of the learning professional throughout the training life cycle.

WHAT TO EXPECT

During this stage, management should expect the reinforcement of the learning among participants post-training, back on the job.

KEY DELIVERABLES

The key deliverables are (1) the implementation of a variety of post-training reinforcement interventions, including coaching by supervisors, colleagues, or the trainer, and performance support tools, for example, e-mail reminders, review tests, job aids, on-site follow-up with participants, and recognition and incentives for good performers, and (2) the collection of evaluation data on application by participants of their learning.

APPROXIMATE TIME

The time required for this stage is contingent on the duration of follow-up activities.

HOW MANAGEMENT CAN HELP

Management can support this stage by allowing sufficient time for participants to practice their new skills and knowledge on the job and by obtaining feedback on the performance of participants since training.

It is critical for management to reinforce the learning transfers in the workplace. According to research, the successful transfer of skills and knowledge through training is closely connected to the direct involvement of management after training, which is almost as important as the involvement of management before training.

Management can also support the transfer of learning by holding staff accountable for implementing what they have learned, measuring their progress, providing coaching and guidance among staff who are having difficulty, and recognizing star performers.

Evaluating learning

PURPOSE

During this stage, the learning professional helps management gauge the effectiveness and impact of the training interventions by monitoring evidence-based results.

WHAT TO EXPECT

Learning evaluation refers to the collection and analysis of data and information to determine how effectively the training intervention has met business, performance, learning, and learner needs.

What is meant by training effectiveness? How does one know if the training went well? Kirkpatrick (1959, 1994) provided the most comprehensive response to this question. He said it boils down to four questions, as follows:

- Did the learners like the training?
- Did they learn the material?
- Did they use it?
- Did it make a difference?

These four questions correspond directly with the four categories of needs of an FBO (figure 5.3).

Figure 5.3 The Links between the Assessment of Needs and the Learning Evaluation

Needs assessment		Evaluation levels
Stage 1: What are the business needs?	⟷	**L4—Business impacts:** Were the business needs met?
Stage 2: What are the performance needs?	⟷	**L3—Job performance:** Can learners perform as required back on the job?
Stage 3: What are the learning needs?	⟷	**L2—Knowledge/skill mastery:** Did they learn what they were supposed to learn?
Stage 4: What are the learner needs?	⟷	**L1—Learner reaction:** Were they satisfied with the learning experience?

The evaluation of each of Kirkpatrick's questions is progressively more complex (table 5.2).

KEY DELIVERABLES

The first key deliverable is a training evaluation report covering the four levels of Kirkpatrick (see table 5.2):

- The reaction of participants to the training intervention
- The comprehension and retention by participants of the learning
- The ability of participants to apply their learning
- The impact of the training on the achievement of FBO goals, which may include a calculation of the return on investment, especially in the case of more complex training projects

Table 5.2 The Four Evaluation Levels

Evaluation level	The question	Why evaluate this level?	When is this level evaluated?	Typical evaluation methods used
1. Reaction	How much did participants like the learning experience?	Checks whether individual learner needs have been accommodated (for instance, correct content level, appropriate content, desirable course duration)	End of training	Participant evaluations at the end of the course, either written or oral
2. Comprehension	How much have participants understood and retained the course contents?	Checks if participants have been able to achieve expected learning outcomes	During training	Course activities and discussion Exams, quizzes, pre- and post-tests for knowledge evaluation Capstone exercises for skill and attitude evaluations, such as individual demonstrations or projects
3. Application	How well are participants applying what they have learned?	Determines if participants are performing as desired and can consistently use their new skills and knowledge back on the job	Three to six months after training	Interviews with participants Focus groups Self-reporting On-site observation
4. Impact[a]	How much has the learning experience affected overall business results?	Determines if business results have been realized and how much of the impact can be attributed to the training versus other environmental factors	Six months to one year after training	Report on the impact of training interventions with regard to agreed key performance indicators

a. Phillips (1997) suggests a fifth level beyond impact—the return on investment—usually in larger scale training projects. Return on investment measures the amount of return relative to an investment and indicates the extent to which the FBO uses its resources efficiently. It is usually expressed as a percentage and defined as net project benefits, divided by project costs. For example, a return of 79 percent means that, for every dollar of investment in a training project, 79 cents is returned after all investment costs have been recovered, such as the cost of developing the training program and the salaries and travel expenses of participants.

In particular, management should examine if the training evaluation addresses each of the performance and training needs in the performance needs assessment (box 5.5). The second key deliverable is represented by recommendations for regular improvements in the training.

Box 5.5 Addressing the Four Training Evaluation Levels

For Ferdinand's training project, Marie addressed the four evaluation levels in the following ways. First, she examined the reaction of participants to the training by surveying participants on their response to the video training course, including the relevance and accuracy of the content, the ease of using the video, and any suggestions for future modifications.

Second, she gauged participant comprehension. She measured the participants' achievements in learning outcomes in two ways. After the video session, the participants took a brief quiz to self-check their understanding of key steps. They were permitted to replay the video as many times as necessary to achieve a score of 90 percent or higher. To receive a certificate for completion of the training, participants were required to demonstrate to their supervisor the sanitation procedure outlined in the job aid.

Third, she studied the application by participants of their learning. Supervisors were required to conduct unannounced checks and observe employees while they were sanitizing equipment. The supervisors also conducted end product testing. They regularly reported the results of employee compliance and tests to management.

Fourth, she analyzed the impact on the FBO. Thanks to the new method of strategic training, Ferdinand's company has eliminated the *Listeria monocytogenes* contamination and has regained market share as one of the most trusted suppliers in the region based on selected key performance indicators.

APPROXIMATE TIME

The length of this stage is contingent on the extent and strategy of the evaluation (see table 5.2).

HOW MANAGEMENT CAN HELP

Management should encourage staff to supply honest feedback about the training course, the training process, and the training content as a way to enhance future training offerings. Management should carefully consider any recommendations forthcoming in reports on the application of learning (evaluation level 3) and the impact of the training on staff performance and, ultimately, business results (evaluation level 4).

For more information, go to the GLC (Grow, Learn, Connect) website, at https://www.growlearnconnect.org/.

What does an auditor seek when assessing food safety training?

An internal or external auditor will not necessarily want to see a certificate from a particular training course, but may request evidence on the delivery of the training and the topics studied. In particular, an auditor will observe hygiene practices and verify the food safety knowledge and skills of staff. The auditor may ask to see food safety records or ask about the nature of the food safety training that has been provided (box 5.6). In essence, merely completing a training program to check off a box is not sufficient. It is critical that management show evidence that the knowledge and skills covered in training have been implemented and are associated with appropriate records. The sample documented standard operating procedure (SOP-014) on training and development provided in chapter 4 may support such a process, while also aiding in the design and implementation of the training program.

Box 5.6 A Word on Budgeting for Training

The cost of a training project is proportional to the complexity of the project. Many variables affect the cost. It is a best practice to pay for training projects in three steps:

- *The performance needs assessment, the report on findings, and the training strategy:* This cost relates to conducting the performance needs assessment and producing a performance needs assessment or strategy report. The cost is based on the materials used and the time taken by the learning professional.
- *Training product development:* Once the training strategy is known, management may obtain a more accurate assessment of the development costs from the learning professional. The development budget is typically included as part of the training strategy. Management can base the cost of contracting on the budget presented in the final training strategy. However, management may want to adjust the implementation approach once there is a better understanding of the budget implications.
- *Training delivery:* Pay for the delivery of the training courses or the training program separately from development because the cost of the former greatly depends on the delivery method, the size of the target audience, the number of trainers needed, the travel costs, and the amount of time.

Some of the following factors will affect the cost of a training project:

- The number and duration of the courses to be developed
- The availability of instructional material on the same subject for adaptation and customization
- The availability and quality of source material
- The types of instructional material to be developed, for example, classroom based, e-learning, and job aid
- The technology needed for the proposed delivery method, for instance, web-based training and videos
- The number of pilot tests involved in the training program
- The translation into multiple languages and adapting the training materials to the local context

The responsibility matrix and training needs analysis

Management should undertake the following as part of the process of developing a training plan and a training program: (1) identify the responsibilities of individuals involved in food safety PRPs and the FSMS; (2) identify the responsibilities of individuals performing tasks that have the potential to generate a significant impact on food safety; (3) determine the proper competencies required of individuals possessing responsibilities in these areas; (4) assess the training needed to link these responsibilities and these competencies; and (5) develop the requisite training program. Tables 5.3 and 5.4 provide guidance in this endeavor. Table 5.5 illustrates a partly completed training agenda.

Table 5.3 Responsibility Matrix and Training Needs Analysis

Role, position, title, position no.	Name	Responsibilities	Qualifications and competency
Managing director	Mike Murphy	• Participate in FSMS management review • Set policy • Review objectives and targets • Allocate resources	Senior business administrator
Food safety manager (management representative)	Joe Bloggs	• Establish, develop, implement, maintain, and improve the FSMS, including food defense • Train the FSMS team members • Implement programs to achieve set objectives and targets • Monitor and measure FSMS performance, including reporting to top management • Maintain awareness of FSMS compliance within the FBO • Liaison with external audit and inspection organizations	BSc Food Science Certificate of Attainment in FSMS FSSC 22000
Internal auditor	Mary Cahill	• Develop internal audit program in liaison with the food safety manager • Conduct internal audits according to schedule • Train other internal auditors	Certified FSSC 22000 Internal Auditor
FSMS team members	A Sullivan J Wright M Brown K Wriggly	• Maintain awareness on policy and FSMS • Implement the program to achieve set objectives and targets • Implement PRP and HACCP plan, and implement, verify, and validate the operational prerequisite program (OPRP) • Help in monitoring and measurement • Train respective staff in implementing FSMS policies and procedures	FSMS, including HACCP principles and practices Planning, operations, management
Department managers	All	• Oversee the context of organization planning, leadership, performance evaluation, improvement of the FSMS	
Laboratory technician	R Harley	• Conduct analytical tests, laboratory equipment maintenance and calibration, p-test, laboratory training	Laboratory management (chemistry/biology)
Associate	All	• Maintain awareness on policy and FSMS	N/A
Transport driver	All	• Maintain awareness on policy and FSMS	Driver's license

Note: FBO = food business operator; FSMS = food safety management system; FSSC = Food Safety System Certification; HAACP = hazard analysis critical control point; LAB = laboratory; N/A = not applicable; OPRP = operational prerequisite program; PRP = prerequisite program.

Training needs	Planned dates	Training details	Remarks
• FSMS orientation (in-house)	April 24, 2018	FSMS-1	Training completed April 24, 2018
• Mandatory FSMS training, reference the training matrix (internal)	March–October 2019	Various	Training to be completed by October 2019
• Emergency preparedness and response/ crisis management (external)	June 2019	FSMS-4	Training to be completed by June 2019
• FSMS orientation (in-house)	April 24, 2018	FSMS-1	Training completed April 24, 2018
• FSSC 22000 lead auditor training (external)	March–April 2019	FSMS-5	Training confirmed
• Mandatory FSMS training, reference the training matrix (internal)	March–October 2019	Various	Training to be completed by October 2019
• Emergency preparedness and response/ crisis management (external)	June 2019	FSMS-4	Training to be completed by June 2019
• FSMS orientation (in-house)	April 24, 2018	FSMS-1	Training completed April 24, 2018
• FSSC 22000 internal auditing course	March–April	FSMS-6	Training confirmed
• FSMS orientation (in-house)	January 19, 2016	FSMS-1	Training completed April 24, 2016
• Mandatory FSMS training, reference the training matrix (internal)	March–October 2019	Various	Training to be completed by October 2019
• Emergency preparedness and response/ crisis management (external)	June 2019	FSMS-4	Training to be completed by June 2019
• FSMS orientation (in-house)	April 24, 2016	FSMS-1	Training completed April 24, 2016
• Emergency preparedness and response/ crisis management (in-house)		FSMS-5	Training to be completed by June 2019
• FSMS orientation (in-house)	April 24, 2016	FSMS-1	Training completed April 24, 2016
• Analytical policies and procedures	September 30, 2016	LAB-1	Training completed September 30, 2016
• Mandatory FSMS training, reference the training matrix (internal)	March–October 2019	Various	Training to be completed by October 2019
• FSMS orientation (in-house)	January 24, 2016	FSMS-1	Training completed January 24, 2016
• Mandatory FSMS training, reference the training matrix (internal)	March–October 2019	Various	Training to be completed by October 2019
• FSMS orientation (in-house)	April 24, 2016	FSMS-1	Training completed April 24, 2016
• Tank cleaning/sanitizing	May 24, 2016	FSMS-10	Training completed April 24, 2016
• Dairy farm raw milk handling/testing	May 24, 2016	FSMS-11	Training completed April 24, 2016
• Mandatory FSMS training, reference the training matrix (internal)	March–October 2019	Various	Training to be completed by October 2019

Table 5.4 Food Safety Training Matrix

Job title	Food safety orientation	Personal hygiene	Pest and waste control	Cleaning	Allergen management	Food delivery	Food processing	Food storage	HACCP	Traceability, recall, and withd
Managing director	M	M	M	M	M	M	M	M	M	M
Food safety manager	M	M	M	M	M	M	M	M	M	M
Hygienist/microbiologist	M	M	M	M	M	M	M	M	M	M
Milk processing manager	M	M	M	M	M	M	M	M	M	M
Laboratory manager	M	M	M	M	M	M	M	M	M	M
Warehousing manager	M	M	M	M	M	M	M	M	M	M
Engineering manager	M	M	M	M	M	M	M	M	M	M
Maintenance manager	M	M	M	M	M	M	M	M	M	M
Logistics manager	M	M	M	M	M	M	M	M	M	M
Laboratory technician	M	M	M	M	M	M	M	M	M	M
Food handler	M	M	M	M	M	M	M	M	M	M
Transport driver	M	M	M	M	M	M	M	M	M	M

Note: M = mandatory; O = optional; N/A = not applicable.

Maintenance	Calibration	FSMS policies and procedures	Analytical policies and procedures	HACCP verification and validation	Document control	Record control	Emergency preparedness and crisis management	Food defense and food fraud	Consumer complaint management	Communications	Internal auditing	FSMS management review
N/A	N/A	M	M	M	M	M	M	M	M	M	N/A	M
N/A	M	M	O	M	M	M	M	M	M	M	M	M
N/A	N/A	M	O	M	M	M	M	M	M	M	M	M
N/A	N/A	M	M	M	M	M	M	M	M	M	N/A	M
M	M	M	N/A	M	M	M	M	M	M	M	M	M
N/A	M	M	N/A	M	M	M	M	M	O	M	O	M
O	M	M	N/A	M	M	M	M	M	O	M	N/A	M
M	M	M	N/A	O	M	M	M	M	O	M	N/A	M
M	O	M	N/A	O	M	M	M	M	M	M	O	M
M	M	M	M	O	M	M	M	O	O	O	O	N/A
M	N/A	M	N/A	N/A	M	M	M	O	O	O	O	N/A
M	O	M	N/A	N/A	M	M	M	M	O	O	O	N/A

Table 5.5 Food Safety Training and Development Matrix

Job title	Food safety orientation	Personal hygiene	Pest and waste control	Cleaning	Allergen management	Food delivery	Food processing	Food storage	HACCP	Traceability, recall, and withdra
Managing director	completed	completed	completed	completed	completed	completed	completed	completed	completed	N/A
Food safety manager	completed	completed	completed	completed	completed	completed	completed	completed	completed	completed
Hygienist/microbiologist	completed	completed	completed	completed	completed	completed	completed	completed	completed	completed
Milk processing manager	completed	completed	completed	completed	completed	completed	completed	completed	completed	completed
Laboratory manager	completed	completed	completed	completed	planned	completed	completed	completed	completed	completed
Warehousing manager	completed	completed	completed	completed	completed	completed	completed	completed	completed	N/A
Engineering manager	completed	completed	completed	completed	completed	completed	completed	completed	completed	completed
Maintenance manager	completed	completed	completed	completed	completed	completed	completed	completed	completed	planned
Logistics manager	completed	completed	completed	completed	completed	completed	completed	completed	completed	N/A
Laboratory technician	completed	completed	completed	completed	completed	completed	completed	completed	completed	completed
Food handler	completed	completed	completed	completed	completed	completed	completed	completed	completed	N/A
Transport driver	completed	completed	completed	completed	completed	completed	completed	completed	planned	N/A

Note: ■ = completed; ■ = planned; ■ = overdue; N/A = not applicable.

CHAPTER 5

Maintenance	Calibration	FSMS policies and procedures	Analytical policies and procedures	HACCP verification and validation	Document control	Record control	Emergency preparedness and crisis management	Food defense and food fraud	Consumer complaint management	Communications	Internal auditing	FSMS management review
			N/A	N/A							N/A	
			N/A								N/A	
			N/A									
			N/A									
			N/A									
			N/A								N/A	
			N/A								N/A	
				N/A								N/A
			N/A	N/A								N/A
			N/A	N/A							N/A	N/A

CHAPTER 5

References

Defelice, Robyn A. 2018. "How Long to Develop One Hour of Training? Updated for 2017." *Insights* (blog), January 9. https://www.td.org/insights/how-long-does-it-take-to-develop-one-hour-of-training-updated-for-2017.

Kirkpatrick, Donald L. 1959. "Techniques for Evaluating Training Programs." *Journal of American Society of Training Directors* 13 (3): 21–26.

———. 1994. *Evaluating Training Programs*. San Francisco: Berrett-Koehler Publishers.

Norris, Thomas D. 2002. "The Seven Attributes of an Effective Records Management Program." Publication 61, Government Records Services, New York State Archives, State Education Department, University of the State of New York, Albany, NY.

Phillips, Jack J. 1997. *Return on Investment in Training and Performance Improvement Programs*. Houston: Gulf Professional Publishing.

CHAPTER 6

Guidance for FBO Management

Introduction

This chapter provides an overview of the responsibilities of the executive management teams of food business operators (FBOs). It covers food safety policy, commitment, and resources, and the management review, including the decisions, actions, and follow-up required to maintain and enhance an effective food safety management system (FSMS).

The chapter supplies information on two important management resources. First is a publication of the International Finance Corporation (IFC), "Investing Wisely in Food Safety: How to Maximize the Benefits and Reduce Costs" (IFC 2015), that outlines the benefits, challenges, and lessons learned by FBO executive management teams in addressing food safety. It is a good store of knowledge and experience for executive management teams that are considering the adoption of an FSMS. Second is the Global Markets Program of the Global Food Safety Initiative (GFSI) and the International Trade Centre, which produces a downloadable checklist that enables FBO executive management teams to gauge the maturity of their FSMSs quickly and use the output to identify the gaps in their systems relative to GFSI standards or other food safety management schemes they may be considering.[1] The IFC Food Safety Foundation Course and the IFC Food Safety Handbook Training Course explain and make these two resources available (chapter 5).

Both resources should be reviewed by any FBO executive management team considering the adoption of an FSMS. If the FBO is targeting Food Safety System Certification (FSSC) 22000, it is best to start with the GFSI Global Markets Program, which implements an approach similar to the FSSC, but which adds two extra initial steps whereby the FBO's FSMS can already be audited and a conformity statement can be issued before engaging in the full FSSC 22000 process.[2]

Also useful are the two toolkits contained in this chapter, a management review toolkit and a cost of nonquality toolkit. The additional guidance they provide can help FBOs determine their strengths and weaknesses in management as well as the cost of failed products and other quality issues that affect an FSMS, thereby confirming the financial and other sorts of benefits deriving from proper FSMS implementation and execution.

Food safety policy

An FBO policy is a statement of intent. It is implemented as a procedure or protocol by an FBO. An FBO food safety policy is generally established and adopted by management, typically on the recommendation of the food safety manager. It is presumably compatible with the strategic direction of the organization. The concept of food safety should be embedded in every organizational feature of an FBO. All aspects of the FBO's FSMS should thus be covered by and consistent with relevant FBO food safety policies and business processes.

International FSMS schemes all require an organization to establish and implement flawlessly relevant food safety policies. The key characteristics these international schemes and food safety assessors and auditors might expect to find on examining the food safety policies of an FBO are as follows: (1) the policies should be established, adopted, and communicated by FBO management; (2) they should clearly set out management's aspirations and expectations in food safety; (3) they should address and satisfy the defined requirements of the food safety scheme adopted by the FBO; and (4) they should be consistent with the FBO's food safety objectives and regulatory and legal obligations.

According to the ISO 22000:2018 food safety standard developed through the International Organization for Standardization (see ISO 2018), top management should establish, implement, and maintain a food safety policy that (1) is appropriate to the purpose and context of the organization; (2) provides a framework for setting and reviewing the FSMS objectives; (3) includes a commitment to satisfy applicable food safety requirements, including statutory and regulatory requirements and mutually agreed customer requirements;

(4) addresses internal and external communication; (5) reflects a commitment to the constant improvement of the FSMS; and (6) ensures adequate organizational knowledge and capacity in food safety.[3]

Management must also ensure that the food safety policy is available and is maintained as documented information (for example, on paper) and that it is communicated, understood, and applied at all levels within the organization. The policy should likewise be available to relevant interested parties, such as customers (product purchasers) or consumers (product users) through the FBO's website.

According to the ISO 22000:2018, the food safety policy needs to be supported by measurable objectives, that is, objectives that are specific, measurable, attainable (or achievable), realistic, and time-bound (SMART). The objectives should also be documented, consistent with the policy, monitored and verified, communicated to the relevant people, and maintained and updated as appropriate. In keeping with the spirit of all food safety schemes, such as BRC Global Standards, FSSC 22000, and the codes of the Safe Quality Food Institute, the primary goal is safe food. Hence, all food safety objectives should involve a reduction or elimination of food safety hazards in FBO products.

Leadership and commitment

As in any other type of management system, an FSMS requires the leadership and commitment of management. The establishment of an environment that encourages food safety is the responsibility of all levels of management, but particularly the highest level. Top management should always be aware that the success of the FBO depends on properly monitoring and enhancing the effectiveness of the organization's risk control measures in guaranteeing the safety of the FBO's products across the food chain. If management is not committed to well-informed, sustained food safety as a primary objective, the goals of the company can easily shift toward other, sometimes conflicting, business objectives, particularly in less mature companies. Leadership and commitment thus imply the direct participation of management, particularly top management, in addressing all food safety issues in an FBO.

According to the ISO 22000:2018, management should exhibit leadership and commitment within an FBO or with respect to an FSMS by (1) ensuring that the food safety policies and objectives of the FSMS are established and are compatible with the strategic direction of the organization; (2) integrating the requirements of food safety management with the FBO's business processes and business decision making; (3) securing adequate resources for the FSMS; (4) communicating within the FBO and elsewhere the importance of effective food safety management and of compliance with the food safety requirements of the FSMS, applicable statutory and regulatory requirements, and mutually agreed customer requirements; (5) evaluating and maintaining the capacity of the FSMS to achieve the intended results; (6) directing and supporting staff and others in the effort to enhance the effectiveness of the FSMS; (7) creating an environment conducive to continuous improvement; and (8) encouraging management and staff to demonstrate leadership in their areas of responsibility.

The goals and outcomes of an FSMS are meant to involve impacts on law, technology, competition, markets, the economy, and the social and cultural environment as well as cybersecurity, food fraud, food defense and intentional contamination, and the knowledge and performance of local, regional, national, or international FBOs. Strong and active leadership is reinforced by the active, visible commitment of management to achieving the appropriate results. FBO management should thus also undertake the following: (1) show passion and interest for food safety; (2) set targets to improve or maintain food safety; (3) maintain the FBO premises and equipment in a good state of repair that enables product production in a food safe manner; (4) ensure that staff, including management, are sufficiently trained and competent to carry out their responsibilities in food safety; (5) establish operational control at all levels of the organization, that is, the hazard analysis critical control point (HACCP) plan, prerequisite programs (PRPs), and operational PRP (OPRP) plans; (6) assess FBO risks thoroughly and maintain appropriate control measures; (7) institute an effective downward and upward communication system and a regular flow of information on food safety, including, for example, performance data, such as consumer complaints and data on waste; (8) consult often on food safety matters with FBO staff and other stakeholders and communicate their concerns to the relevant actors; (9) evaluate and review the FSMS in light of audits or assessments and initiate management reviews of the FSMS (see the toolkit below); and (10) promote a culture of food safety throughout the FBO.[4]

To create effective management structures, it is important that any change in the FSMS, including working arrangements and personnel changes, must be evaluated for the implications for the performance of the FSMS, the availability of resources, and the reallocation of responsibilities within the FBO.

Resources

Resources has a reserved meaning within an FSMS. The term refers to the four types of FSMS resources in an FBO, namely, financial resources, human resources, infrastructure, and the work environment. These resources are typically controlled and supervised by management. This section examines two types of resources: financial resources and human resources.

The IFC has developed a useful executive management guide, "Investing Wisely in Food Safety: How to Maximize the Benefits and Reduce Costs" (IFC 2015) as an aid for FBO executive management teams that are weighing whether to design and implement an FSMS based on an HACCP plan and a relevant food safety scheme. The guide illustrates the benefits and challenges involved in this process, especially in five areas: planning, financing, changing behaviors, outsourcing wisely, and typical mistakes and how to avoid them.

The Global Markets Program of the GFSI and the International Trade Centre provides a downloadable checklist that enables teams to gauge the maturity of their FSMS systems quickly and use the output to identify the gaps in their systems relative to the GFSI standards or other food safety management schemes they may be considering.[5]

Financial resources

Finance is clearly a key input at any organization. Because management controls the finances of an FBO, management is responsible for ensuring the FBO has sufficient financial resources to realize its food safety policies, achieve its objectives and, especially, prevent any shortage of financing from becoming a factor in hindering the safe supply of food along the food chain.

A food safety auditor examining the financing of an FBO would typically look at the role of capital expenditure investment in preventing food safety risks across the FBO food chain and take into consideration the requirements of the relevant FSMS certification scheme. Such an auditor would also investigate the financial investment of the FBO in food safety training among all individuals working for or on behalf of the FBO. Note that a financial risk assessment does not, by itself, guarantee food safety along an FBO food chain.

Human resources

The most difficult challenge facing most FBOs is the need to encourage the personnel involved in the FSMS to take responsibility for their performance in ensuring food safety. Management must believe in the value of the FSMS and communicate this confidence to all staff and others engaged on behalf of the FBO. Management must demonstrate that all these people can contribute to the success of the FSMS and have the responsibility to guarantee food safety at all times. Management must seek to empower all to express their views on ways to improve the FBO's efforts and take action if they spot a problem in operations that could compromise food safety.

Management should thus strive to ensure that all personnel are provided with sufficient training in food safety to understand the purpose of the FBO's food safety policies and practices and the role of each FBO staff member in supporting food safety.

Management should also single out individual staff members to be responsible for ensuring that the FSMS conforms to the requirements of the FBO, reporting to management on the performance of the FSMS, forming the food safety team and nominating the food safety team leader, and designating staff with responsibility for implementing and documenting relevant initiatives.

The food safety team leader is responsible for overseeing the establishment, launch, maintenance, and updating of the FSMS; organizing and managing the work of the food safety team; arranging for relevant training and capacity building among the food safety team; and reporting to management on the effectiveness and suitability of the FSMS.

FBO managers have additional responsibilities. Thus, line managers are required to respond to the food safety concerns of personnel in a timely manner, welcome their ideas on ways to improve the FSMS, and pass these along to top management. Line managers are also responsible for helping create a culture of food safety across the FBO. Top management must provide line managers with sufficient resources to maintain a robust FSMS and to comply fully with food safety regulations, standards, and the expectations set by the FBO, regulators and customers. Executive management and line managers must lead by example and display proper food safety practices.

For most food safety auditors, the focus is training and the effectiveness of training. From the perspective of an FBO, the relevant food safety regulations and standards set the requirements, and the regulators and standards organizations provide practical advice to FBOs. The management of FBOs should therefore analyze the recommendations of these regulators and standards bodies on training in food safety. Most food safety regulators require FBOs to provide training to personnel in food safety or, if they are small operations, accept supervision. The training must be appropriate and adequate. For example, in an FBO, line managers and employees who process or serve food will need different types of training.

In some cases, government authorities provide guidance on training requirements. For instance, the Food Safety Authority of Ireland has produced guides to assist FBOs with training personnel in the workplace. One of the guides supplies information on basic food safety skills that staff should be able to demonstrate within the first month of employment and information on the additional food safety skills that staff should be able to demonstrate within 3–12 months of commencing employment in an FBO (FSAI 2015). Another guide offers information on the food safety skills that should be demonstrated by managers and supervisors in food operations (FSAI 2016).[6]

Toolkits

The two toolkits below are useful in gauging and guiding the management of an FBO. The first covers the management review, and the second supplies tools that may be used to reduce the costs of out-of-specification or nonquality products.

Toolkit 1: Results of a sample management review

Readers may benefit from an examination of the sample standard operating procedure, SOP-021, "Management Review," in chapter 4 before proceeding through this toolkit.

AGENDA

The agenda or sequence of steps in the management review below includes 13 main topics, as follows: (1) actions identified during the previous management review, (2) changes affecting FSMSs, (3) performance review (site objectives 2018, human resources, supplier performance, key process performance indicators, downtime, good manufacturing practices [GMPs] and glass inspections, verification activities, consumer complaints, overview: corrective and preventive actions, and customer satisfaction), (4) effectiveness of the food safety team, (5) new statutory and regulatory requirements, (6) emergency situations and withdrawals, (7) adequacy of resources, (8) risk and opportunity management, (9) proposed improvement activities, (10) electronic document management system (electronic DMS) performance, (11) food safety policy and documentation, (12) proposed objectives for 2019, and (13) any other business.

AGENDA ITEM 1: ACTIONS IDENTIFIED DURING THE PREVIOUS MANAGEMENT REVIEW

Table 6.1 reproduces the sample results on the actions and decisions of a previous management review.

Table 6.1 Actions Identified during the Previous Management Review

Action or decision	Agenda point	Who	Due date	Status
All process owners to verify access to the electronic DMS and be able to locate relevant controlled documents	1	All department managers	Immediate	In process
Establishment and confirmation of factory objectives for 2018	3	Joe	End February 2019	In process
Establish key performance indicator scorecard 2018 and align with factory objectives 2018 and key process description performance indictors	3	Joe	End February 2019	In process
Align factory objectives 2018, and key performance indicator scorecard	3	Joe, Mary, and Natia	End February 2019	In process
Industrial performance coordinator to publish status of factory objectives, key performance indicator scorecard, months	3	Natia to discuss and agree on process with Lisa	End February 2019	In process
Factory risk register to be established to manage risks associated with the attainment of factory objectives and key performance indicators	3	Natia to discuss and agree on process with Lisa	End February 2019	In process

continued

Table 6.1 ***(Continued)***

Action or decision	Agenda point	Who	Due date	Status
FSMS PRP development project plan to be created and published following completion of the gap analysis	3	Mary	End January 2019	In process
Improve the quality of management review data and presentation	3	Mary and Mike	End February 2019	In process
Ownership of process for collection and analysis of management review presentation pack (this document) to be managed by industrial performance coordinator	3	Natia to discuss and agree on process with Lisa	End February 2019	In process
Improvement plan to be developed for downtime (total)	3	Joe and Henry	End February 2019	In process
Improvement plan to be defined and documented to improve downtime (total)	3	Mike and Frank	End February 2019	In process
Improvement plan to be developed to reduce consumer complaints associated with packaging and maintenance	3	Joe	End February 2019	In process
Improvement plan to be developed to reduce variation in GMP inspection results	3	Joe	End February 2019	In process
Improve cycle time for closure of nonconforming and corrective and preventive action to under 30 days	3	Department managers and Joe	End February 2019	In process
Manager and supervisor development program to be introduced to ensure supervisors actively coach, mentor, and supervise associates with operational control FSMS	3	Jack, Sheila, Mary, and Joe	End February 2019	In process
Development plan, including hiring of food safety resources to be implemented to ensure the sustainability of FSMS	3	Sheila	End January 2019	In process
Implementation plan to be developed for FSMS following initial stage 1 audit	3	Natia to discuss and agree on process with Lisa	End February 2019	In process
Implementation plan to be developed for FSMS following initial stage 1 audit	6	Joe and Mary	End January 2019	In process
FSMS visualization performance to be introduced in all operational areas	6	Process owners, Joe, and Mary	End February 2019	In process
Schedule February and May 2019 management review meetings in leadership team diaries	7	Joe and Mary	End January 2019	In process
Microlaboratory construction and upgrade to be completed	7	Jack and Sheila	End February 2019	In process

AGENDA ITEM 2: CHANGES AFFECTING FSMSs

The following are highlights of the findings on any changes that may affect the FSMS. The findings are the results of the analysis during the sample management review.

- The core discipline standard, namely, FSSC 22000, will be undergoing a major change in September 2019. The timetable for the changes will not begin until March 2019 when the draft international standard versions will be published.
- Planned changes as a result of changes in the U.S. Food Safety Modernization Act (FSMA) will require key changes in the FSMS DMS.
- The trees at the rear of the factory are becoming so big that it is easy for people to climb a tree and get over the fence. The FBO should look into trimming or cutting the trees.
- The predicted quality of this year's sunflower oil is lower than normal. This might give rise to adulteration issues.
- The FBO is receiving a growing number of phishing emails. It is strongly suggested that the cybersecurity awareness program be relaunched among all employees.

AGENDA ITEM 3: PERFORMANCE REVIEW

Site objectives 2018

Table 6.2 supplies sample results of the analysis of the site objectives of an FBO.

Table 6.2 Performance Review: Site Objectives, 2018

FG	Indicator	Jan	Feb	Mar	Apr	May	Jun	Jul
1	Cases produced, number	428,157	377,311	442,489	443,016	444,102	419,861	454,705
2	Cases produced, % of budget	121.7	109.7	107.2	81.3	88.0	96.3	97.0
3	Cases produced, per employee hour	32.82	32.86	34.70	31.14	31.81	33.07	33.37
4	Operational downtime, hours	18.64	21.05	20.26	20.96	19.25	19.69	19.28
5	Mechanical downtime, hours	7.41	10.14	7.88	6.77	9.49	9.06	9.45
6	3 liter line efficiency, tonnes	596	571	608	624	582	605	610
7	Overhead, cost per case, $	3.68	3.71	3.32	3.22	3.49	3.10	3.06
8	Raw material yields, %	97.73	99.00	98.40	99.06	98.56	99.46	98.32
9	Warehouse cases shipped, per employee hour	127.92	152.54	157.57	170.25	154.03	152.36	164.08
10	On-time and complete shipments, %	96.08	96.88	98.80	95.87	95.66	97.17	96.11
11	Obsolete, damage, defects, rework, packed product	(35,060)	(23,215)	(18,123)	(47,731)	(36,845)	(23,754)	(31,871)
12	Injury frequency, individuals	3	0	0	0	0	0	0
13	Sanitation score, %	83.0	85.0	82.0	90.0	85.0	90.0	92.0
14	Customer complaints, number	2	7	3	7	24	2	11
Performance matrix index								
	Base	300	300	300	300	300	300	300
	Score, month	454	623	730	714	605	707	702

Note: FG = factory goal; YTD = year to date.

Human resources

Table 6.3 illustrates selected results on employees that were produced during the sample management review.

Table 6.3 Performance Review: Employees, 2018

Indicator	Unit	Target	Nov	Dec
Training completion	% of employees	80	63	59
Employee turnover	Number of employee terminations, %	<5	—	—
Employee engagement index	% of employees responding favorably (4 or 5 on a 5-point scale) to the engagement dimension questions in the company survey	81	—	—

Note: — = not available.

Supplier performance

The following illustrate the findings on supplier performance produced through the analysis during the sample management review.

- Among the suppliers, 85 percent have caused no issues and have delivered in full and on time. The packaging supplier is responsible for 12 percent of the issues, followed by the supplier of whey powder. The packaging supplier is still struggling to deliver in full. On average, this supplier provides 10 percent less input because of production issues on their end. The FBO aims to cope in

Aug	Sep	Oct	Nov	Dec	YTD	Budget	Base	Goal	Change, YTD, %
493,307	440,600	371,886	254,463	0	4,569,897	4,844,254			
93.9	100.2	94.8	59.9	0.0	94.3				
33.01	31.65	29.24			32.36		33.99	35.00	4.80
20.80	17.58	19.00	20.02	20.28	19.71		19.67	15.00	0.20
6.19	8.57	8.74	11.24	9.04	8.52		5.78	5.45	47.40
616	594	558			597		550	600	8.55
3.48	3.38	3.98	4.13		3.43	3.93	3.93	3.85	12.72
99.78	99.26	96.82			98.69	98.00	98.17	98.50	0.53
138.06	138.45	136.44			148.62		143.68	158.00	3.44
97.72	95.78	97.12			96.70		87.77	97.00	10.17
(137,756)	(131,741)	(111,398)			(59,749)		(59,545)	(53,590)	0.34
1	0	2	0	1	0.58		0.67	0.47	12.94
81.0	89.0	86.0	90.0		86.64		86.25	90.00	0.45
9	3	7	4		7.18		13.50	12.15	46.80
300	300	300	300	300	300				
458	563	250	2412	519	565			1,000	

the short term by maintaining a larger amount of stock. The same supplier also exhibits a strong seasonal drop in the top-load strength of the container cases because of high moisture levels.

- The FBO is currently investigating the opportunity to use a summer version and a winter version of the container cases to ensure that collapsing cases are no longer an issue.
- The whey powder supplier has had problems with *Clostridium botulinum*. FBO management believes this is linked to structural issues in their spray drying tower. Hence, FBO management is in the process of engaging an alternative supplier.

Key process performance indicators

Table 6.4 illustrates selected results on key process performance indicators that were gauged during the sample management review.

Table 6.4 Performance Review: Key Process Performance Indicators, 2018

Indicators	Definition	Unit responsible for measurement	Target value	Nov value	YTD value
Master schedule attainment, %	Planned orders vs. actual production	Planning	80	72	72
Case fill rate, %	Cases delivered vs. cases ordered	Warehouse	99	99	99.61
Stock cover	Number of calendar days for finished goods at month end vs. the demand plan for the following months	Planning	3.5 weeks	4 weeks	4.5 weeks
Obsolescence, value, €	Value of material past the shelf life date and due to be written off, related only to plants	Planning	<31,000, per month	—	59,749
Downtime because of lack of supplies, %	Unplanned stoppages, % of net production hours	Production manager	0.00	0.00	0.00
Line performance, %	Stoppage, downtime, % of net production hours	Production manager	20	20	19.67
Case fill rate, %	Cases shipped vs. cases ordered	Warehouse	99	99	99.61
Inventory count accuracy, %	Physical count vs. actual count x 100	Warehouse	92	—	—
Order fill rate, %	Order filled vs. total orders	Warehouse	95	—	97

Note: — = not available; YTD = year to date.

Downtime

Figure 6.1 shows the results on monthly downtime that were revealed during the sample management review.

Figure 6.1 Performance Review: Monthly Downtime, by Type, 2018

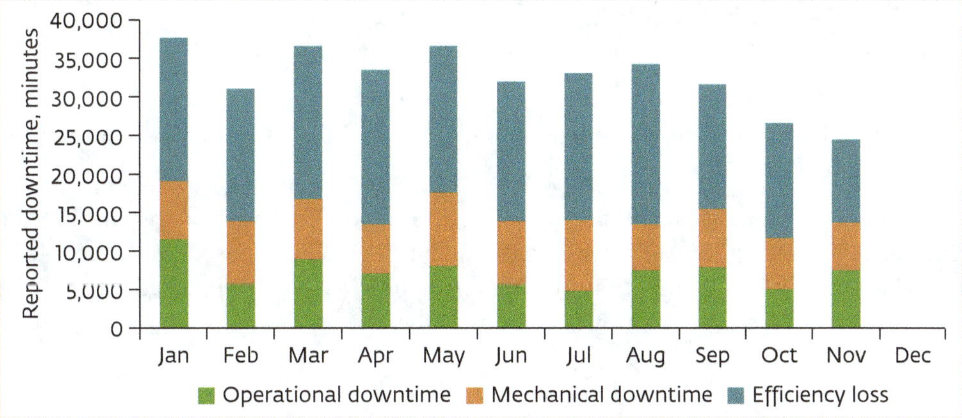

GMPs and glass inspections

The main observations during the GMP rounds include the following (figures 6.2 and 6.3):

- No broken glass was found during the glass inspection rounds in 2018.
- Two sets of labels were observed at the line at the same time.
- The cleaning performance of the dosing systems was improper.
- Hand hygiene performance in high care was under par.
- The level of mosquito infestation was high in the warehouse because of standing water at the back.
- Maintenance staff leave behind a substantial amount of debris after production activities.

Figure 6.2 Performance Review: GMP Findings, by Type, January–November 2018

percentage of nonconformities

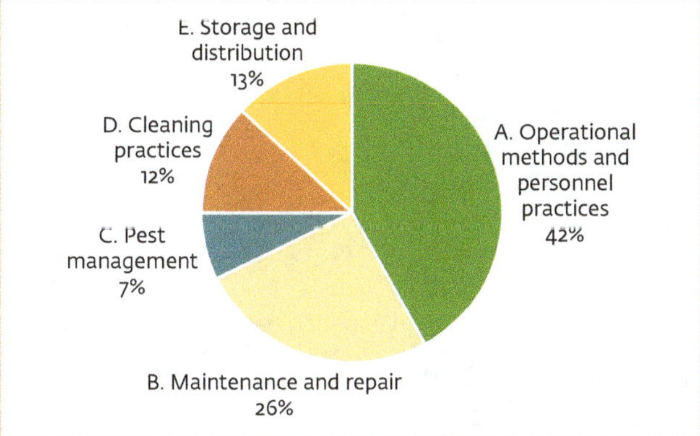

Figure 6.3 GMP Inspection Scores, January–November 2018

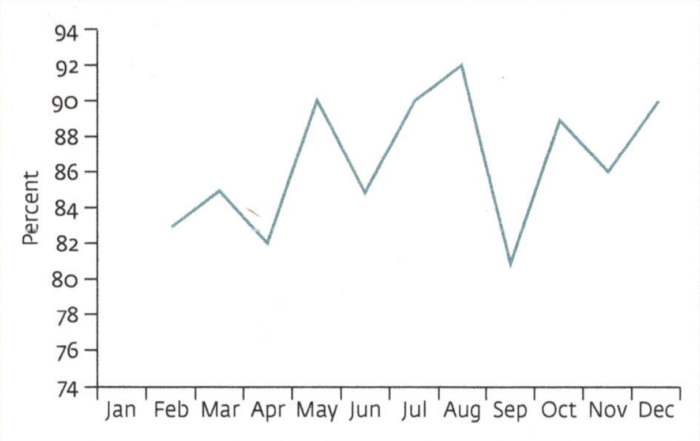

Verification activities

The following are selected findings on the performance in verification activities produced during the sample management review.

Findings on HACCP:

- Scheduled annual review performed on July 10, 2018

- Unscheduled HACCP verification review because of an OPRP failure on March 9 (metal detector) and October 10, 2018
- New products added: super milk, semiskim milk, organic milk, orange juice

Findings on new projects:

- FBO ongoing initiative to add ethylenediaminetetraacetic acid to formulas containing sodium benzoate.
- New hazards: A warning was issued by the U.S. Food and Drug Administration (FDA) on carbendazim in orange juice; there was a subsequent FDA release based on the conclusions of the preliminary risk assessment of the U.S. Environmental Protection Agency that the consumption of orange juice with carbendazim at the low levels that had been reported did not raise safety concerns.

Finding on PRPs:

- The transfer to a format approved by the FBO is in progress; the due date is January 31, 2019.

Finding on the FSMS:

- The FSMS verification checklist has been assessed. The closure of gaps is in progress; the due date is to be announced.

Consumer complaints

Figure 6.4 describes the number and the share of consumer complaints by type that were found during the sample management review.

Figure 6.4 Performance Review: Consumer Complaints, Number and Share, by Type, 2018

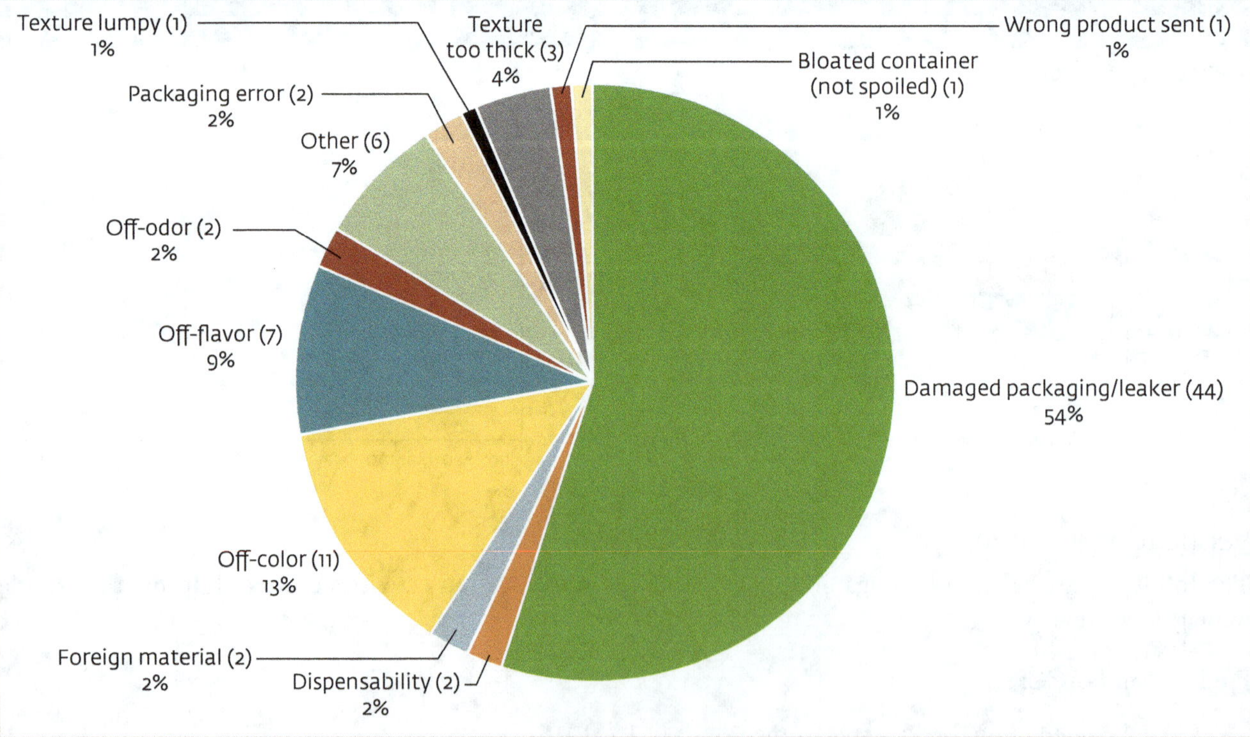

Overview: Corrective and preventive actions

The sample management review involved an examination of the quality of the data on corrective and preventive actions. It revealed that there were 17 corrective and preventive actions year-to-date (November 2018). Six were complaint related, and 11 were process or compliance related. There were five open corrective and preventive actions. Two were complaint related, and three were micro related. The average duration of corrective and preventive actions on micro-issues was five months. The average cycle time to close corrective and preventive actions was 30 to 45 days.

Customer satisfaction

The sample management review investigated customer satisfaction and found that the net promoter sales score was 32 percent. This means that, compared with its competitors in the same sector, the FBO was considered average as a supplier in the view of customers. Especially in the area of understanding customer requirements, the review concluded there was room for improvement. The FBO therefore launched a project focusing on its three largest customers. It adopted a multidisciplinary approach to managing these accounts. This involved the following: (1) sales would lead the overall account strategy and process; (2) the net promoter sales score would be used as an aid in defining new concepts in terms of approaching the customer; (3) quality assurance would become part of the monthly analysis of product quality performance; (4) logistics would be a center of the monthly discussion on the performance in product delivery; and (5) all the departments participating in these issues will become part of a wider internal team, including production, purchasing, and product maintenance, to enhance the customer experience.

AGENDA ITEM 4: EFFECTIVENESS OF THE FOOD SAFETY TEAM

An analysis of the effectiveness of the food safety team was undertaken as part of the sample management review. The analysis concluded that the management of change process was functioning well and that the food safety team had been involved in all changes occurring on-site. It was expected that the Food Safety Modernization Act (FSMA) in the United States would introduce two important rules during 2019, namely, rules on food defense and rules on the importation of raw materials, ingredients, and packaging from countries outside the United States. The main priority within the FBO currently is to establish a formal and systematic system for managing compliance with client technical standards. The plan is to introduce a new software-as-a-service distribution model to allow a third-party provider to host applications and offer them to customers over the Internet. The service would be provided as a tool to assist with the management of statutory and regulatory compliance.

AGENDA ITEM 5: NEW STATUTORY AND REGULATORY REQUIREMENTS

The national food safety authorities have sharpened the rules for the management of *Listeria monocytogenes* in ready-to-eat food. They now require the following: (1) more tests on certain classes of finished products, including five tests per 25 grams of product instead of one test on 25 grams; (2) new procedures for determining product shelf life, including storage at different temperatures during the shelf life test to reflect temperatures across the entire cold chain (warehouse–retail–consumer); and (3) the food safety authorities were to start actively checking on this in mid-2019. In the European Union (EU), meanwhile, a new regulation on official controls came into force on December 14, 2019; however, there was no immediate impact on the processes of the FBO.

AGENDA ITEM 6: EMERGENCY SITUATIONS AND WITHDRAWALS

In 2018, the FBO experienced one product withdrawal because of the presence of *Escherichia coli* in finished products. In total, 118 tons of product was retrieved from the market. The retrieval cost $47,000 in additional shipping costs, and the FBO lost a total sales value of $523,000 because of the issue. If the problem would have been detected sooner, the expenses might have been lower in terms of lost product and costs. The root cause was a nonproper execution of the manual cleaning process of the (manual) product dosing valve. As a solution, the manual valve was replaced by an automatic valve, which is connected to the clean-in-place system. The new cleaning step has been properly validated.

As a side benefit, additional exploration revealed that two other locations on site relied on similar valves. These will be replaced in the short term.

AGENDA ITEM 7: ADEQUACY OF RESOURCES

FBO management needs to hire a new food safety team leader. This is a mandatory position. The annual salary will be $15,000–$25,000 depending on experience.

The roles of manager and supervisor and the relevant levels of active engagement with the FSMS need to be clarified. To accomplish this, the following steps will be undertaken: (1) one full day of training on the corresponding issues among all managers and supervisors, (2) inclusion of a detailed list of responsibilities in the job descriptions of managers and supervisors, and (3) the institution of a requirement that each manager and supervisor must register an additional hour of FSMS-related activity per day.

Given the rate of change in the FSMS, the FBO had to invest in product information management software at a cost of $10,000.

The metal detectors were old and often broke down. The FBO therefore needed to invest in three new metal detectors in 2019. This represented a total investment of $120,000.

The roof of the headquarters building needed maintenance. There were several leaks during heavy rains. The estimated repair costs were $17,000.

AGENDA ITEM 8: RISK AND OPPORTUNITY MANAGEMENT

The FBO anticipated there would be strikes among cleaning staff because of the negotiations on a new union contract. To avoid downtime, the FBO could engage a third-party cleaning company to bridge the gap. This would cost an additional $5,000 per week, which is low compared with the $17,500 per day cost of downtime.

The FBO has been given the opportunity to make a pitch for a sales volume of $7.5 million with the biggest customer in the U.S. market. A dedicated team was installed in preparation. If the FBO were able to secure the account, it would be operating at close to maximum capacity. So, it had to establish a second team to evaluate the opportunity to raise capacity in the short to medium term and ensure it could meet the new customer's requirements if it succeeded in making the deal. The initial estimate was that the FBO would need to install a new production line, at a total cost of $457,000.

AGENDA ITEM 9: PROPOSED IMPROVEMENT ACTIVITIES

The FBO proposed the establishment of two fresh FSMS management reviews during 2019 to ensure it remained on top of all the planned changes. The reviews would occur in February 2019 (prior to the FSSC 22000 stage 1 audit) and May 2019 (prior to the FSMS stage 2 audit). This would involve one day of extra commitment by management.

More active communication on GMP topics would be called for during monthly staff sessions to increase GMP awareness. This would require no additional staff time.

Added mandatory discussions on the GMP was to be included among safety issues during each team meeting in all departments. This would require no additional staff time.

As the FBO prepares for the initial stage 2 audit, the active participation of and communication with managers, supervisors, and associates will be critical. This would call for one hour per week for six weeks among all managers and supervisors.

Training will be required among internal auditors on the provision of feedback and input to executive management and department managers. This is a key routine requirement that forms part of the FSMS improvement process. The estimated cost will be $3,000.

AGENDA ITEM 10: ELECTRONIC DMS PERFORMANCE

The performance of the electronic DMS and related training is within an acceptable average. In 2018, 124 of 379 documents were reviewed. The process was on time in the case of 80 percent of the documents. The delay was less than one month in the case of 10 percent of the documents, one to two months for 6 percent of the documents, and over two months for 4 percent of the documents.

Table 6.5 illustrates compliance with the training program on the updated and new document procedures among staff, by department.

Table 6.5 Training Compliance, New and Updated Document Procedures, by Duration of the Training
percent

Department	< 1 month	1–2 months	2–3 months	3–6 months	6–12 months
Warehouse	19	33	53	67	89
Production	5	22	23	33	44
Maintenance	17	31	43	61	78
Quality assurance	17	40	63	71	78
Human resources	2	13	28	31	53
Finance	3	4	22	25	50
Planning	24	48	58	65	69
Sales	9	27	35	53	78
Information technology	15	16	21	33	48

AGENDA ITEM 11: FOOD SAFETY POLICY AND DOCUMENTATION

The first draft of the FSMS manual was produced based on Annex SL (see ISO and IEC 2018) and the publicly available specification PAS 99:2012 (BSI Group 2012). It covers the FSMS, with the option for an extension to the ISO/IEC 17025 standard on laboratories (see ISO and IEC 2017). In particular, it addresses all FSMS process descriptions in the electronic and the published DMS, all core FSMS core procedures in the electronic DMS, and all current job profiles in the electronic DMS.

All FSMS process owners need to become masters in the use of the electronic DMS. As the FBO rolls out the implementation of the FSMS, awareness sessions for managers, supervisors, and associates will become essential.

The FBO will also need to review and upload all other FSMS documentation into the electronic DMS during the first half of 2019.

No changes have been proposed in FSMS policies.

AGENDA ITEM 12: PROPOSED OBJECTIVES FOR 2019

Table 6.6 summarizes the achievements of the FSMS in 2018 and the targets for 2019.

Table 6.6 Achievements of the FSMS in 2018 and Targets for 2019

Item	2018 actual	2019 target
Sales volume, up to 650,000 units	$49,500,000	$59,000,000
Cases produced, per employee hour	36	40
Operational downtime, %	15	<10
Overhead cost, per case	$4.25	<$3.75
Raw material yields, %	92	>98.50
Warehouse cases shipped, per employee hour	141.32	>165.00
On-time and complete shipments, %	87	>98
Obsolete, damage, defects, rework, value lost	$412,132	<$40,000
Injury frequency, % of personnel	0.73	<0.45
Sanitation score, %	92	>95
Consumer complaints, per ton	16	<10
Net profit	$3,562,456	>$6,500,000

AGENDA ITEM 13: ANY OTHER BUSINESS

The quality assurance manager is taking a long leave in the summer of 2019. FBO management plans to assign the assistant quality manager to the manager position temporarily to ease any disruptions in FSMS operations. This will be communicated during all monthly departmental meetings in May 2019.

Toolkit 2: The cost of nonquality

The second toolkit presented in this chapter is designed to assist FBO management to gain a broader grasp of the costs involved in nonquality. The management and personnel at many organizations do not possess a sufficiently deep understanding of the drivers of the costs of nonquality and thus base their views on the initiatives required to achieve progress and measure effectiveness on limited information.

Among the several factors involved in determining the costs of nonquality, four main issues may be highlighted: (1) internal problems, such as blocked stock, products that are to be reworked, scrapped products, and degraded products; (2) complaints, such as customer complaints or consumer complaints; there is a distinction between complaints related to food safety and other complaints; (3) severe issues, such as complaints related to food safety that are escalated by the customer or consumer and lead to intervention by food safety authorities; while these are still complaints, they may well result in a recall; and (4) recalls and withdrawals.

Explanations and descriptions of improvement activities routinely place areas with the highest incidence or frequency at the top of any list of statistics or priorities. For instance, consumer complaints related to product flavor might be at the top of such a list because these sorts of complaints are typically the most numerous. However, the incidence or frequency of issues does not necessarily mean that the corresponding financial impact on an FBO is of the same order of magnitude as the financial impact of other issue characteristics. Thus, complaints related to product flavor might be the most numerous and frequent, but addressing them costs far less than dealing with microbiological complaints related to illnesses among consumers. The latter might even result in a product recall. To make the best use of FBO resources, management should base decisions on data on the costs of nonquality, thereby obtaining valuable insight into the returns on the related investments in improvements and facilitating clearer and more transparent decision making on improvement projects.

To gain a more precise understanding of the costs of nonquality, a costing model based on standard costs, historical performance, and financial data broken down by issue categories should be more revealing than an elaborate activity-based costing model. The former model could be structured according to the following costs: (1) the average cost for not first time right products, (2) the average cost per complaint, (3) the average cost to right a serious issue, and (4) the average cost of a recall (table 6.7).

Table 6.7 The Cost of Nonquality Tool, by Cost Area, Cost Item, Month, and Quarter

Cost area	Cost item	Unit	Owner	Jan	Feb
Not first time right	Production runs, batches	Number	Production		
	Still good product	kg	QA		
	Rework	kg	QA		
	Degraded product	kg	QA		
	Scrapped product	kg	QA		
	Product: no decision taken (on hold)	kg	QA		
Complaints	Faults detected by producer	Number	QA		
	Customer complaints	Number	QA		
	Consumer complaints	Number	QA		
	Still good product	kg	QA		
	Rework	kg	QA		
	Degraded product	kg	QA		
	Scrapped product	kg	QA		
	Product: no decision taken (on hold)	kg	QA		
	Return shipments	Number	Operations		
	Cost of return shipments (transport, plus handling)	$	Operations		
	Replacement shipments	Number	Operations		
	Product in replacement shipments	kg	Operations		
	Shipping cost, replacement (excluding product costs)	$	Operations		
	Credit notes	Number	Operations		
	Value of credit notes	$	Operations		
	Rebates, penalties	Number	Operations		
	Value of rebates, penalties	$	Operations		
	Additional audits	Number	QA		
	QA travel costs for the last year in relation to customer complaints	$	Finance		
	QA travel time for the last year in relation to customer complaints	Hours	QA		
	Sales-related travel costs for the last year in relation to customer complaints	$	Finance		
	Sales travel time for the last year in relation to customer complaints	Hours	Sales		
	Other mitigation costs	$	Sales		
Severe issues	Severe issues	Number	QA		
	Still good product	kg	QA		
	Rework	kg	QA		
	Degraded product	kg	QA		

Mar	Apr	May	Jun	Jul	Aug	Sep	Oct	Nov	Dec	Q1	Q2	Q3	Q4	Total

continued

Table 6.7 The Cost of Nonquality Tool, by Cost Area, Cost Item, Month, and Quarter *(Continued)*

Cost area	Cost item	Unit	Owner	Jan	Feb
Severe issues *(continued)*	Scrapped product	kg	QA		
	Product: no decision taken (on hold)	kg	QA		
	Return shipments	Number	Operations		
	Cost of return shipments (transport, plus handling)	$	Operations		
	Replacement shipments	Number	Operations		
	Product in replacement shipments	kg	Operations		
	Shipping cost, replacement (excluding product costs)	$	Operations		
	Credit notes	Number	Operations		
	Value of credit notes	$	Operations		
	Rebates, penalties	Number	Operations		
	Value of rebates, penalties	$	Operations		
	Additional audits	Number	QA		
	Rescheduled production runs	Number	Production		
	Added production costs because of suboptimal planning	$	Production		
	Compensation paid to customer as margin compensation	$	Sales		
	Loss of turnover (until year end) due to customers leaving	$	Sales		
	Cost of external experts (laboratories and so on)	$	QA		
	QA travel costs for the last year in relation to customer severe issues	$	Finance		
	QA travel time for the last year in relation to customer severe issues	Hours	QA		
	Sales-related travel costs for the last year in relation to customer severe issues	$	Finance		
	Sales travel time for the last year in relation to customer severe issues	Hours	Sales		
	Other mitigation costs	$	Sales		
Recalls	Recalls	Number	QA		
	Still good product	kg	QA		
	Rework	kg	QA		
	Degraded product	kg	QA		
	Scrapped product	kg	QA		
	Product: no decision taken (on hold)	kg	QA		
	Return shipments	Number	Operations		
	Cost of return shipments (transport, plus handling)	$	Operations		
	Replacement shipments	Number	Operations		

Mar	Apr	May	Jun	Jul	Aug	Sep	Oct	Nov	Dec	Q1	Q2	Q3	Q4	Total

continued

Table 6.7 The Cost of Nonquality Tool, by Cost Area, Cost Item, Month, and Quarter *(Continued)*

Cost area	Cost item	Unit	Owner	Jan	Feb
Recalls *(continued)*	Product in replacement shipments	kg	Operations		
	Shipping cost, replacement (excluding product costs)	$	Operations		
	Credit notes	Number	Operations		
	Value of credit notes	$	Operations		
	Rebates, penalties	Number	Operations		
	Value of rebates, penalties	$	Operations		
	Additional audits	Number	QA		
	Rescheduled production runs	Number	Production		
	Additional production costs because of suboptimal planning	$	Production		
	Compensation paid to customer as margin compensation	$	Sales		
	Loss of turnover (until year end) due to customers leaving	$	Sales		
	Cost of external experts (laboratories and so on)	$	QA		
	QA travel costs in relation to customer recalls	$	Finance		
	QA travel time in relation to customer recalls	Hours	QA		
	Sales-related travel costs in relation to customer recalls	$	Finance		
	Sales travel time in relation to customer recalls	Hours	Sales		
	Other mitigation costs	$	Sales		
	Additional advertising costs due to public recalls	$	Sales		
	Margin loss due to (temporary) stop of sale of product(s)	$	Sales		
	Margin loss due to (temporary) production pause (due to the authorities)	$	Sales		
	Margin loss due to customers leaving	$	Sales		
	Margin loss due to market access restrictions in other countries	$	Sales		

Note: QA = quality assurance. Editable work sheets and templates can be found at http://www.ifc.org/foodsafety/handbook/template

The subsections below supply an overview of the typical building blocks in related costs that should be factored in based on the last one to three years of financial information available to the FBO to derive a standard cost for each type of cost. To determine priorities and assess the impact of improvement activities, FBO management would then multiply the standard cost by cost or issue type by the associated incidence rate to gauge the approximate cost of resolving an issue.

NOT FIRST TIME RIGHT PRODUCT

A not first time right product is a product that has a quality defect after production that (1) can still be corrected or reworked to make it a quality product after all; (2) has been degraded to, for example, animal feed; or (3) has been scrapped because it cannot be sold or fixed.

Mar	Apr	May	Jun	Jul	Aug	Sep	Oct	Nov	Dec	Q1	Q2	Q3	Q4	Total

Data on the following should be gathered to derive a typical cost for a not first time right product: total amount of product produced (kg or liter); total product cost, consisting of raw material and packaging material costs and manufacturing costs (labor, machine depreciation, and indirect manufacturing costs); number of production runs or batches per year; total number of hours spent by quality assurance on product release per year; total number of hours spent by quality control on the product release per year; total quality assurance expenses for product release per year; total rework costs; average price difference between a normal product and a degraded product; total scrapping costs (excluding product); warehouse storage costs per pallet per day (if pallets are not used, use warehouse product storage locations); average weight or volume per pallet (if pallets are not used, use warehouse product storage locations); total amount of rework last year (if not available, take 5 percent of the overall product volume); total amount of degraded product last year (if not available, take 0 percent if degradation does not occur and take 5 percent if it does occur); total amount of scrapped product last year (if not available, take 1 percent of the overall product volume); total quality control expenses for the last year (excluding personnel cost); average hourly rate quality assurance department; number of hours per year spent by quality assurance on product release; average hourly rate, quality control department; number of hours per year spent by quality control on product sampling, analysis, and data entry; and average amount of product on hold per year over the last three years.

COMPLAINTS

Data on the following should be collected to derive a typical cost for complaints: total amount of rework last year due to complaints; total amount of degraded product last year due to complaints; total amount of scrapped product last year due to complaints; average hourly rate, sales department; number of hours per year spent by sales on complaint-related issues; average hourly rate, operations department; number of hours per year spent by operations on complaint-related issues; quality assurance travel costs for the last year in relation to consumer complaints; sales-related travel costs for the last year in relation to consumer complaints; return shipment costs last year; replacement product value in relation to complaints over the last year; credit notes sent last year; rebates given and penalties received last year; number of additional audits per customer in relation to complaints last year; consumer-related costs (restitution, mail costs, and additional goods delivered); and average number of complaints per year over the last three years.

SEVERE ISSUES

Data on the following should be assembled to gauge a typical cost for severe issues: total amount of rework last year due to severe issues; total amount of degraded product last year due to severe issues; total amount of scrapped product last year due to severe issues; number of hours per year spent by sales on severe issues; number of hours per year spent by operations on severe issues; quality assurance travel costs for the last year in relation to severe issues; sales-related travel costs for the last year because of severe issues; return shipment costs last year; replacement product value in relation to severe issues over the last year; credit notes sent last year; rebates given and penalties received last year; number of additional audits per customer in relation to severe issues last year; consumer-related costs (restitution, mail costs, and additional goods delivered); number of additional audits received last year due to severe issues; estimate of costs because of rescheduling caused by severe issues last year; estimate of additional costs due to suboptimal scheduling last year; loss of turnover and related margin last year; expert advice costs in relation to severe issues last year; and average number of severe issues per year over the last three years.

RECALLS

Data on the following items should be gathered to come to a typical cost for recalls: total amount of rework last year due to recalls; total amount of degraded product last year due to recalls; total amount of scrapped product last year due to recalls; number of hours per year spent by sales on recalls; number of hours per year spent by operations on recalls; quality assurance travel costs for the last year in relation to recalls; sales-related travel costs for the last year in relation to recalls; return shipment costs last year; replacement product value in relation to recalls over the last year; credit notes sent last year; rebates given and penalties received last year; number of additional audits per customer in relation to recalls last year; consumer-related costs (restitution, mail costs, additional goods delivered; number of additional audits received last year due to recalls; estimate of costs because of rescheduling caused by recalls last year; estimate of additional costs due to suboptimal scheduling last year; loss of turnover and related margin last year; expert advice costs in relation to recalls last year; additional advertising costs due to public recalls; margin loss due to (temporary) halt in sale of product(s); margin loss due to (temporary) closure of production (due to the authorities); margin loss due to customers walking out; margin loss due to market access restrictions in other countries; and average number of recalls per year over the last three years.

Notes

1. See "Global Markets: A Pathway to Certification," Global Food Safety Initiative, Consumer Goods Forum, Levallois-Perret, France, https://mygfsi.com/how-to-implement/global-markets.
2. For more information on the link between the GFSI Global Markets Program and the FSSC 22000 process, see "Global Markets Program," FSSC 22000, Food Safety System Certification, Global Food Safety Initiative, Consumer Goods Forum, Levallois-Perret, France, https://www.fssc22000.com/developmentprogram/.
3. Top management is the person or group who directs and controls the FBO that is being certified. If the scope of the management system covers only part of an organization, then top management refers to those who direct and control that part of the organization. The term appropriate refers to the scope of the products, food chain activities, and markets of the FBO's FSMS.
4. These issues are discussed in detail in a report on the culture of food safety that also includes a comprehensive reading list (see GFSI 2018).
5. See "Global Markets: A Pathway to Certification," Global Food Safety Initiative, Consumer Goods Forum, Levallois-Perret, France, https://mygfsi.com/how-to-implement/global-markets.
6. A wide range of training resources may be found at BiaBiz: Empowering Those Who Feed the World (database), Biabiz Limited, Garrycloyne, Blarney, County Cork, Ireland, https://www.bia-biz.com/.

References

BSI Group. 2012. "Publicly Available Specification: Specification of Common Management System Requirements as a Framework for Integration." PAS 99:2012 (October), BSI Group, London.

FSAI (Food Safety Authority of Ireland). 2015. "Guide to Food Safety Training, Level 1: Induction Skills, and Level 2: Additional Skills; for Food and Non-food Handlers (Food Service, Retail, and Manufacturing Sectors)." FSAI, Dublin. https://www.fsai.ie/WorkArea/DownloadAsset.aspx?id=8310.

———. 2016. "Guide to Food Safety Training, Level 3: Food Safety Skills for Management; Food Service, Retail, and Manufacturing Sectors." FSAI, Dublin. https://www.fsai.ie/WorkArea/DownloadAsset.aspx?id=784.

GFSI (Global Food Safety Initiative). 2018. "A Culture of Food Safety: A Position Paper from the Global Food Safety Initiative (GFSI)." Version 1.0 (April 11), GFSI, Consumer Goods Forum, Levallois-Perret, France.

IFC (International Finance Corporation). 2015. "Investing Wisely in Food Safety: How to Maximize the Benefits and Reduce Costs." IFC, Washington, DC. https://www.ifc.org/wps/wcm/connect/industry_ext_content/ifc_external_corporate_site/agribusiness/resources/investing+wisely+in+food+safety.

ISO (International Organization for Standardization). 2018. "Food Safety Management Systems: Requirements for Any Organization in the Food Chain." International Standard ISO 22000:2018(E), ISO, Geneva.

ISO (International Organization for Standardization) and IEC (International Electrotechnical Commission). 2017. "General Requirements for the Competence of Testing and Calibration Laboratories." International Standard ISO/IEC 17025:2017(E), 3rd ed. (November), ISO and IEC, Geneva.

———. 2018. *ISO/IEC Directives, Part 1: Consolidated ISO Supplement, Procedures Specific to ISO*, 9th ed. Geneva: ISO and IEC.

CHAPTER 6

APPENDIX A

Institutions and Other Entities Involved in Food Safety

Table A.1 Institutions and Other Entities Involved in Food Safety

State institution or other entity	Internet address
Asia Pacific Food Industry	http://www.apfoodonline.com/
Austrian Agency for Health and Food Safety	https://www.ages.at/en/ages/basics/
Austrian Federal Ministry of Agriculture, Forestry, Environment and Water Management – Food	http://www.lebensministerium.at/lebensmittel.html
Belgian Federal Agency for the Safety of the Food Chain (FASFC)	http://www.favv-afsca.fgov.be/home-en/
Belgian Federal Public Service for Health, Food Chain Safety and Environment – Food Safety	http://www.health.belgium.be/eportal/foodsafety/index.htm
BRC Global Standards	http://www.brcglobalstandards.com
Bulgarian Food Safety Agency	http://www.babh.government.bg/en/
CanadaGap	http://www.canadagap.ca
CHINA HACCP	http://www.gbstandards.org/index/standards_search.asp?word=HACCP
Croatian Food Agency	https://www.hah.hr/en/
Cyprus Ministry of Agriculture, Rural Development and Environment	http://www.moa.gov.cy/moa/agriculture.nsf/All/9638239B67CB5B93C22578A200307D00?OpenDocument
Cyprus Ministry of Health – Food Safety Council	https://www.moh.gov.cy/MOH/fsc/fsc.nsf/index_en/index_en?OpenDocument
Cyprus State General Laboratory	https://www.moh.gov.cy/moh/sgl/sgl.nsf/index_en/index_en?OpenDocument
Czech Republic Ministry of Agriculture	http://eagri.cz/public/web/en/mze/
Danish Ministry of Environment and Food	https://en.mfvm.dk/the-ministry/
Danish National Food Institute	http://www.food.dtu.dk/english
Danish Veterinary and Food Administration – Food	http://www.foedevarestyrelsen.dk/english/Food/Pages/default.aspx
Estonian Ministry of Agriculture	http://www.agri.ee/food-safety/
EUR-Lex - Direct free access to European Union law with full search facility	http://eur-lex.europa.eu/en/index.htm
EUROPA Summaries of EU legislation – Food Safety	https://europa.eu/european-union/topics/food-safety_en
European Commission – food hygiene legislation page	https://ec.europa.eu/food/safety/biosafety/food_hygiene/legislation_en
European Commission Directorate-General for Health and Food Safety (DG SANTE)	https://ec.europa.eu/knowledge4policy/organisation/dg-sante-dg-health-food-safety_en
European Commission Health EU Portal – Food Safety	https://ec.europa.eu/food/safety_en

continued

Table A.1 *(Continued)*

State institution or other entity	Internet address
European Food Information Council (EUFIC) – Food Safety	https://www.eufic.org/en/food-safety/
European Food Safety Authority	http://www.efsa.europa.eu/
FAO/WHO (Food and Agriculture Organization of the United Nations/World Health Organization) – Codex Alimentarius	http://www.fao.org/fao-who-codexalimentarius/home/en/
Finnish Food Authority	https://www.ruokavirasto.fi/en/
Finnish Ministry of Agriculture and Forestry – Food and Agriculture	https://mmm.fi/en/food-and-agriculture
Food and Agriculture Organization (FAO) of the United Nations – Food Safety and Quality	http://www.fao.org/food/food-safety-quality/en/
Food and Drink Technology	http://www.foodanddrinktechnology.com/
Food Engineering	http://www.foodengineeringmag.com/
Food Processing	http://www.foodprocessing.com
FoodRisk.org	https://www.foodrisk.org/
French Agency for Food, Environment, and Occupational Health and Safety	http://www.anses.fr/
French Ministry of Agriculture and Food	https://agriculture.gouv.fr /french-ministry-agriculture-and-food
FSSC 22000	http://www.fssc22000.com
German Federal Institute for Risk Assessment (BFR) – Food Safety	https://www.bfr.bund.de/en/food_safety-737.html
German Federal Ministry of Food and Agriculture – Safe Food and a Healthy Diet	https://www.bmel.de/EN/Food/food_node.html
German Federal Office for Agriculture and Food	https://www.ble.de/EN/Home/home_node.html
Global Aquaculture Alliance	https://www.aquaculturealliance.org/
Global Food Safety Initiative	http://www.mygfsi.com/
GlobalG.A.P.	https://www.globalgap.org/uk_en/
Global Red Meat Standard	https://grms.org/
GMP+ International	http://www.gmpplus.org/
Hellenic Food Authority	http://www.efet.gr/
Hellenic Ministry of Agriculture and Food	http://www.efet.gr/
Hungarian National Food Chain Safety Office	https://portal.nebih.gov.hu/
Icelandic Food and Veterinary Authority (MAST)	http://www.mast.is/english/frontpage/about-mast/
International Featured Standards (IFS)	https://www.ifs-certification.com/index.php/en/

continued

Table A.1 Institutions and Other Entities Involved in Food Safety *(Continued)*

State institution or other entity	Internet address
Irish Department of Agriculture, Food and the Marine	http://www.agriculture.gov.ie/
Irish Food Safety Authority	http://www.fsai.ie/links.html
ISO, the International Organization for Standardization	http://www.iso.org
Italian Istituto Superiore di Sanità (ISS)	http://old.iss.it/index.php
Latvian Veterinary and Food Department	https://www.zm.gov.lv/en/statiskas-lapas/zemkopibas-ministrija/statiskas-lapas/veterinary-and-food-department?id=4226#jump
Lithuanian State Food and Veterinary Service	https://vmvt.lt/kontaktai/state-food-and-veterinary-service?language=en
Luxembourg Ministry of Health	http://www.ms.public.lu/fr/
Malta Competition and Consumer Affairs Authority (MCCAA)	https://mccaa.org.mt/
Maltese Environment and Resources Authority	https://era.org.mt/en/Themes/Pages/Welcome.aspx
Maltese Ministry for Health	http://ehealth.gov.mt/HealthPortal/others/foodsafety_week/food_safety_week.aspx
Netherlands Food and Consumer Product Safety Authority	https://english.nvwa.nl/
Netherlands Ministry of Agriculture, Nature and Food Quality	https://www.government.nl/ministries/ministry-of-agriculture-nature-and-food-quality
Norwegian Food Safety Authority	http://www.regjeringen.no/en/dep/hod/About-the-Ministry/Subordinateinstitutions/Norwegian-Food-Safety-Authority.html?id=279765
Norwegian Ministry of Agriculture and Food	http://www.regjeringen.no/en/dep/lmd.html?id=627
Polish Chief Sanitary Inspectorate	http://www.gis.gov.pl/?lang=en&go=content&id=10
Portuguese Economy and Food Safety Authority	http://www.asae.gov.pt/
PrimusGFS	http://www.primusgfs.com
Romanian National Sanitary Veterinary and Food Safety Authority	http://www.ansvsa.ro/
Slovak Republic Ministry of Agriculture and Rural Development	http://www.mpsr.sk/en/index.php?navID=1
Slovenian Ministry of Agriculture, Forestry and Food	http://www.arhiv.mkgp.gov.si/en/areas_of_work/food_safety/
Spanish Agency on Food Safety and Nutrition	http://www.aecosan.msssi.gob.es/en/AECOSAN/web/home/aecosan_inicio.htm
SQF Institute	https://www.sqfi.com/
Standards and Trade Development Facility (STDF)	https://www.standardsfacility.org/
Swedish Food Agency	https://www.livsmedelsverket.se/en

continued

Table A.1 *(Continued)*

State institution or other entity	Internet address
Swedish Ministry of Rural Affairs	https://www.government.se/government-policy/rural-affairs/
Swiss Federal Food Safety and Veterinary Office	https://www.blv.admin.ch/blv/en/home.html
U.K. Department for Environment, Food and Rural Affairs	http://www.defra.gov.uk/
U.K. Food Standards Agency	http://www.food.gov.uk/
U.S. Department of Agriculture – Food Safety and Inspection Service	http://www.fsis.usda.gov/
U.S. Department of Agriculture – Food Safety Topics	https://www.nal.usda.gov/fsrio/food-safety-topics
U.S. Department of Agriculture – Food Security	https://www.usda.gov/topics/food-and-nutrition/food-security
U.S. Environmental Protection Agency – Agriculture	http://www.epa.gov/agriculture/tfsy.html
U.S. Food and Drug Administration – Animal and Veterinary	http://www.fda.gov/AnimalVeterinary/default.htm
U.S. Food and Drug Administration – Food	http://www.fda.gov/Food/default.htm
U.S. Food and Drug Administration – Food Defense	https://www.fda.gov/Food/FoodDefense/default.htm
U.S. Juice HACCP	https://www.fda.gov/food/hazard-analysis-critical-control-point-haccp/juice-haccp
U.S. Meat and Poultry HACCP	https://www.nal.usda.gov/fsrio/meat-and-poultry-haccp
U.S. Seafood HACCP	https://www.fda.gov/food/hazard-analysis-critical-control-point-haccp/seafood-haccp
World Health Organization (WHO) – Food Safety	http://www.who.int/foodsafety/en/

www.ingramcontent.com/pod-product-compliance
Lightning Source LLC
LaVergne TN
LVHW080012210926
848481LV00011B/2997